Electromagnetic Reverberation Chambers

Related titles on electromagnetic waves:

Dielectric Resonators, 2nd Edition Kajfez and Guillon
Electronic Applications of the Smith Chart Smith
Fiber Optic Technology Jha
Filtering in the Time and Frequency Domains Blinchikoff and Zverev
HF Filter Design and Computer Simulation Rhea
HF Radio Systems and Circuits Sabin
Microwave Field-Effect Transistors: Theory, design and application, 3rd Edition Pengelly
Microwave Semiconductor Engineering White
Microwave Transmission Line Impedance Data Gunston
Optical Fibers and RF: A natural combination Romeiser
Oscillator Design and Computer Simulation Rhea
Radio-Electronic Transmission Fundamentals, 2nd Edition Griffith, Jr
RF and Microwave Modeling and Measurement Techniques for Field Effect Transistors Gao
RF Power Amplifiers Albulet
Small Signal Microwave Amplifier Design Grosch
Small Signal Microwave Amplifier Design: Solutions Grosch
2008+ Solved Problems in Electromagnetics Nasar
Antennas: Fundamentals, design, measurement, 3rd Edition Blake and Long
Designing Electronic Systems for EMC Duff
Electromagnetic Measurements in the Near Field, 2nd Edition Bienkowski and Trzaska
Fundamentals of Electromagnetics with MATLAB®, 2nd Edition Lonngren et al.
Fundamentals of Wave Phenomena, 2nd Edition Hirose and Lonngren
Integral Equation Methods for Electromagnetics Volakis and Sertel
Introduction to Adaptive Arrays, 2nd Edition Monzingo et al.
Microstrip and Printed Antenna Design, 2nd Edition Bancroft
Numerical Methods for Engineering: An introduction using MATLAB® and computational electromagnetics Warnick
Return of the Ether Deutsch
The Finite Difference Time Domain Method for Electromagnetics: With MATLAB® simulations Elsherbeni and Demir
Theory of Edge Diffraction in Electromagnetics Ufimtsev
Scattering of Wedges and Cones with Impedance Boundary Conditions Lyalinov and Zhu
Circuit Modeling for Electromagnetic Compatibility Darney
The Wiener–Hopf Method in Electromagnetics Daniele and Zich
Microwave and RF Design: A systems approach, 2nd Edition Steer
Spectrum and Network Measurements, 2nd Edition Witte
EMI Troubleshooting Cookbook for Product Designers Andre and Wyatt
Transmission Line Transformers Mack and Sevick
Electromagnetic Field Standards and Exposure Systems Grudzinski and Trzaska
Practical Communication Theory, 2nd Edition Adamy
Complex Space Source Theory of Spatially Localized Electromagnetic Waves Seshadri
Electromagnetic Compatibility Pocket Guide: Key EMC facts, equations and data Wyatt and Jost
Antenna Analysis and Design Using FEKO Electromagnetic Simulation Software Elsherbeni, Nayeri and Reddy
Scattering of Electromagnetic Waves by Obstacles Kristensson
Adjoint Sensitivity Analysis of High Frequency Structures with MATLAB® Bakr, Elsherbeni and Demir
Developments in Antenna Analysis and Synthesis Vol. 1 and Vol. 2 Mittra
Advances in Planar Filters Design Hong
Post-Processing Techniques in Antenna Measurement Castañer and Foged
Nano-Electromagnetic Communication at Terahertz and Optical Frequencies: Principles and applications Alomainy, Yang, Imran, Yao and Abbasi
Nanoantennas and Plasmonics: Modelling, design and fabrication Werner, Campbell and Kang

Electromagnetic Reverberation Chambers

Recent advances and innovative applications

Edited by
Guillaume Andrieu

The Institution of Engineering and Technology

Published by SciTech Publishing, an imprint of The Institution of Engineering and Technology, London, United Kingdom

The Institution of Engineering and Technology is registered as a Charity in England & Wales (no. 211014) and Scotland (no. SC038698).

First published 2020

The Institution of Engineering and Technology
Michael Faraday House
Six Hills Way, Stevenage
Herts, SG1 2AY, United Kingdom

www.theiet.org

British Library Cataloguing in Publication Data
A catalogue record for this product is available from the British Library

ISBN 978-1-78561-931-1 (hardback)
ISBN 978-1-78561-932-8 (PDF)

Typeset in India by MPS Limited

Cover image designed by Dr Guillaume Andrieu

Contents

About the editor   xi

Introduction   1
Guillaume Andrieu

  I.1 Motivation and goal   1
  I.2 Organization   2
  I.3 Usage   4
  References   5

1 Performance of reverberation chambers using the "well-stirred condition method"   7
Guillaume Andrieu

  1.1 The well-stirred condition method   8
     1.1.1 Definition of the well-stirred condition of an RC   8
     1.1.2 Principle of the method   9
     1.1.3 Required experimental setup (for a mechanically stirred RC)   10
     1.1.4 On S_{11} measurements in RC   10
     1.1.5 Analysis of the EM field distributions   13
     1.1.6 Analysis of the sample correlation   15
     1.1.7 Determination of f_{wsc} 20
  1.2 Detailed analysis of the method performance   20
     1.2.1 Description of the measurements   20
     1.2.2 Sensitivity of the method related to the antenna position and orientation   21
     1.2.3 Sensitivity of the method related to the antenna type   26
     1.2.4 On the uncertainty of the method related to the PCF process   28
     1.2.5 On the effect of a heavily lossy EUT   30
  1.3 Examples of parametric analysis   31
     1.3.1 Determination of the "optimal loading"   33
     1.3.2 Effect of the number of metallic blades of a mode stirrer   34
  1.4 An alternative calibration procedure   36
  1.5 Conclusion   41
  References   41

2 **Review of the stirring techniques proposed in the literature – performance assessment of the most common ones using the well-stirred condition method** 45
Guillaume Andrieu

2.1 Mechanical stirring techniques 46
    2.1.1 Overview 46
    2.1.2 Dual-mode stirrer 47
    2.1.3 Reference results 48
2.2 Source signal stirring techniques 49
    2.2.1 The frequency stepping stirring technique 50
    2.2.2 Variants 61
2.3 Source position stirring techniques 63
    2.3.1 Moving antenna stirring technique 63
    2.3.2 Antenna network stirring technique 67
    2.3.3 Hybrid stirring technique combining multiple antennas and frequency stepping 68
2.4 Other mode stirring techniques 75
    2.4.1 Reactively loaded antenna stirring technique 75
    2.4.2 Switched stirred reverberation chamber 75
2.5 Conclusion 76
References 77

3 **Frequency and time-domain performance assessment of vibrating intrinsic reverberation chambers** 81
Guillaume Andrieu

3.1 Description of the VIRC used 82
    3.1.1 General characteristics 82
    3.1.2 Stirring strategy 83
    3.1.3 Q-factor and average mode bandwidth Δ_f 85
3.2 Frequency-domain characterization 86
3.3 Time-domain characterization 92
    3.3.1 Decorrelation time and detection of periodicity 92
    3.3.2 Rayleigh distribution in the time-domain 94
    3.3.3 Discussion 96
3.4 Conclusion 98
References 99

4 **Probabilistic model about the influence of the number of stirring conditions considered during a radiated susceptibility test** 101
Guillaume Andrieu

4.1 About RS tests in RC 102
    4.1.1 Principle of an RS test 102
    4.1.2 Values of n recommended by the standard 103

		4.1.3	EM field distributions in a "well stirred" RC	103
		4.1.4	On the choice of the metric representing the EM field strength	104
	4.2	Description of the probabilistic model		106
		4.2.1	Assumptions and definitions	106
		4.2.2	Uncertainty on the estimation of $\langle |E_\mathrm{T}| \rangle_m^s$	106
		4.2.3	Demonstration of the method applicability for any EUT	108
		4.2.4	Computation of the risk r	109
		4.2.5	Experimental results	111
	4.3	Overtesting		116
		4.3.1	Principle	116
		4.3.2	Example of numerical results	117
		4.3.3	A two-step approach using overtesting	118
	4.4	Determination of the minimum susceptibility level of the EUT (after a successful test)		119
	4.5	Conclusion		120
	References			121

| **5** | **Over-the-air testing of wireless devices in heavily loaded reverberation chambers** | | | **123** |

Kate A. Remley, Chih-Ming Wang and Robert D. Horansky

	5.1	Chamber characterization for OTA tests		125
		5.1.1	Configuring a reverberation chamber for wireless device testing	127
		5.1.2	Distribution of mode-stirring samples in loaded chambers	131
		5.1.3	Coherence bandwidth	135
		5.1.4	Spatial correlation and the mode-stirring sequence	139
		5.1.5	Lack of spatial uniformity due to loading	144
		5.1.6	Isotropy for loaded chambers	146
		5.1.7	K-factor	149
		5.1.8	Chamber characterization: a summary	150
	5.2	Over-the-air tests for radiated power and receiver sensitivity		151
		5.2.1	Total radiated power	151
		5.2.2	Total isotropic sensitivity	152
	5.3	Replicating specific multipath channels for OTA test		156
		5.3.1	Highly reflective power delay profiles	157
		5.3.2	Longer power delay profiles	158
		5.3.3	Emulating spatial channels	158
	5.4	Uncertainty in OTA measurements with heavily loaded reverberation chambers		160
		5.4.1	Number of samples vs. spatial uniformity: which uncertainty mechanism dominates?	160
		5.4.2	Relative uncertainty due to the type of stirring mechanisms	162

5.4.3   Uncertainty due to difference in reference and DUT
          antennas   164
5.4.4   Combined and expanded uncertainties   167
5.5   Conclusion   168
5.6   Acknowledgments   169
References   169

**6   From material absorption to dosimetry for exposure of animals in
      reverberation chambers   175**
Philippe Besnier

6.1   Losses and Q-factor   176
      6.1.1   Role of losses in a RC   176
      6.1.2   Definition of the Q-factor   177
      6.1.3   Origin of losses and their contribution to the Q-factor 178
      6.1.4 Q-Factor measurements   179
6.2   The average absorption effective area of an object in a RC   186
      6.2.1   The Q-factor of an object in a RC and its average
              absorption cross section   186
      6.2.2   Theoretical absorbing cross section of a rectangular piece
              of absorbing material   187
      6.2.3   Measurement examples   189
6.3   Dosimetry for animals   191
      6.3.1   Using RC for animal exposure   191
      6.3.2   Theory about heat transfer in a RC   193
      6.3.3   Experimental validation   199
6.4   Discussion   200
References   201

7   Characterization of antenna efficiency in reverberation chambers   205
Wei Xue and Xiaoming Chen

7.1   Antenna radiation efficiency measurement methods   206
      7.1.1   Standard reference antenna method   206
      7.1.2   One-antenna method   208
      7.1.3   Two-antenna method   212
      7.1.4   Three-antenna method   214
      7.1.5   Time reversal method   216
      7.1.6   Discussion on antenna efficiency measurement methods   221
7.2   Statistical analysis of antenna efficiency measurement uncertainty 224
      7.2.1   Statistics of measured antenna efficiency   225
      7.2.2   Simulations   228
      7.2.3   Measurements   229
7.3   Conclusion   235
References   235

**8   Characterization of antenna radiation pattern in
     reverberating enclosures**   **239**
     Guillaume Andrieu

     8.1   *K*-factor method   240
     8.2   Doppler spectrum method   242
     8.3   Time-gating method   245
           8.3.1   Principle of the method   245
           8.3.2   Example of results   247
     8.4   Comparison of the methods   256
     References   258

9   Radar cross-section estimation in reverberation chambers   **261**
     Philippe Besnier

     9.1   Definition of radar cross section and the radar equation   262
     9.2   Radar cross-section measurements   264
     9.3   Scattered cross section and antenna patterns in reverberation
           chambers   264
     9.4   Theory of radar cross-section pattern measurements in
           reverberation chambers   266
           9.4.1   Measurement setup   267
           9.4.2   Backscattered field in an empty RC   268
           9.4.3   Backscattered field from the target   269
           9.4.4   The backscattered coefficient   269
           9.4.5   RCS equation of the target   270
           9.4.6   Extraction of RCS   271
     9.5   Validation on simple targets   274
           9.5.1   Preliminary measurements in anechoic chamber   274
           9.5.2   Calibration in RC   276
           9.5.3   Measurement of the metal plate   277
           9.5.4   Measurement of a dihedral target   278
           9.5.5   The role of the stirrer   278
     9.6   Discussion   282
     References   283

Conclusion   **285**
     Guillaume Andrieu

     C.1   OTA test for new protocols   285
     C.2   "Anechoic" measurements   285
     C.3   Chaotic cavities   286
     C.4   VIRCs   287
     C.5   Future EMC strategy in the transportation industries   288

Index   **289**

About the editor

Guillaume Andrieu was born in Limoges, France, in 1980. He received the Master's degree in Radiofrequencies and Optical Communications from the University of Limoges in 2003, and the Ph.D. degree in Electronics from the IEMN Laboratory at the University of Lille, Villeneuve d'Ascq, France, in 2006 (with also the Renault Technocentre, Guyancourt, France). In 2006, he joined the XLIM Laboratory, University of Limoges, as a Postdoctoral Fellow, where he has been an Associate Professor since 2009. His current research interests include coupling on cables and electromagnetic compatibility testing, including reverberation chambers and bulk current injection tests.

Introduction

Guillaume Andrieu

I.1 Motivation and goal

Pioneer works about electromagnetic reverberation chambers (RC) have been made at the university of Naples in the 1970s under the direction of Paolo Corona [?] and in parallel in the United States at the National Bureau of Standard (which is right now the National Institute of Standards and Technology (NIST)) [?]. At the origin, the main purpose was to perform electromagnetic compatibility (EMC) measurements, in particular shielding effectiveness of cables and connectors, radiated susceptibility tests and total radiated power of systems or subsystems (a brief history of the work made during the first 20 years (roughly) after the appearance of RCs can be found in [?]).

The last 20 years have shown a spectacular diversification of the RC applications in parallel with a spectacular growth of the number of RCs installed all around the world and of people working with this facility.

Many excellent general books fully or partially devoted to RCs are already available ([?,?,?,?] for instance) where the most fundamental theoretical concepts, the methods, and metrics used to characterize their functioning and their main applications are generally detailed. Therefore, to avoid repetition with these books, the readers will not find generalities about RCs here and whose who are not familiar with the basics of RCs may refer first to a more general book as for instance the ones mentioned before.

Probably partly in reason of the diversification of the applications over the last years, the worldwide research about RCs reaches a new level of maturity. The purpose of this book is to provide, in a compact volume, a thorough presentation by some of the most world renowned RC researchers of a few "recent advances and innovative applications." However, this book is not exhaustive and I offer in advance my apologies to the authors having published recently interesting works! It is worth noting that, even some material in this book is new, much of it is a restatement of results already published (recently of course) in the literature.

I hope that this book can contribute, at its level, to the increase of interest about this fascinating facility and will convince more and more persons, laboratories and companies to work on it!

[1]XLIM Laboratory, UMR7252, University of Limoges, Limoges, France

I.2   Organization

This book is composed of nine different chapters which can be sorted in three categories.

The first three chapters deal with the fundamental topic of characterization (or calibration) of an RC, an important task to be performed before any measurement or test.

Chapter 1 provides a general method (the so-called well-stirred condition method) to characterize the quality of the field generated in an RC. The method, which is fast and accurate, aims at defining the minimum frequency where an RC can be considered "well-stirred." After giving a clear definition of this important concept, the method is illustrated on the stirring technique the most used nowadays, a mechanically stirred RC, in other terms when the enclosure is equipped with a rotating mode stirrer. In addition, MATLAB® programs related to this method are available on the IET website.

Chapter 2 consists in presenting the mode stirring techniques proposed, to date, for RCs, including a critical discussion on each of them. In this chapter, the extension of the well-stirred method is shown on the most common techniques, i.e. all the techniques able to be implemented in the same parallelepipedic enclosure. This gives the possibility to compare the efficiency of each technique while avoiding the bias of using enclosures of different sizes. The most important parameters to play with in order to improve the performance of each stirring technique are, therefore, identified.

Chapter 3 concerns the extension of the well-stirred method to the particular case of vibrating intrinsic reverberation chambers (the so-called VIRCs) where the enclosure is made of a metallized tent mounted in a rigid structure. The movement of the tent as a function of the time, which justifies to treat this kind of RC in a dedicated chapter, leads also to explore the possibility to characterize the VIRC behavior in the time-domain.

In these first three chapters, the effect of the quality factor Q (modified through the insertion of small amount of absorbers) on the RC performance is investigated for each stirring technique. It is shown that this parameter is of great importance and strongly drives the RC behavior.

The next four chapters present recent advances on RC fundamental applications.

Chapter 4 presents a probabilistic model (the corresponding MATLAB programs being also provided on the IET website) intending to determine the influence of the number of fixed independent configurations considered successively during a radiated susceptibility test. This model, applicable for any system under test regardless of its directivity, computes the risk to declare EMC compliant a faulty system. One understands of course that this risk increases if the number of configurations is reduced, a choice generally made in the purpose of reducing the testing time.

Chapter 5, written by Kate A. Remley, Chih-Ming Wang and Robert D. Horansky from NIST (Boulder, United States), presents the latest advances in the field of "over-the-air" testing of wireless devices currently performed in (loaded) RCs. Two main kinds of tests are discussed: how to assess through standardized tests specific metrics associated with device performance such as total radiated power or total isotropic sensitivity and how to use the reflective properties of the RC to replicate specific multipath conditions. On both aspects, metrics and techniques used to assess and to reduce the uncertainty budget of such measurements (which is fundamental for those kind of tests) are discussed.

Chapter 6, written by Philippe Besnier from IETR Laboratory in Rennes (France), deals with energy absorption in RC. Two specific but related applications are discussed. The first application deals with material characterization and particularly the absorbing area cross section of any object provided that a correct estimation of the average composite Q-factor of the RC is known. The second application deals with bioelectromagnetics and particularly the potential use of RC for experimental dosimetry. It is specifically attractive to provide calibrated dosimetry dedicated to animal exposure since they are allowed to move freely in their cage, under electromagnetic stress. Such a test setup makes it possible to assess the risks of exposure to various sources of electromagnetic fields. This recent application of RC appears particularly relevant in the context of the growth of the frequencies of operation of future communication standards (5G and beyond).

Chapter 7, written by Wei Xue and Xiaoming Chen of Xi'an Jiaotong University (China), details the main methods already proposed in the literature in order to assess the efficiency of antennas in RCs, this environment being cost-effective and time efficient with respect to anechoic chamber measurements. A special emphasis is made on frequency domain methods using or not a reference antenna of known efficiency but a time-reversal method is also presented with many details. As in the previous chapter, methods used to characterize the uncertainty of such measurements as a function of the number of uncorrelated samples collected during the measurements are discussed.

Finally, the last two chapters deal with an original use of metallic oversized (with respect to the wavelength) enclosures, stirred or not. Indeed, an increasing interest appears recently (roughly since 2010) to perform free space measurements generally performed in an anechoic chamber within such enclosures. This, which may appear counter-intuitive to perform in such a highly multipath environment, presents some non-negligible advantages, the first being the low cost of a metallic enclosure with respect to an anechoic chamber of same size (in reason of the absence of absorbers).

Chapter 8 presents the recent advances on the measurement of antenna radiation pattern within reverberating enclosures. Different methods are described briefly and a special emphasis is done on the time-gating method supposed to provide the best performance.

Chapter 9, written by Philippe Besnier, describes a recent method proposed for the measurement of radar cross-section of targets within reverberating enclosures. The detection of the target is made possible by observing the modification of the reflection coefficient of a measurement antenna. After a description of the related theory, the method is applied on canonical targets. In particular, radar cross-sectional pattern in the azimuthal plan is shown to demonstrate the feasibility of such approach in an RC.

I.3 Usage

This book is designed to be a textbook for people familiar with RCs, for instance using or studying it on a regular basis, whatever they are academics (researchers, PhD students, etc.) or industrials (practicing engineers, technicians, etc.). The aim is to help them to pursue their academic research and/or their product development using this facility. The required background concerns the basic theoretical concepts underlying the RC behavior. However, in each chapter (all the chapters can be read separately), the specific theoretical background required to ensure a deep understanding of the chapter is given. It can happen in some locations of the book that some theoretical descriptions are repeated, especially when these chapters are written by different contributors. I have considered that this practice improves the readability of each chapter.

Acknowledgments

I would like first to thank my former PhD director Bernard Démoulin, emeritus professor, as well as Lamine Koné, CNRS researcher, from the university of Lille (France), for introducing me, during my PhD (between 2004 and 2006), in the fascinating field of RCs. I had the privilege to use at this time the first RC installed in France (in the 1990s). I could not reach my current stage of career without the foundation they helped me build during this period.

I need also to express my gratitude to Alain Reineix, CNRS researcher at the university of Limoges, who gave me the opportunity, by his constant support, to obtain a permanent position as an associate professor at the XLIM Laboratory.

I am of course grateful to all the researchers who have accepted my proposal to write a chapter in this book (special thanks to Kate, Xiaoming and Philippe!).

I want to thank also all my former coauthors on RC papers, among them the PhD students I have supervised (Aziz Adardour, Ayoub Soltane and right now Guillaume Alberto and Sahand Rasm), the colleagues working with industrial partners as Fabrice Tristant (Dassault Aviation), Charles Jullien (Safran Electrical & Power) (thanks for letting me use your VIRC!), Frédéric Lescoat (Airbus Defense & Space) and Laurent Trougnou (European Space Agency, ESA). I want also to express a special thanks to my main coauthor Nicolas Ticaud, a former postdoc fellow at the XLIM Laboratory and working now at Cisteme company. Through my interactions with each of them, I was able to better polish my understanding of RCs.

I am indebted to my colleagues and students at the university of Limoges. The dynamic, stimulating and friendly environment greatly helped foster this book. My appreciation goes particularly to my colleagues of the XLIM Laboratory, of the Physics and of the ICT Departments at the Science Faculty.

I want to express of course my gratitude to the IET with a special acknowledgment to M. Nicki Dennis, the commissioning editor, who has identified me as the potential editor of a book and helped me to build the proposal. I also thank the anonymous reviewers who have been interested by this proposal. The valuable support of Ms.

Olivia Wilkins from IET and of Ms. Sujatha Subramaniane from MPS Limited during the preparation of the book have been also very much appreciated!

Finally, I am deeply grateful to my family, probably because they had to bear my bad mood when I was working on the book.

References

[1] Corona P, Latmiral G, Paolini E, *et al.* Use of a reverberating enclosure for measurements of radiated power in the microwave range. IEEE Transactions on Electromagnetic Compatibility. 1976;EMC-18(2):54–59.

[2] Crawford ML and Koepke GH. Design, evaluation and use of a reverberation chamber for performing electromagnetic susceptibility/vulnerability measurements. NBS Technical Note; 1986.

[3] Corona P, Ladbury J, and Latmiral G. Reverberation-chamber research-then and now: a review of early work and comparison with current understanding. IEEE Transactions on Electromagnetic Compatibility. 2002;44(1):87–94.

[4] Hill DA. Electromagnetic Fields in Cavities: Deterministic and Statistical Theories. Wiley-IEEE Press, Hoboken, NJ, USA; 2009.

[5] Besnier P and Démoulin B. Electromagnetic Reverberation Chambers. Wiley-ISTE, Paris, France; 2013.

[6] Boyes SJ and Huang Y. Reverberation Chambers: Theory and Applications to EMC and Antenna Measurements. Wiley, Chichester, West Sussex, UK; 2015.

[7] Xu Q and Huang Y. Anechoic and Reverberation Chambers: Theory, Design, and Measurements. Wiley-IEEE Press; 2018.

Chapter 1

Performance of reverberation chambers using the "well-stirred condition method"

Guillaume Andrieu[1]

Before performing any test or measurement in an RC, it is of fundamental importance to validate its performance from the point of view of the quality of the electromagnetic (EM) field generated inside. It is in particular the occasion to assess the efficiency of the stirring process used.

This is a difficult task to handle with the reference IEC 61000-4-21 standard [1] examining the field uniformity obtained in a volume defined as the "working volume," this standard procedure being more a pass/fail method in order to validate or not an RC. Indeed, it is difficult to perform parametric studies with this procedure, for instance on the effect of the average composite quality factor Q of the facility after inserting different amount of absorbers, in reason of the measurement time required by the method (more or less 1 or 2 working days for each configuration). In addition to this drawback, it is commonly assumed in the literature that this procedure suffers from other disadvantages; among them are a relative lack of sensitivity of the proposed metrics as well as the fact that some threshold values are empirical (or in other terms not justified by any theoretical development).

In this context, a large diversity of criteria (many of them being presented in [2]) has been proposed in the literature to reach the objective of characterizing the quality of the field generated in an RC. We can cite, for instance, metrics related to

- the field inhomogeneity or anisotropy [1,3];
- the EM field distributions [4];
- the sample correlation [1,5];
- the relative modification of scattering parameters (stirring ratio, K-factor [6,7] or enhanced backscatter coefficient [8–10] for instance);
- the time-domain behavior of the RC (power-delay profile [11] or coherence bandwidth [12] for instance).

All these criteria are of course more or less powerful as discussed extensively in the literature. Moreover, the threshold values (the value used to decide if the corresponding criteria is satisfied or not) of some indicators is sometimes selected empirically based on the experience of the RC users (for instance, the 20 dB criteria for the stirring ratio).

[1]XLIM Laboratory, UMR7252, University of Limoges, Limoges, France

In this context, the first chapter of this book presents with many details a complete method recently proposed [13] and called the "well-stirred condition method". This method aims at defining from which frequency (called f_{wsc}) the RC can be considered "well stirred." This condition implying isotropy and uniformity of the EM field is in many papers the assumption underlying any theoretical development. As far as the author knowledge, no unquestionable definition of the well-stirred condition was available in the literature before [13]. Indeed, the well-stirred condition (also often referred to as the overmoded condition [14]) was generally focused only on the EM field distributions [15–18] resulting in an incomplete definition.

The aim of the method is to propose to the community of RC users a fast, accurate and easy to implement experimental method based on a clear definition of the well-stirred condition of an RC. In the case of a mechanically stirred RC, the method requires the S_{11} scattering parameter of an antenna located in a fixed position to be measured for a given number N of stirring conditions (or samples), a value of $N = 50$ samples being considered in this chapter. This simple measurement protocol, requiring a few minutes of measurements, leads to determine f_{wsc} accurately. These performances are achieved in reason of the thousands of frequencies (even correlated) available from vector network analyzer (VNA) measurements. It is worth noting here that fast electric field probes able to separate stirred and unstirred contributions on thousands of frequencies per second would also be usable in the method (with small easy adjustments), even if these expensive probes are not often present in an EMC laboratory.

The chapter is organized as follows: the proposed method is described in Section 1.1 while the method performances are shown from experimental results in Section 1.2. Section 1.3 presents two parametric analysis defining the optimal loading of an RC and investigating the effect of the mode stirrer geometry. Finally, the chapter ends in Section 1.4 with a discussion on the use of the method as an alternative RC calibration procedure.

1.1 The well-stirred condition method

1.1.1 Definition of the well-stirred condition of an RC

The proposed method requires a clear and unquestionable definition of the well-stirred condition, which is given in this subsection.

On the one hand, the EM field distributions in such regime are known for years thanks to pioneering works published some decades ago [15–18]. They are summarized in Table 1.1. Under these ideal conditions, the EM field in any point can be described as the summation of plane waves of random angles of arrival, polarization and amplitude [18]. This isotropic diffusion regime results in a homogeneous spatial distribution of energy. Therefore, these ideal distributions can be verified from measurements made in N different points sufficiently spaced [19] (typically $\lambda/2$, with λ being the wavelength) in the RC for a given position of the mode stirrer. The same distributions are also obtained in any point of the RC (located sufficiently far from any

Table 1.1   *Ideal distributions in a well-stirred RC (E_r is a rectangular component of the electric field, E_t the total electric field and P the induced power on an antenna)*

Variable	Distribution
$Re(E_r)$, $Im(E_r)$	Normal (or Gaussian) distribution
$\|E_r\|$	χ with 2 degrees of freedom (or Rayleigh)
$\|E_t\|$	χ with 6 degrees of freedom
P	χ^2 with 6 degrees of freedom (or exponential)

object [20]) if collected over N different stirring conditions, this latter verification being easier to implement from a measurement point of view.

On the other hand, when the well-stirred condition is achieved, the EM environment generated in the RC has to lead to independent realizations over the N measured samples. In other terms, each sample has to be uncorrelated with respect to the other samples. This fundamental point means that each collected sample brings an entirely new information about the RC behavior. In the case of correlated samples, the new measured samples give a partially new information that traduces a waste of time and a potential bias during the analysis of the results. One understands that this situation has to be avoided as much as possible during tests and measurements made in an RC.

1.1.2   Principle of the method

According to both aspects of the well-stirred definition given in Section 1.1.1, the determination of f_{wsc} requires the simultaneous analysis of the empirical EM field distribution (with respect to the ideal ones) and of the sample correlation. Each criteria leads to a specific frequency (i.e. f_1 and f_2 as defined hereafter) confirming that the criteria is matched above. These frequencies are defined after the comparison of the results with a critical value theoretically justified. It is understood here that f_{wsc} is determined as a function of the number N of stirring configurations considered during the measurements.

In reason of the thousands of frequencies available using S_{11} measurements, it is assumed that these measurements can be performed at only one position of the antenna, which allows the measurement time to be reasonable (i.e. a few minutes). As proposed for the first time in [21], this high number of frequency samples compensates the fact that the measurements are not collected in different positions of the RC.

The EM field properties are proven in [20] to be similar for any point of an RC located sufficiently far from any object. As a consequence, the position and the orientation of the antenna is free in the method provided that a minimum distance greater than $\lambda/2$ is respected between the antenna and any object such as the RC walls, the mode stirrer or the equipment under test (EUT) for instance (if applicable).

1.1.3   Required experimental setup (for a mechanically stirred RC)

In the case of a mechanically stirred RC, which is the most popular stirring technique, the typical experimental setup required by the method is depicted in Figure 1.1. As shown in the figure, the method consists in measuring the S_{11} scattering parameter of a broadband antenna located in the RC for N different mode stirrer positions.

Two measurement guidelines are required concerning the antenna:

- As mentioned in Section 1.1.2, the antenna has to be located sufficiently far from any object (RC walls or the mode stirrer for instance), typically $\lambda/2$.
- The antenna has to be sufficiently matched over the whole frequency range of interest, typically by respecting the condition $|S_{11}| < -10$ dB.

1.1.4   On S_{11} measurements in RC

In an RC, the S_{11} parameter of an antenna can be decomposed as follows for any collected sample i

$$S_{11,i} = S_{11}^{fs} + S_{11}^{uns} + S_{11,i}^{sti}, \tag{1.1}$$

where

- S_{11}^{fs} is the free space reflection coefficient of the antenna (measured for instance in anechoic chamber);
- S_{11}^{uns} corresponds to the contribution of the unstirred paths, i.e. paths without any interaction with the stirrer paddles;
- $S_{11}^{sti,i}$ represents the contribution of the stirred paths, i.e. paths interacting with the stirrer paddles, for a given mode stirrer position i.

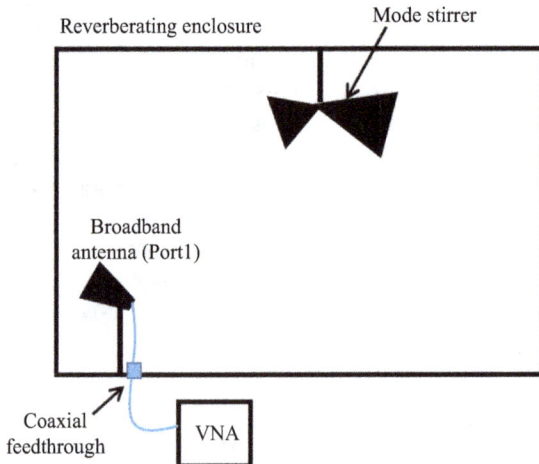

Figure 1.1   Schematic description of the required experimental setup (for a mechanically stirred RC)

$S_{11}^{\text{sti},i}$ is easily computed for each sample i by removing the average value of S_{11}:

$$S_{11,i}^{\text{sti}} = S_{11,i} - S_{11}^{\text{fs}} - S_{11}^{\text{uns}} = S_{11,i} - \langle S_{11} \rangle, \tag{1.2}$$

where the operator $\langle . \rangle$ denotes averaging on all the stirring conditions at a particular frequency. S_{11}^{uns} is also estimated easily from (1.1) and (1.2):

$$S_{11}^{\text{uns}} = \langle S_{11} \rangle - S_{11}^{\text{fs}}. \tag{1.3}$$

Figure 1.2 presents the magnitude of S_{11} measured in the unloaded RC we have used (see Section 1.2.1 for a complete description of the experimental setup) for each mode stirrer position, as well as the magnitude of this parameter measured in free space and the magnitude of the unstirred contribution, i.e. S_{11}^{uns}. We can see that the amplitude of the stirred contribution is clearly in average greater than the amplitude of the unstirred contribution, which makes us confident in the efficiency of our mode stirrer.

The method focuses on the analysis of the stirred part of S_{11}. This aims at removing the contribution of the unstirred paths and of the free space reflection coefficient of the antenna. The idea is not to artificially improve the RC performance but to assess

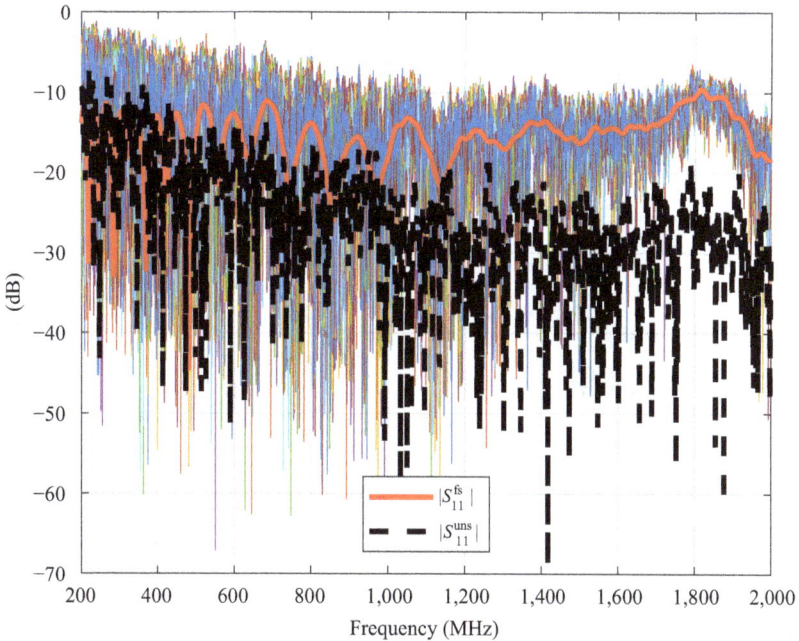

Figure 1.2   Magnitude of S_{11} for $N = 50$ mode stirrer positions (all the curves without legend), S_{11}^{uns} and S_{11}^{fs} in the unloaded RC

the ability of the RC to create stirred contributions having properties matching with the ones of a well-stirred RC. Therefore, the method has the great advantage to be independent of the antenna position and orientation as discussed further down. The magnitude of S_{11}^{sti} obtained for each mode stirrer position is plotted in Figure 1.3.

The method is based on the fact that all the information required to characterize the RC behavior is contained in all these curves showing a large dynamic of around 50 dB.

One can ask why the method uses the S_{11} parameter instead of S_{21}, traditionally more used in RC studies. In reality, S_{11} (and subsequently S_{11}^{sti}) allows the RC to be characterized exactly in the same conditions than a subsequent test (see Section 1.2.5 for the application of the method in presence of an EUT). Indeed, as an S_{21} measurement implies to insert a second antenna (not present later during the test, if we consider for instance a radiated susceptibility test), the insertion of this additional antenna has the effect to modify the average composite quality factor Q particularly in the lowest frequency range of the RC and, therefore, the overall RC behavior.

From the stirred contributions point of view, the theoretical background behind the definition of the enhanced backscatter coefficient [8–10] validates that the statistics of S_{11} and S_{21} are similar in a well-stirred RC and in particular that $|S_{11}^{sti}|$ or $|S_{21}^{sti}|$ both follow the Rayleigh distribution [6,22], similarly to the magnitude of a rectangular component of the electric field.

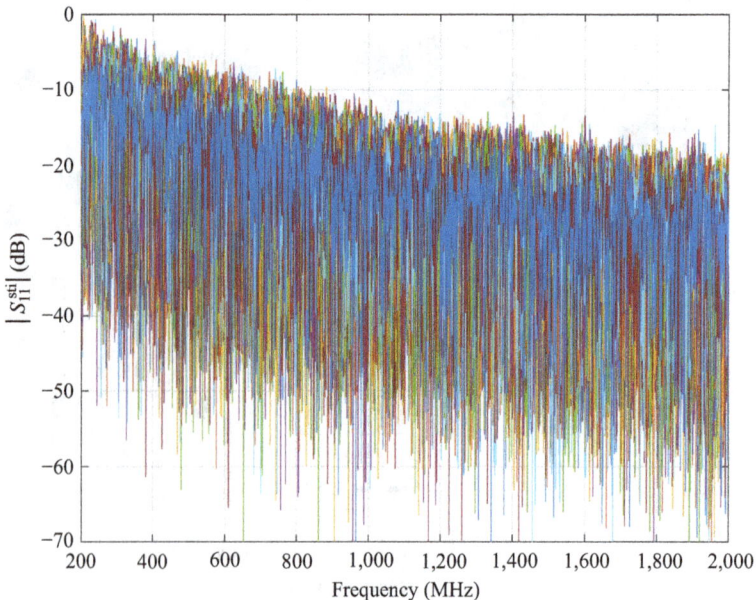

Figure 1.3   Magnitude of S_{11}^{sti} for $N = 50$ mode stirrer positions in the unloaded XLIM RC

1.1.5 Analysis of the EM field distributions

The first task of the method consists in assessing the EM field distributions obtained in the RC in order to define a first frequency f_1. Above this frequency, the EM field distributions are assumed to match the ideal distributions obtained in a well-stirred RC. This is made with the Anderson–Darling (AD) goodness-of-fit (GOF) test.

1.1.5.1   Principle of a goodness-of-fit test

GOF tests are used in statistics in order to test if a set of samples follows a given distribution. In RCs, GOF tests have been introduced in order to compare the empirical EM field distributions (i.e. computed using the measured values) to the ideal ones.

The first step of a GOF test consists in claiming a hypothesis H_0: "these N samples are from this distribution." It should be mentioned here that, rigorously, mathematicians prefer to use the terminology "a sample of size N." For the sake of clarity, this is not the choice we have made in this book.

In the second step, the empirical and the theoretical (i.e. using the theoretical formula of the distribution) cumulative density functions (CDFs) have to be calculated. The theoretical CDF of the Rayleigh distribution writes as follows:

$$f(x) = \frac{2x}{\theta} \cdot e^{-(x^2/\theta)}. \tag{1.4}$$

The computation of the theoretical CDF requires the parameter θ to be computed. Using the maximum likelihood method, θ in the case of the Rayleigh distribution corresponds to the quadratic mean of the samples [4]:

$$\theta = \frac{1}{N} \cdot \sum_{i=1}^{N} x_i^2. \tag{1.5}$$

The third step consists in computing a parameter specific to the test and giving the information about the difference between both CDFs previously computed. An example of theoretical and empirical CDF is shown in Figure 1.4.

This parameter is then compared in a fourth step to a critical value depending of the chosen distribution, of the number N of samples and of the level of significance α. When the test statistic is lower than the critical value, the hypothesis is accepted. Otherwise, the hypothesis is rejected.

The level of significance α of a GOF test equals the probability that the test claims that the samples are not from the tested distribution when the samples follow in reality this distribution. In common practice, the level of significance $\alpha = 5\%$ is generally considered.

1.1.5.2   The Anderson–Darling goodness-of-fit test

The specific parameter related to the AD GOF test, i.e. A^2, is computed from the following relationship

$$A^2 = -\frac{\sum_{i=1}^{N} (2i-1) \left[\ln F(x_i) + \ln (1 - F(x_{N+1-i})) \right]}{N} - N, \tag{1.6}$$

where $F_0(x_i)$ is the theoretical CDF and x_i corresponds to the experimental value of rank i sorted in ascendant order. To take into account the number N of samples, a modified statistic A_m^2 is computed as a function of A^2 and N

$$A_m^2 = A^2 \left(1 + \frac{0.6}{N}\right). \tag{1.7}$$

A_m^2 is then compared to the appropriate critical value c. In [4], it is proved that the most suitable critical values have been given by Stephens [23]. Table 1.2 provides these critical values in the case of the Rayleigh distribution for different levels of significance α.

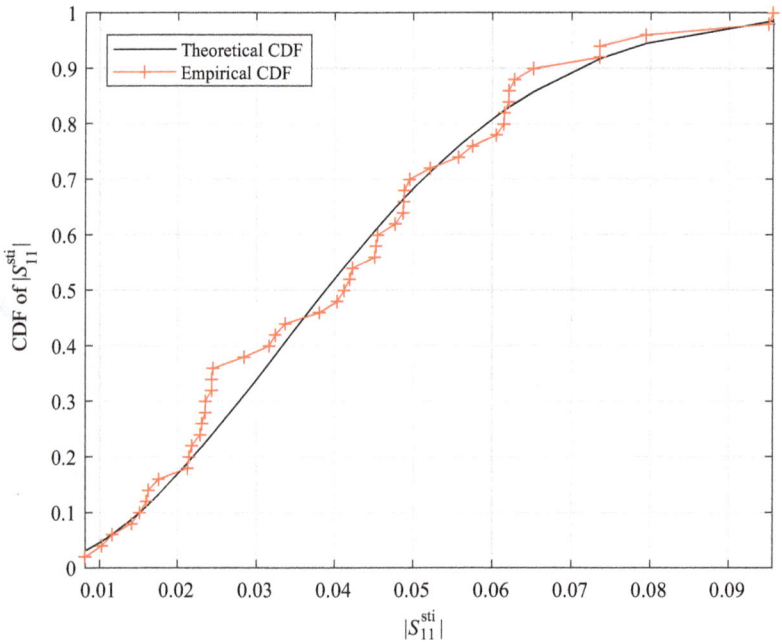

Figure 1.4   Example of theoretical and empirical CDFs required by a GOF test (obtained in our RC from N = 50 samples of S_{11}^{sti})

Table 1.2   Suitable critical values of the AD GOF test for the Rayleigh distribution (from [23])

Level of significance α (%)	15	10	5	1
Critical value	0.922	1.078	1.341	1.957

In [4], the AD GOF test is shown to be the most relevant GOF test to assess the EM field distributions in an RC even other tests have already been used in the literature. Indeed, the popular chi-square (i.e. χ^2) test is not well adapted to the case of continuous distributions, which is the case of RC measurements. The Kolmogorov–Smirnov GOF test is also used. However, the KS statistic, being the absolute maximum deviation between both CDFs, is less sensitive to the extreme values of the distribution than the AD GOF test, the AD statistic being more a "cumulated" error as shown by (1.6).

1.1.5.3   Determination of a first frequency f_1

In the method, the AD GOF test is performed at each frequency of interest on the N S_{11}^{sti} measured samples in order to analyze if their magnitudes, i.e. $|S_{11}^{\text{sti}}|$, are from the Rayleigh distribution. In the case of the Rayleigh distribution and a level of significance α of 5%, c equals 1.341 for $N = 50$ samples (the number of samples considered in this chapter) according to Table 1.2.

As an illustrative example, A_m^2 is plotted on Figure 1.5 according to the frequency for our unloaded RC (when no additional loading is inserted in, the complete experimental setup being described later in Section 1.2.1).

In order to focus the analysis on the general trend of A_m^2 with the frequency, a polynomial curve fit (PCF) is computed (with functions provided by MATLAB® or GNU Octave software packages) as also shown in the figure. A sixth-order PCF is used in all the results shown in this chapter unless something else is explicitly stated.

The particular frequency f_1 is defined as the frequency when the fitted data A_m^2 equals the threshold value t_1 which is the critical value of the AD GOF test c itself. Consequently, this first frequency f_1 is derived rigorously with a threshold value theoretically justified.

In [4], it has been shown that below the frequency range where the ideal distributions are matched, the agreement of the empirical measurements with the Weibull distribution is better. In particular, the parameters of the Weibull distribution tends toward the ones related to the Rayleigh distribution when frequency grows, i.e. in the "transition" region.

It is important to emphasize that the Rayleigh distribution is a particular case of the Weibull distribution. However, it is not relevant to test the Weibull distributions in the well-stirred condition method for two reasons:

- The aim of this part of the method is to determine a specific frequency f_1 but not to observe a transition.
- The Rayleigh distribution has a clear physical meaning which corresponds to the sum of a large number of plane waves of any incidence and polarization. The physical meaning of the Weibull distribution is unclear.

Therefore, it is preferred to consider the ideal case and try to see when the results tends satisfyingly toward it.

1.1.6   Analysis of the sample correlation

The second task of the method consists in assessing the sample correlation obtained in the RC in order to define a second frequency f_2. Above this frequency, the collected samples are assumed to be independent (or uncorrelated). This is made with the well-known first-order autocorrelation coefficient.

1.1.6.1 The first-order autocorrelation coefficient

The autocorrelation coefficient $r(n)$ is used for years in the field of RCs to quantify the sample correlation [1]. The correlation of a set of data x is computed with the same set of data shifted of n ranks, n being the order of the autocorrelation coefficient. This coefficient is included in the range between -1 and 1, 0 meaning that the data are uncorrelated, and an absolute value of 1 meaning that the data are perfectly correlated.

The first-order autocorrelation coefficient $r(1)$ is therefore used to assess the sample correlation of two successive mode stirrer positions. In the case of a set of N data, $r(1)$ is calculated as follows for a set of data x

$$r(1) = \frac{\sum_{k=1}^{N} (x(k) - \langle x \rangle)(x(k+1) - \langle x \rangle)}{\sum_{k=1}^{N} (x(k) - \langle x \rangle)^2}. \tag{1.8}$$

An example of the autocorrelation coefficient calculated as a function of the order n is plotted in Figure 1.6 from $N = 50$ samples. It should first be noticed that the plot is symmetric. This is logical as, for instance $r(n)$ and $r(N - n)$, corresponds

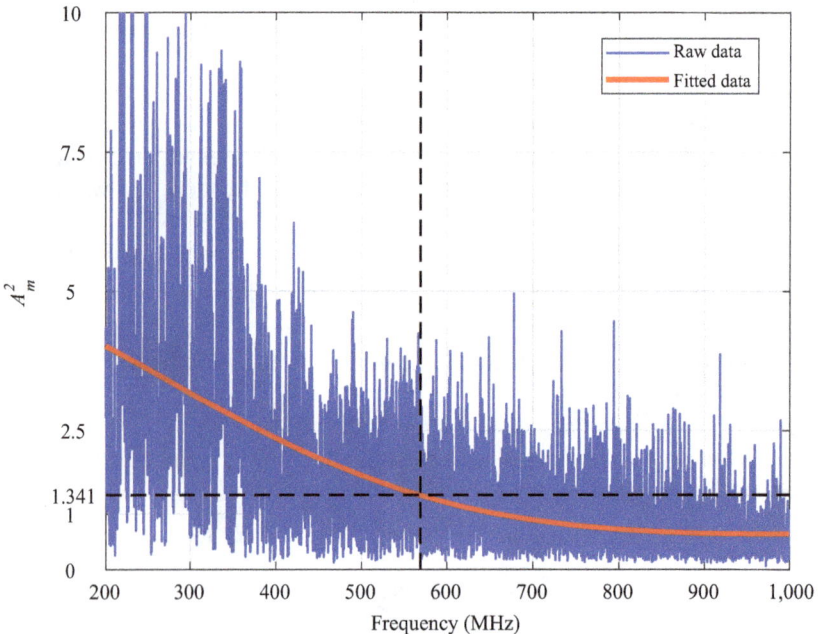

Figure 1.5   Principle of determination of f_1 using a sixth-order PCF. Results obtained in the unloaded RC. In this example, $f_1 = 569$ MHz

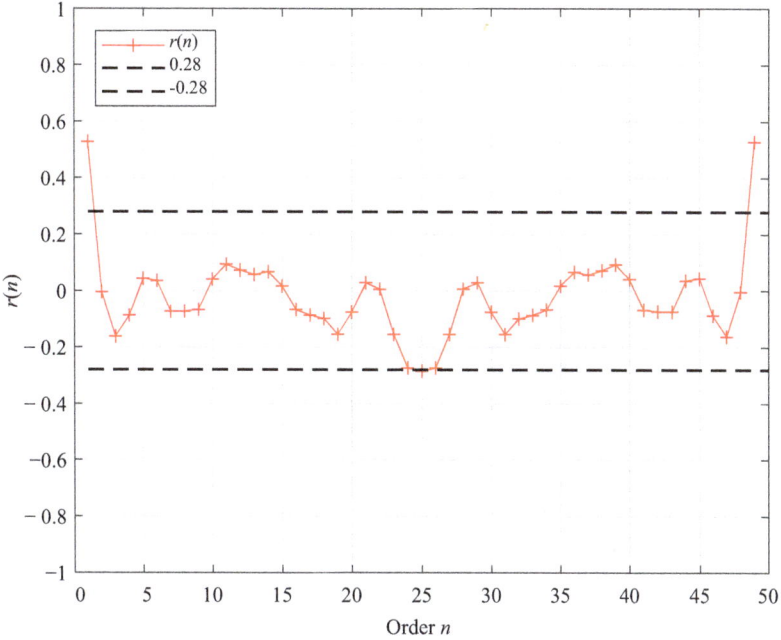

Figure 1.6   Example of an autocorrelation coefficient r(n) obtained at different orders n (obtained from N = 50 samples of S_{11}^{sti})

to the same shift. In this illustrative example, only $r(1)$ is greater than the threshold value defined later in Section 1.1.6.2.

It is also straightforward to convert the order of the autocorrelation coefficient in the angle of rotation of the mode stirrer. As an example, for $N = 50$, $r(1)$ corresponds to the rotation of the mode stirrer of 7.2° (i.e. 360/50). In this figure, it can, therefore, be concluded that a rotation of the mode stirrer of 7.2° is insufficient to have uncorrelated data unlike a rotation of 14.4°. Therefore, 25 independent samples are at least available over one entire mode stirrer rotation in this simple example.

1.1.6.2   Determination of a second frequency f_2

In the method, the first-order autocorrelation coefficient $r(1)$ is calculated at each frequency of interest from the magnitude of the S_{11}^{sti} samples. As previously done for the determination of f_1, a frequency-dependent PCF of $r(1)$ is computed in order to determine accurately a second frequency f_2 when the interpolation of $r(1)$ equals a threshold value t_2. An example of result obtained for the unloaded RC is shown in Figure 1.7.

In the field of RCs, data are often considered as uncorrelated when $r(1)$ is lower than $1/e$, i.e. 0.37. However, it has been demonstrated in the literature that this threshold value neglecting the number N of samples of the set of data is not adapted to the

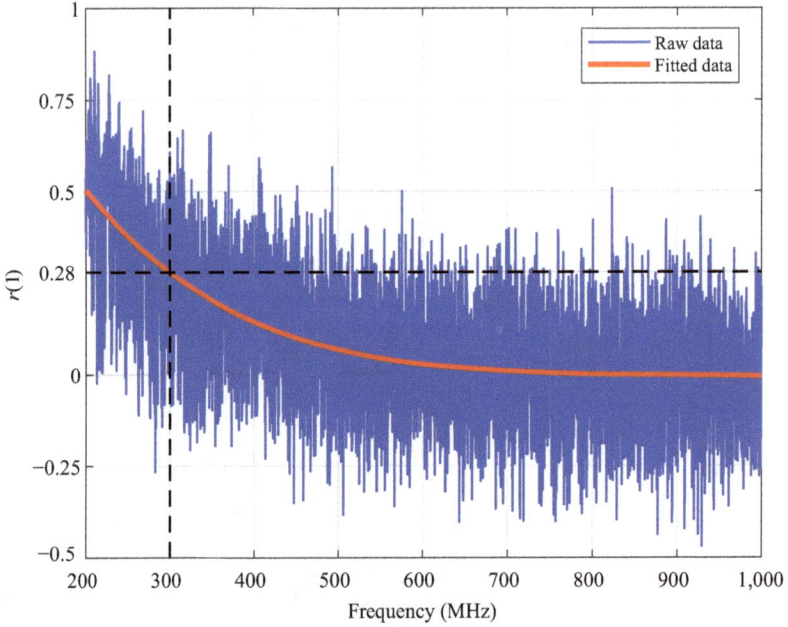

Figure 1.7   Principle of determination of f_2. Results obtained in the unloaded RC. In this example, $f_2 = 300$ MHz

problem [24]. Indeed, the autocorrelation function is a random variable having a CDF $\psi(r)$ depending of the number of samples N

$$\psi(r) = \frac{N-2}{\sqrt{2\pi}} \cdot \frac{\Gamma(N-1)}{\Gamma(N-1/2)} \cdot \frac{\left(1-\rho^2\right)^{(N-1)/2} \cdot \left(1-r^2\right)^{(N-4)/2}}{(1-\rho r)^{N-(3/2)}} \cdot A(r),$$

with

$$A(r) = 1 + \frac{1+\rho r}{4(2N-1)} + \cdots, \tag{1.9}$$

where r is the autocorrelation coefficient calculated from a set of N samples at a given order, ρ the real autocorrelation coefficient which would be obtained for an infinite number of samples and Γ the mathematical function. It is worth noting that this probability density function (PDF) is independent of the underlying distribution of samples.

The threshold value t_2 considered in the well-stirred condition method is determined with the method given in [5] and summarized as follows. As mentioned before, $r(1)$ is a random variable. The method considers the PDF of $r(1)$ in the case of uncorrelated data (i.e. $r(1) = 0$ for an infinite value of N). Then, for a given N and a given level of confidence γ (in %), the method computes the interval containing γ% of the computed $r(1)$ values. Therefore, a Monte Carlo approach yields easily to the threshold value corresponding to any (N, γ) combination. As t_1, t_2 is, therefore, also a theoretically justified threshold value.

Following this approach, t_2 equals 0.28 for $N = 50$ samples and $\gamma = 95\%$ (the level of confidence generally considered). In other terms, $r(1)$ is lower than 0.28 for 95% of the sets of $N = 50$ uncorrelated samples. Other values of t_2 according to N are given in Table 1.3 and in Figure 1.8 according to the number N of samples and of the level of confidence γ.

One understands here that f_2 (and by extension f_{wsc}) is related to the number of samples N considered, a larger N leading to a greater value of f_2. Indeed, in the case of a rotating mode stirrer, the minimum frequency when 360 positions of the

Table 1.3 t_2 according to N from [5] for a level of confidence γ of 95%

N	10	20	30	40	50	75	100
t_2	0.64	0.45	0.37	0.32	0.28	0.23	0.2

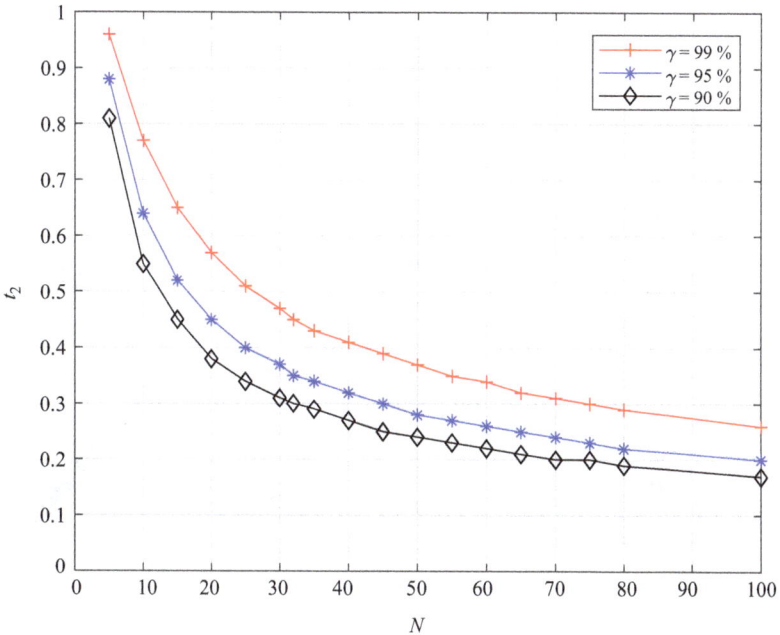

Figure 1.8   Values of t_2 as a function of N and of the level of confidence γ (from [5])

mode stirrer (1° step) are independent is necessarily higher with respect to another configuration considering only 60 positions (6° step) of the same stirrer.

1.1.7   Determination of f_{wsc}

In order to be sure that both criteria are matched above, the frequency f_{wsc} which constitutes the main output of the method is the maximum value between both frequencies f_1 and f_2. For instance, in the results shown in Figures 1.5 and 1.7, f_{wsc} equals 569 MHz.

1.2   Detailed analysis of the method performance

1.2.1   Description of the measurements

The RC considered in all the experiments described in this chapter is the RC of the XLIM laboratory (3.75 m length, 2.45 m width, 2.46 m height, volume $V \approx 22.6\ \text{m}^3$, theoretical fundamental resonance $f_0 \approx 74$ MHz). The experimental conditions described hereafter are the general measurement conditions, unless something else is explicitly stated (which is the case in some specific subsections of this chapter).

The mode stirrer is made of eight rectangular metallic blades of $60 \times 40\ \text{cm}^2$ dimensions rotating around a metallic mast as shown in Figure 1.9.

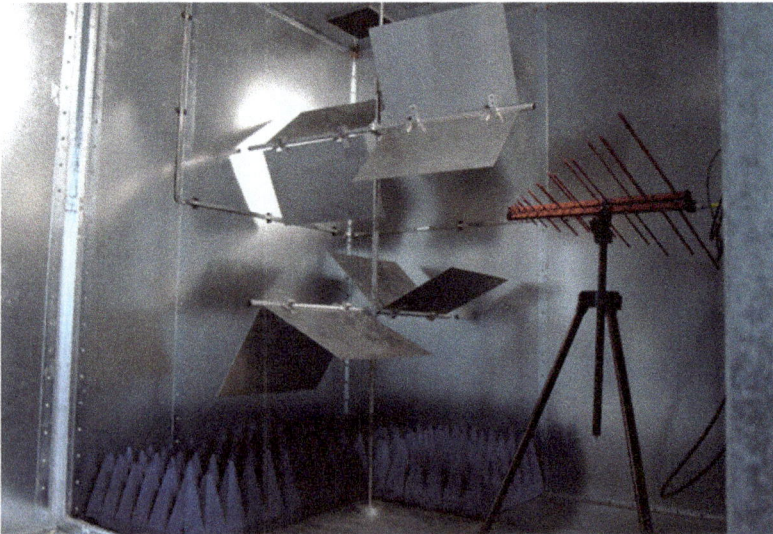

Figure 1.9   Picture of the XLIM RC showing the eight-blade rotating mode stirrer, four blocks of absorbers located below and the log-periodic antenna

The S_{11} parameter of the emitting broadband log-periodic antenna is measured with a VNA for $N = 50$ positions of the mode stirrer (rotation of 7.2° between two successive positions) over 18,001 frequencies linearly spaced between 0.2 and 2 GHz ($\delta f = 100$ kHz). In those measurement conditions, the total measurement time equals 10 min. For each position of the mode stirrer, the total sweep time of the VNA equals 2.4 s for a resolution bandwidth of 10 kHz. Consequently, even after considering the time required by the VNA to switch from a frequency to the following one, the total time allowed for each frequency (133 µs) is sufficient to reach a steady state and then obtain reliable measurements. Indeed, this time is greater than the root-mean-square (RMS) delay spread generally observed in RC [11,12], typically a few microseconds.

In addition to the configuration when the RC is unloaded, three other different loading conditions (with absorbers inserted below the mode stirrer) are used. A block of 9 pyramidal absorbers (PAs), one block and four blocks of absorber are also considered, each block containing 36 PAs. Each pyramid has a squared ground surface of 100 cm² and a height of 30 cm. Table 1.4 gives the total volume of the absorbers for each loading configuration as well as the ratio (in %) between this volume and the total volume of the RC. It allows this small ratio to be emphasized, the largest ratio being of only 0.53%.

Finally, to quantify the sensitivity of the different outputs of the method (i.e. f_1, f_2 and f_{wsc}) in the case of a parametric analysis (for instance related to different positions of the antenna in the RC), a deviation parameter d is introduced in order to assess the maximum deviation (in %) of each metric m_i (with i being the configuration number) with respect to the average value $\langle m \rangle$ obtained for all the different configurations

$$d = 100 \left(\frac{|m_i - \langle m \rangle|_{\max}}{\langle m \rangle} \right). \tag{1.10}$$

1.2.2 Sensitivity of the method to the antenna position and orientation

The well-stirred condition method is claimed to be independent of the position and of the orientation of the antenna used provided that the antenna impedance mismatch is reasonable, i.e. $|S_{11}| < -10$ dB. This is investigated in this subsection.

To validate the inherent uncertainty related to the antenna position and orientation within the RC, four different positions (with also different orientations) of the antenna are considered for each loading configuration, each position being separated from the

Table 1.4   *Ratio between the volume of the absorbers and the RC volume for the considered loading conditions*

	9 PA	One block	Four blocks
V_{abs} (m³)	0.009	0.03	0.12
V_{abs}/V_{RC} (%)	0.04	0.13	0.53

others with a minimum distance of $\lambda/2$ [19] in order to make sure that each position of the antenna is independent.

The obtained results for the 9 PA loading configuration are presented in Figures 1.10 and 1.11, while Table 1.5 collects the results obtained for each loading configuration. As expected, low deviations of f_1, f_2 and f_{wsc} are obtained when the antenna is moved in the RC for a given loading condition. The maximum deviation d obtained on f_1 is greater with respect to f_2 (respectively 7.6% and 0.7% for a 9 PA loading). This is due to the higher sensitivity of the AD GOF test. The maximum deviation d related to f_{wsc} is satisfying (lower than 4%). This deviation of the results is attributed to the inherent uncertainties of any RC measurement due to the finite number N of samples considered for each antenna position.

These results confirm the relevance of the choice to consider only one position of the antenna, the information obtained for each position and orientation of the antenna being similar. Therefore, f_{wsc} is assumed to be only driven by the ability of the RC to create stirred contributions matching with the ideal properties of an RC, independently of the antenna position and orientation in the RC.

When analyzing the raw data shown in Figures 1.5 and 1.7, it is clear that both metrics of interest in the method (i.e. A_m^2 and $r(1)$) are not lower than the corresponding threshold at all the frequencies. This is due to the level of significance $\alpha = 5\%$ for the AD GOF test and to the level of confidence $\gamma = 95\%$ for $r(1)$.

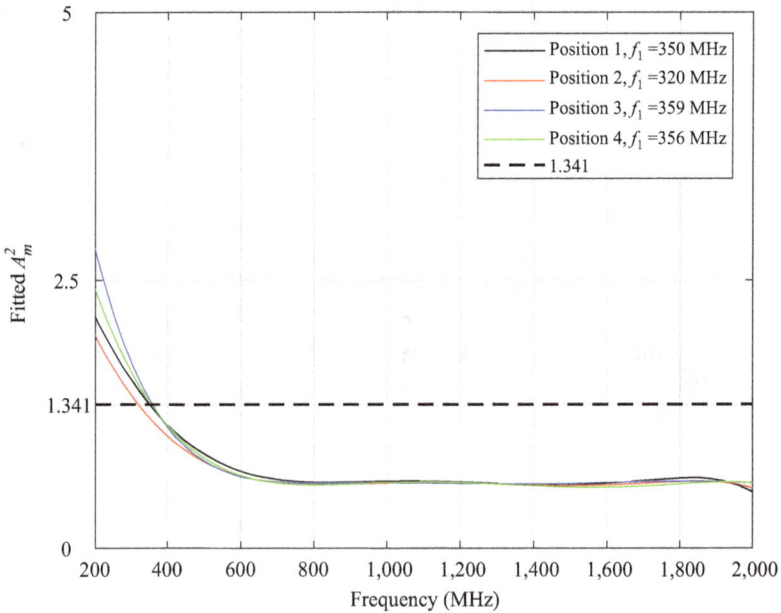

Figure 1.10   *Fitted modified AD statistic A_m^2 for the RC loaded with 9 PA obtained for four different positions (and orientations) of the log-periodic antenna*

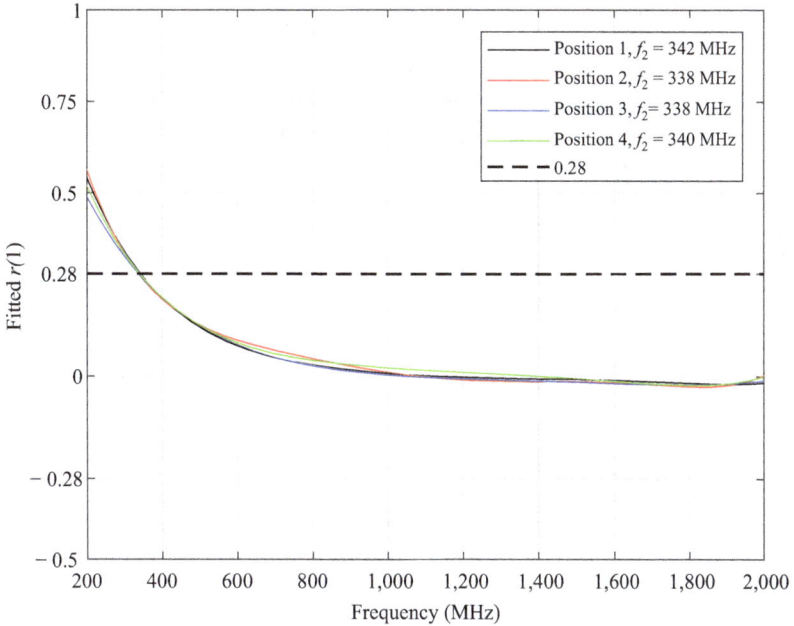

Figure 1.11   Fitted first-order autocorrelation coefficient r(1) for the RC loaded
             with 9 PA obtained for four different positions (and orientations) of
             the log-periodic antenna

Table 1.5 f_1, f_2, f_{wsc} and the deviation parameter d obtained according to the
           log-periodic antenna position (and orientation) for different loading
           conditions

		Unloaded	9 PA	One block	Four blocks
f_1 (MHz)	Pos. 1	569	350	267	<200
	Pos. 2	557	320	262	<200
	Pos. 3	553	358	301	<200
	Pos. 4	559	356	290	<200
	d (%)	1.7	7.6	7.4	N/A
f_2 (MHz)	Pos. 1	300	342	402	679
	Pos. 2	267	338	384	669
	Pos. 3	289	338	406	649
	Pos. 4	299	340	396	656
	d (%)	7.5	0.7	3.2	2.3
f_{wsc} (MHz)	Pos. 1	569	350	402	679
	Pos. 2	557	338	384	669
	Pos. 3	553	358	406	649
	Pos. 4	559	356	396	656
	d (%)	1.7	3.6	3.2	2.3

"<200" means that f_1 is lower than 200 MHz; "N/A" means not applicable.

Table 1.6 contains the percentage of frequencies when the conditions $A_m^2 > 1.341$ and $|r(1)| > 0.28$ are observed for each loading configuration between 1.5 and 2 GHz, i.e. well above f_{wsc} so when the well-stirred condition is reached without any doubt. As expected, this percentage tends toward 5% and shows that the rejection of these criteria is inherent to the metrics used and not to bad performance of the RC at some particular frequencies.

To illustrate the influence of the loading condition on the RC behavior, Figures 1.12–1.13 present a comparison of the results obtained for each loading configuration.

Table 1.6   *Frequency of appearance [in %] of the mentioned condition from results obtained for the four loading configurations and averaged over four positions of the antenna (between 1.5 and 2 GHz)*

	Unloaded	9 PA	One block	Four blocks
$A_m^2 > 1.341$	4.9	5.1	4.7	4.6
$|r(1)| > 0.28$	4.5	4.4	4.1	5

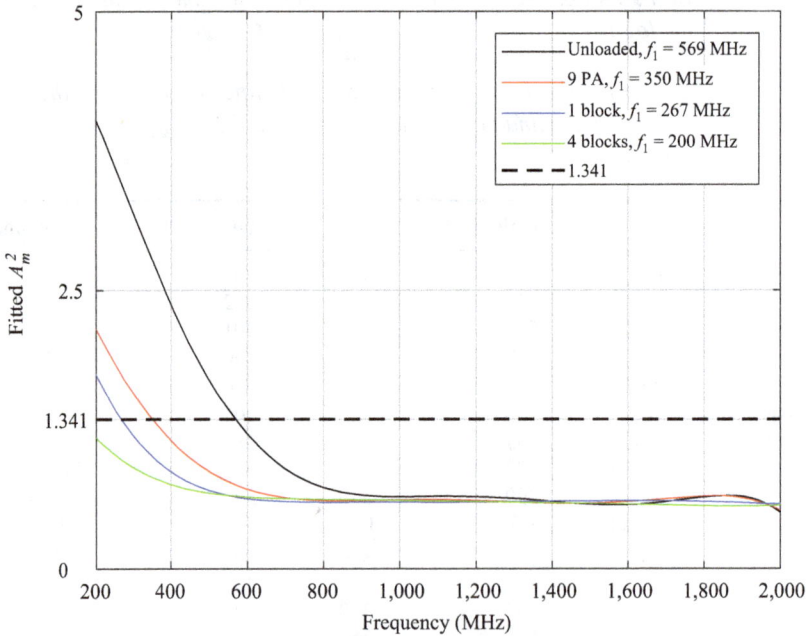

Figure 1.12   *Fitted modified AD statistic A_m^2 in the RC as a function of the amount of loading*

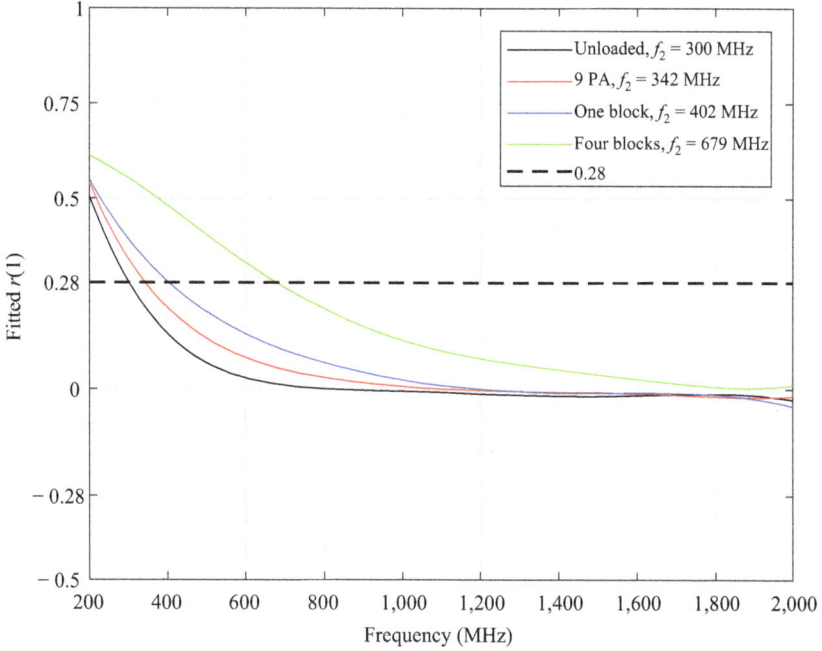

Figure 1.13   Fitted first-order autocorrelation coefficient r(1) in the RC as a function of the amount of loading

The decrease of Q obtained when absorbers are inserted in the RC has an opposite effect on both metrics. Indeed, a better agreement of the EM field distributions with the ideal ones are observed as well as a higher sample correlation. Indeed, the fact to insert absorbers in a mechanically stirred RC [25] increases the modal overlapping, each mode being "significant" [26] on a larger bandwidth.

This has a double effect:

- Improvement of the EM field distributions as more modes having a significant weight are excited at each frequency. When the Q-factor is too high and each mode is narrowband, the total field distributions at a given frequency has more chance to be dominated by a few numbers of modes which leads to significantly rejects the ideal distributions.
- Increase of the sample correlation as each mode has a significant effect on a larger bandwidth.

In Section 1.3.1, a parametric analysis is performed in order to determine the "optimal loading" to insert in the RC leading to the lowest value of f_{wsc}.

The definition of the optimal loading of an RC makes sense as

- an RC having an extremely large Q-factor is inadequate to work correctly because of the insufficient modal overlap;

- an RC of extremely low Q-factor makes impossible to generate a high EM field strength, one fundamental advantage of the RCs.

1.2.3   Sensitivity of the method to the antenna type

The effect of the type of antenna used is shown in this subsection. In addition to the log-periodic antenna, two other antennas have been used (one position being considered for each additional antenna)

- a "biconilog" antenna matched on the frequency range from 80 MHz to 6 GHz;
- a monopole antenna of 15 cm long lying on a-70 cm squared metallic ground plane.

Figure 1.14 presenting the magnitude of the free space reflection coefficient, i.e. $|S_{11}^{fs}|$, of each antenna on the considered frequency range highlights clearly the strong impedance mismatch of the monopole on most of the frequency range.

On the one hand, the effect of this impedance mismatch is shown in Figure 1.15 to be important on the modified AD statistic. Indeed, the highest value of $|S_{11}^{fs}|$ reduces the accuracy of the measurement of the unstirred contribution of S_{11}, i.e. S_{11}^{sti}. On the contrary, the result obtained with the biconilog antenna is similar to the reference results obtained with the log-periodic antenna.

On the other hand, no effect related to the type of the antenna is visible in Figure 1.16 concerning $r(1)$, which is a more robust metric insensitive to the

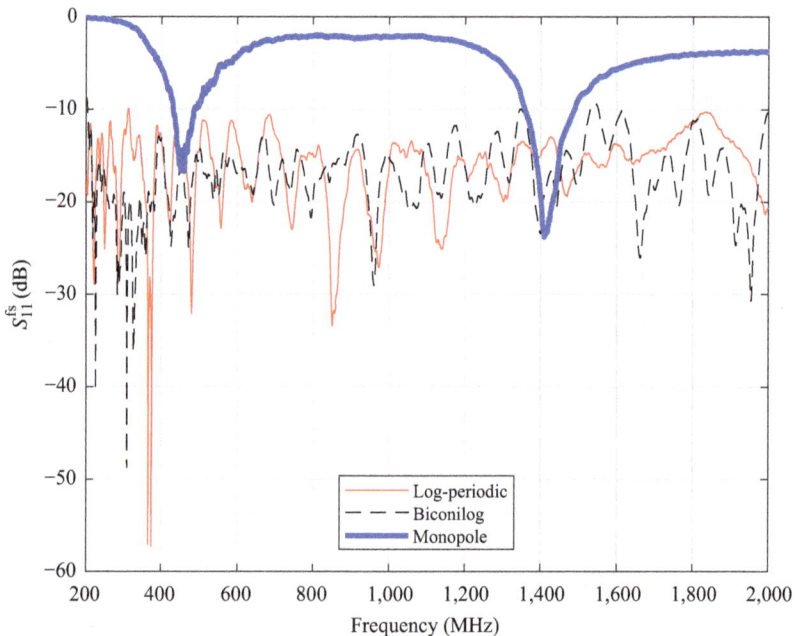

Figure 1.14   Magnitude of S_{11} measured in free space for the three antennas used

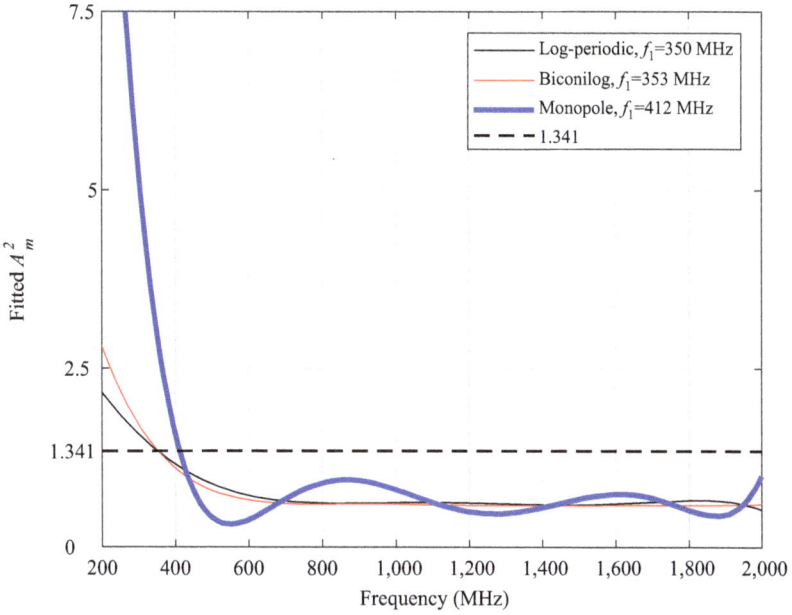

Figure 1.15   Fitted modified AD statistic A_m^2 for the RC loaded with a 9 PA absorber for three different emitting antennas

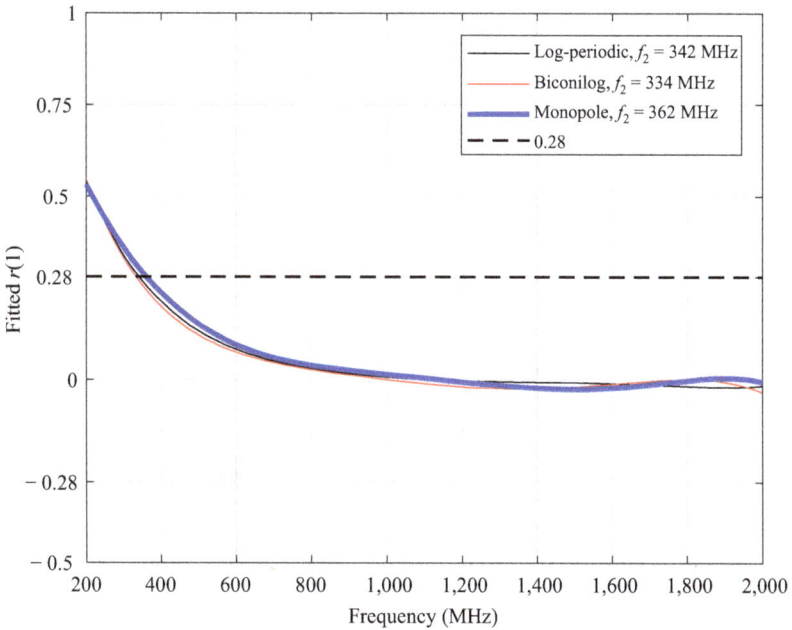

Figure 1.16   Fitted first-order autocorrelation coefficient r(1) for the RC loaded with a 9 PA absorber for three different emitting antennas

impedance mismatch of the antenna. It should be noticed that the same conclusions can also be drawn from the three other loading configurations.

As a conclusion, the antenna used has to be satisfyingly matched, i.e. $|S_{11}| < -10$ dB on the whole frequency range of the S_{11} measurements, in order to be sure to obtain reliable results, particularly on f_1.

1.2.4 On the uncertainty of the method related to the PCF process

The influence of the PCF process on the determination of f_1 and f_2 (and subsequently f_{wsc}) is examined carefully in this subsection.

As a good practice, it is recommended to measure S_{11} on a large bandwidth, typically from $f_{\text{min}} = 3 f_0$ to a frequency f_{max} equals to 10 or 15 f_0 with thousands of collected frequencies (a number easily achievable for any VNA).

The inherent uncertainty of the PCF process depends in major part of the polynomial order used during the PCF process. Figures 1.17 and 1.18 present the PCF of A_m^2 and $r(1)$ obtained for the unloaded RC for different polynomial orders. It can be concluded that low orders (two but also three) are generally not sufficient to reproduce faithfully the frequency dependence of both metrics, while on the contrary, orders greater than six often introduce nonphysical ripples. It is, therefore, recommended to compute the PCF process with an order included in the range from 4 to 6.

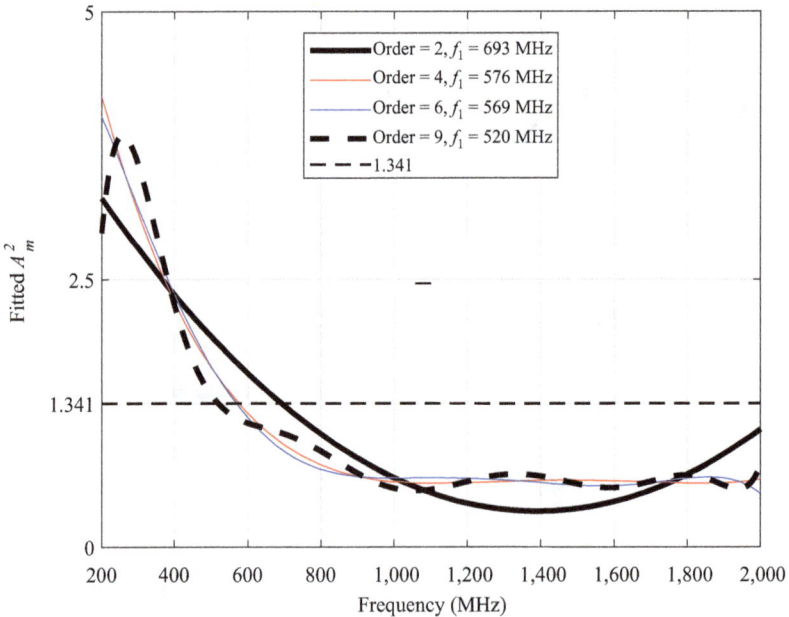

Figure 1.17   Fitted modified AD statistic A_m^2 for the unloaded RC according to the polynomial order of the PCF process

In addition, Table 1.7 collects the values of f_1 and f_2 obtained for such orders for one position of the log-periodic antenna in the unloaded RC. The maximum value of the deviation parameter d reaches 1.4% when comparing the results obtained for these different orders. Therefore, when following all these recommendations, the

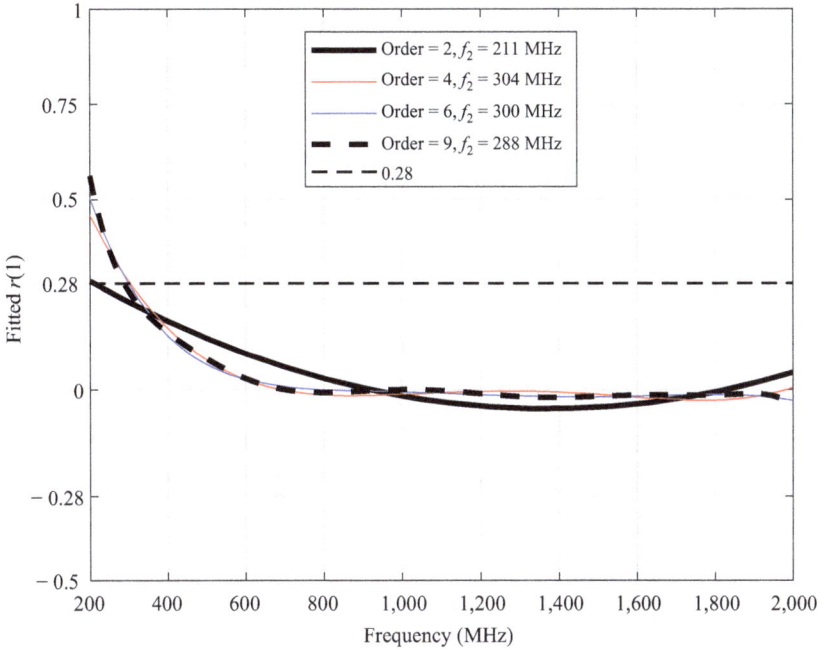

Figure 1.18   Fitted first-order autocorrelation coefficient r(1) for the unloaded RC according to the polynomial order of the PCF process

Table 1.7 f_1, f_2 and the deviation parameter d obtained for the unloaded RC according to the order of the polynomial fitting process

	Loading	Unloaded	9 PA	One block	Four blocks
f_1 (MHz)	Order 4	576	353	264	<200
	Order 5	577	345	267	<200
	Order 6	569	350	267	<200
	d (%)	0.9	1.2	0.8	N/A
f_2 (MHz)	Order 4	304	350	411	684
	Order 5	301	345	403	679
	Order 6	300	342	401	679
	d (%)	0.8	1.4	1.4	0.6

uncertainties related to the order of the PCF process can reasonably be assumed to be lower than 2%.

1.2.5   On the effect of a heavily lossy EUT

Due to the simplicity of the experimental setup required and its short inherent measurement time, the method can be used in order to characterize the RC behavior in presence of an EUT. This constitutes an important advantage as the RC performance can be assessed in real testing conditions. It is of course recommended to make sure that the antenna is located sufficiently far ($\lambda/2$) from the EUT. Indeed, it is known in the literature that in the presence of a heavily lossy EUT, the isotropy of the field can be disturbed [27] at the vicinity of the EUT.

To investigate what happens in the presence of an EUT, measurements have been carried out by considering one block of absorber acting as a (heavily) lossy EUT. This block of absorber has been inserted in the center of the RC, i.e. far from any walls, as shown in Figure 1.19. It is worth noting that this is an extreme case as it is weakly probable to work with such a lossy EUT.

The measurements are repeated for several positions and orientations of the antenna, the antenna directly pointing toward the absorber (i.e. line-of sight (LOS) configuration) or not (i.e. non-LOS configuration (NLOS)).

As shown in Figures 1.20 and 1.21 and in Table 1.8, the result are, as expected, independent of the orientation and the position of the antenna whatever if the antenna is directly oriented toward the EUT or not. The values of f_1 and f_2 collected in Table 1.8 obtained for 3 LOS and 5 NLOS configurations confirm the relevance of the method even in the presence of a lossy EUT.

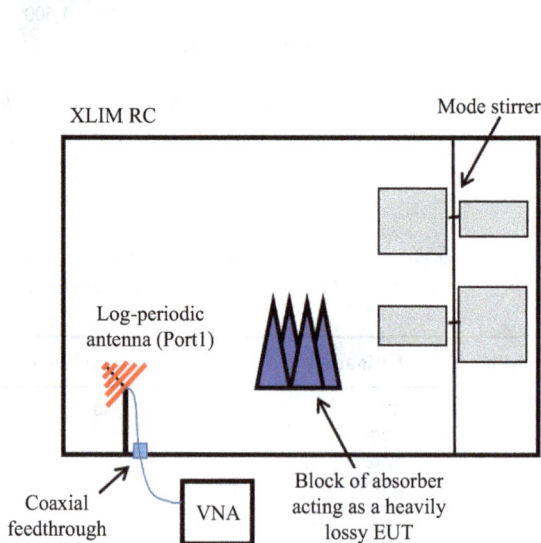

Figure 1.19   Schematic description of the experimental setup in the XLIM RC in presence of the lossy EUT, being represented by one block of absorber

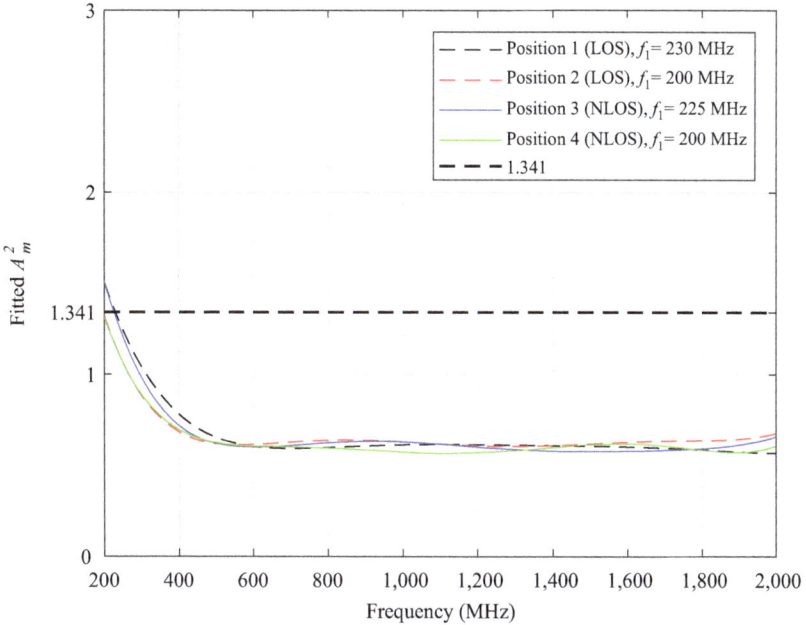

Figure 1.20 *Fitted modified AD statistic A_m^2 obtained for four positions and orientations of the antenna in presence of a highly dissipative EUT located in the center of the RC. In positions 1 and 2 (dotted curves), the antenna points directly toward the absorber*

The knowledge of the "chamber loading factor" [1] related to the EUT is, therefore, useless as the average total electric field $\langle|E_T|\rangle$ obtained in the RC for a given injected power to the antenna can be predicted from (1.13) as developed later in this chapter, in Section 1.4. During the subsequent tests or measurements, it is of course highly recommended to avoid a direct LOS between the antenna and the EUT, a commonly accepted measurement guideline in the field of RC measurements.

1.3 Examples of parametric analysis

This section presents two examples of parametric analysis performed in our RC. These analyses take advantage of

- the accuracy of the method that allows small modifications of the RC performance to be highlighted;
- the short measurement time required which allows different configurations to be tested in a reasonable amount of time.

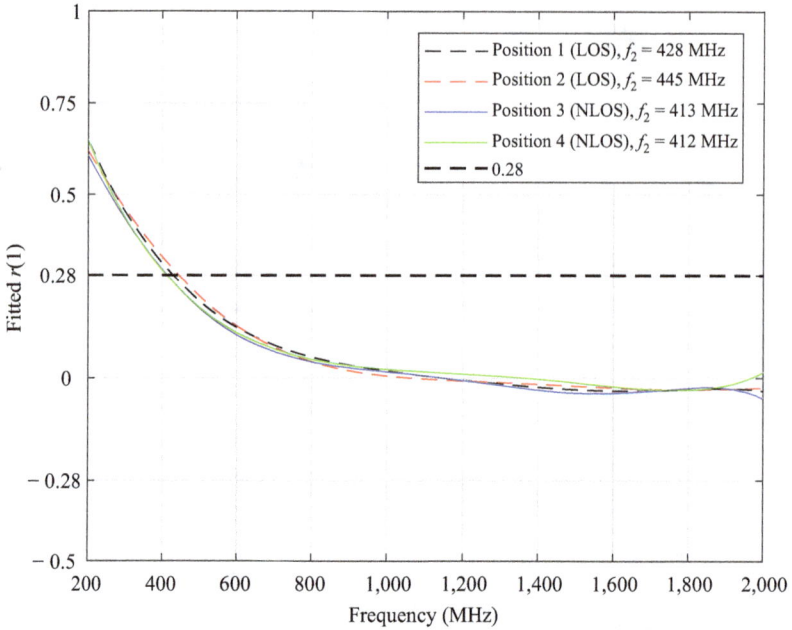

*Figure 1.21   Fitted first-order autocorrelation coefficient r(1) obtained for four
positions and orientations of the antenna in the presence of a highly
dissipative EUT located in the center of the RC. In positions 1 and 2
(dotted curves), the antenna points directly toward the absorber*

*Table 1.8 f_1, f_2 and the deviation parameter d obtained with $|S_{11}^{sti}|$
in the presence of a lossy EUT for different orientations
(and positions) of the antenna with respect to the EUT*

Configuration	f_1 (MHz)	f_2 (MHz)
Pos. 1 (LOS)	229	428
Pos. 2 (LOS)	<200	445
Pos. 3 (LOS)	<200	439
Pos. 4 (NLOS)	225	413
Pos. 5 (NLOS)	<200	412
Pos. 6 (NLOS)	228	428
Pos. 7 (NLOS)	<200	431
Pos. 8 (NLOS)	229	432
d (%)	N/A	3.8

1.3.1 Determination of the "optimal loading"

The first parametric analysis concerns the determination of the "optimal loading" of our RC, i.e. the amount of absorbers to insert in the enclosure in order to obtain the lowest value of f_{wsc}. Indeed, among the four loading configurations tested previously, the "9 PA" condition exhibits the best performance but it is not sure that this is the optimal loading.

A new measurement campaign respecting the following procedure has been carried out:

- Step 1: The antenna is positioned in one particular location of the RC. For this antenna position, the measurements are repeated for 10 different loading conditions of the RC close to each other and included in the range from 0 (i.e. "unloaded") to 15 PAs. The idea is to consider numerous loading condition around the 9 PA condition. The absorbers are located always in the same place within the RC, i.e. below the mode stirrer.
- Step 2: The measurements described in the first step are repeated for four different positions of the antenna.

Figures 1.22 and 1.23 present the PCF of both metrics of interest (i.e. A_m^2 and $r(1)$) for one particular position of the antenna.

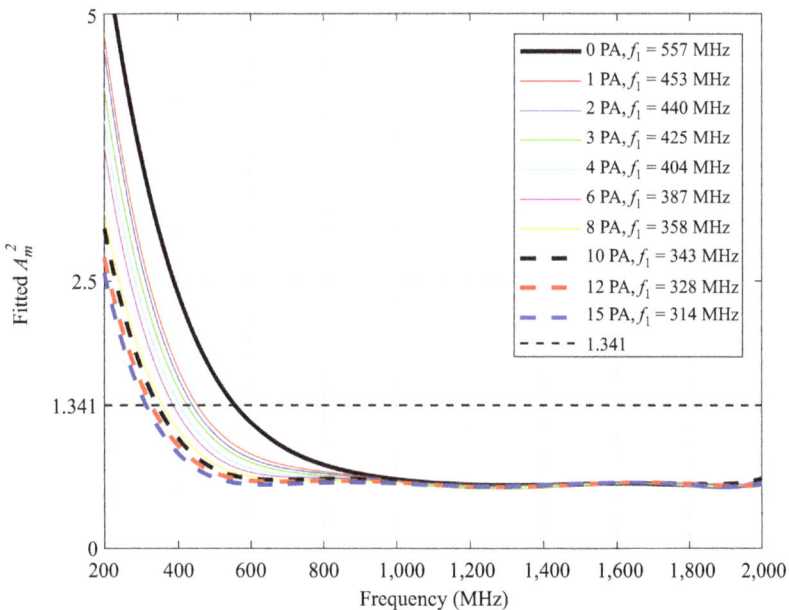

Figure 1.22   Fitted modified AD statistic A_m^2 obtained in the RC for 10 different loading conditions in order to define the optimal loading (fixed antenna position for all the measurements)

The effect of the modification of the loading, even extremely small (in particular with respect to the RC total volume), is particularly clear. Working with a fixed position of the antenna helps to emphasize the effect of loading on the results. As already shown in Section 1.2.2, f_1 decreases while f_2 increases with the amount of absorbers inserted in the RC.

Table 1.9 compiles the values of f_1, f_2 and f_{wsc} averaged for the four positions of the antenna considered. The "8 PA" loading is, among all the amount of absorbers we have tested, the "optimal loading" of our RC. This is also shown graphically in Figure 1.24.

1.3.2   Effect of the number of metallic blades of a mode stirrer

The second parametric analysis shown in this chapter concerns the influence of the mode stirrer geometry on the RC behavior, i.e. on f_{wsc}. Different number of plates of the mode stirrer have been considered by removing successively some plates. The study has been repeated for two loading configurations, i.e., for the RC unloaded and loaded with one block of PAs.

On the one hand, Figure 1.25 illustrates the absence of influence of the mode stirrer geometry on the quality of the EM field distributions generated within the RC. It means that the geometry of the mode stirrer is not critical from the point of view of

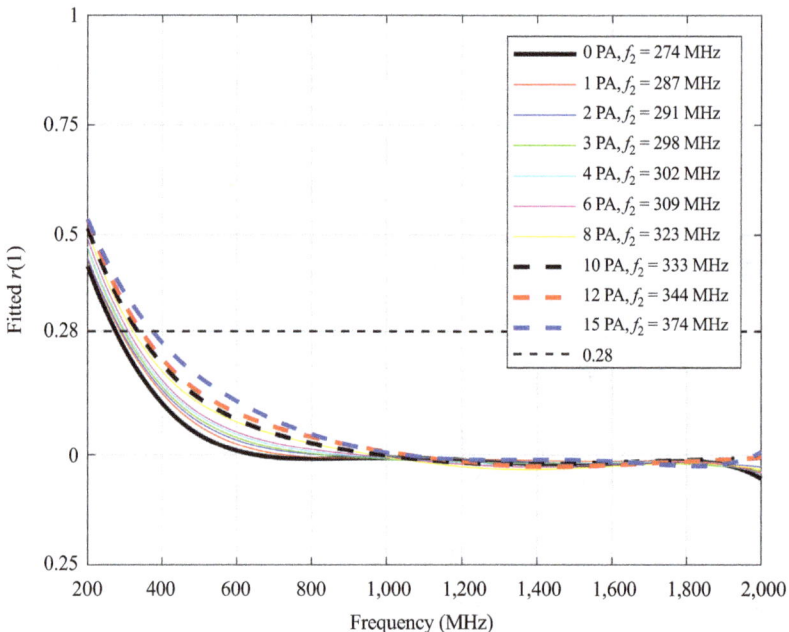

Figure 1.23   Fitted first-order autocorrelation coefficient r(1) obtained in the RC for 10 different loading conditions in order to define the optimal loading (fixed antenna position for all the measurements)

Table 1.9 $\langle f_1 \rangle$, $\langle f_2 \rangle$ and $\langle f_{wsc} \rangle$ *(in MHz) (averaged over four antenna positions) obtained as a function of the loading inserted in the RC in order to define the optimal loading condition*

Number of PA	V_{abs}/V_{RC} (%)	$\langle f_1 \rangle$ (MHz)	$\langle f_2 \rangle$ (MHz)	$\langle f_{wsc} \rangle$ (MHz)
0	0	551	284	551
1	0.004	424	298	424
2	0.009	412	305	412
3	0.013	402	309	402
4	0.018	383	315	383
6	0.027	356	320	361
8	0.035	326	336	345
10	0.044	313	348	351
12	0.053	289	362	362
15	0.066	282	383	383

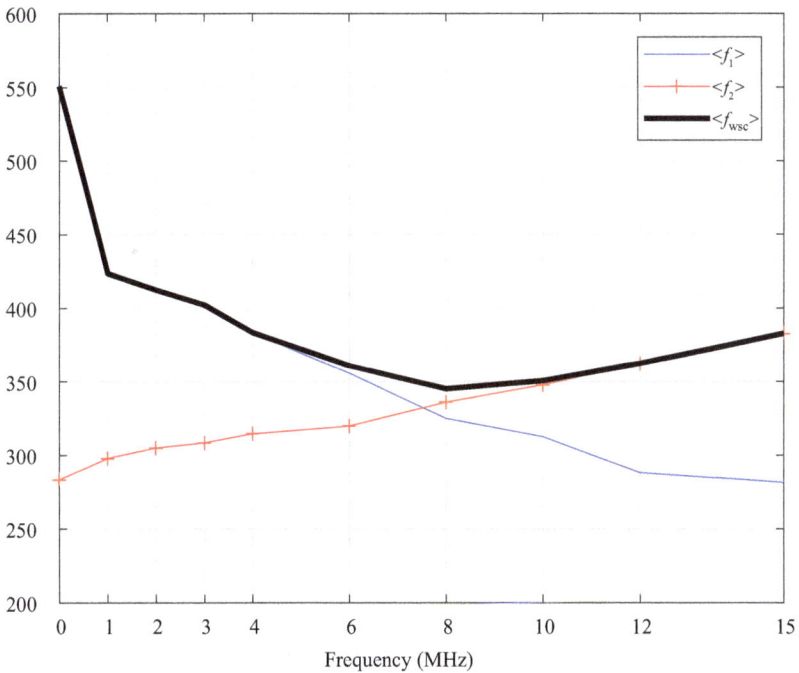

Figure 1.24   *Values of* $\langle f_1 \rangle$, $\langle f_2 \rangle$ *and* $\langle f_{wsc} \rangle$ *(in MHz) (averaged over four antenna positions) obtained as a function of the loading inserted in the RC. The optimal loading leading to the lowest value of* f_{wsc} *is the "8 PA" loading*

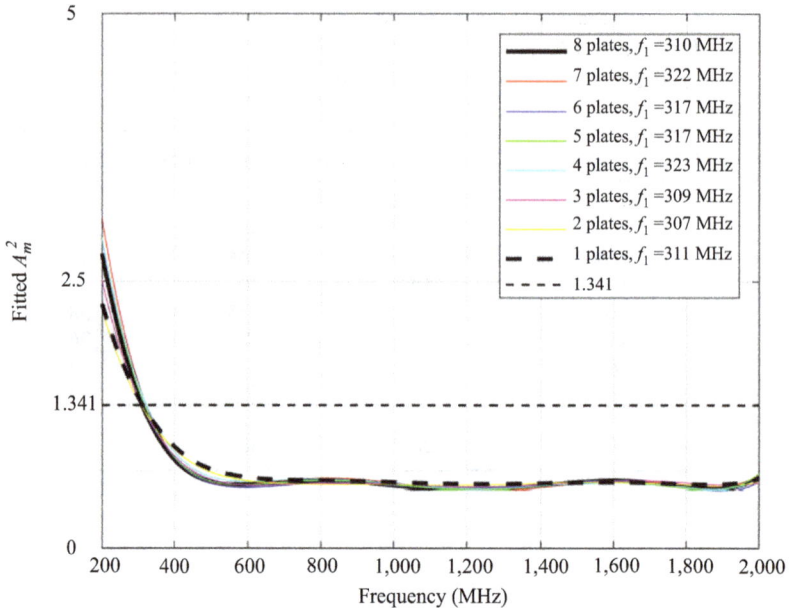

Figure 1.25   Fitted modified AD statistic A_m^2 obtained in the RC loaded with one block of absorbers as a function of the number of metallic plates of the rotating mode stirrer

the EM field distributions provided that its size is sufficient with respect to the total volume of the RC.

On the other hand, the influence of the stirrer geometry on the sample correlation appears clearly in Figure 1.26. Indeed, each mode stirrer plate removed involves a (slight) increase of the sample correlation.

Figure 1.27 showing the evolution of f_1, f_2 and f_{wsc} as a function of the number of mode stirrer plates confirms that these conclusions are valid for both loading configurations.

1.4   An alternative calibration procedure

The IEC 61000-4-21 procedure [1] defines the lowest usable frequency (LUF) of an RC from the analysis of the field uniformity obtained in a volume called the working volume. This procedure has been implemented in the case of our unloaded RC. In this case, as depicted in Figure 1.28, the 3-dB limit is respected on the whole frequency range. Consequently, the LUF is estimated to be lower than 200 MHz, while f_{wsc} for this loading configuration is around 560 MHz as shown in Table 1.5.

It is well-known in the literature (for instance in [4,14,28,29]) that the well-stirred condition is not reached around the LUF. This is potentially an important problem as the well-stirred condition is often the assumption underlying any theoretical

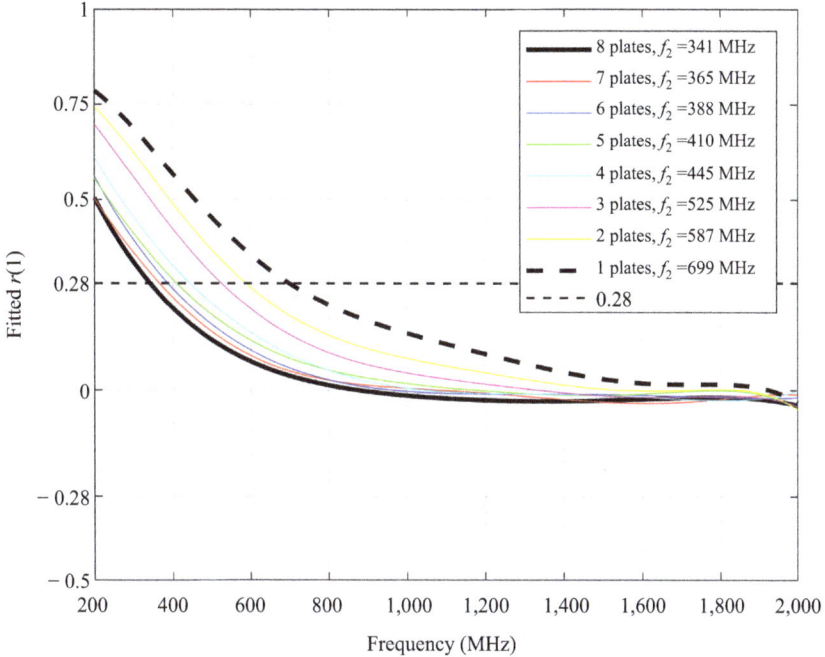

Figure 1.26   Fitted first-order autocorrelation coefficient r(1) obtained in the RC loaded with one block of absorbers as a function of the number of metallic plates of the rotating mode stirrer

development. In reason of the measurement time required by the IEC standard (more or less 1 or 2 working days for each RC configuration), we also know that it is difficult to perform parametric studies with this procedure, for instance in order to define the optimal absorber quantity to insert in the RC as done in Section 1.3.1.

Considering the advantages of the well-stirred condition method in terms of rapidity, accuracy and ease of implementation (simple experimental setup), the method is ready to be used as an alternative (and more stringent) RC calibration procedure. The IEC measurement setup for an RC calibration requires an RF synthesizer, an RF amplifier (optional depending on the required electric field strength), a power-meter, an RF coupler and a triaxial electric field probe while only a VNA is required in the well-stirred condition method presented in this chapter. The measurement setup is consequently simpler even if all these equipment would still be required later to perform a radiated susceptibility EMC test. However, as described hereafter, the use of a triaxial field probe is no longer necessary.

Three fundamental points have to be evocated at this stage.

The first one concerns the fact that the well-stirred condition method does not deal with the assessment of the uniformity and the isotropy of the field generated within the RC whereas these metrics are the most important metrics of interest in the IEC procedure. In reality, an indirect verification of the field uniformity and isotropy

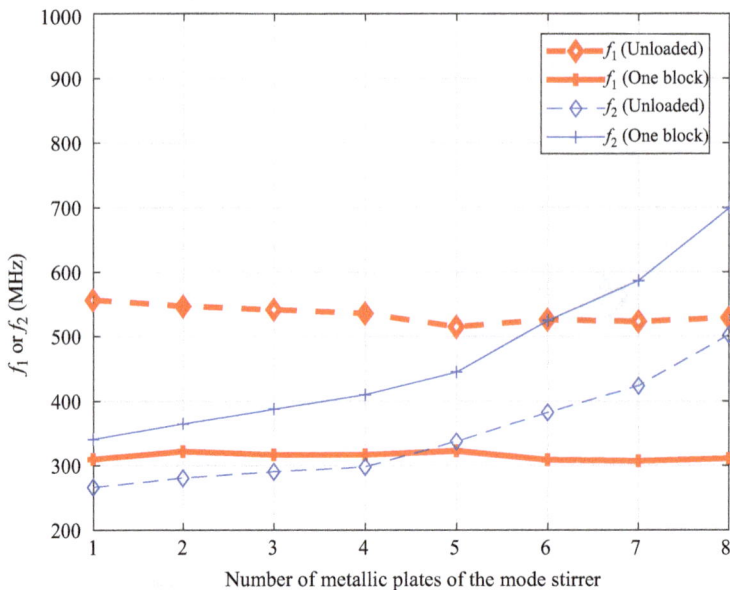

Figure 1.27   Comparison of f_1 and f_2 obtained in the RC for two loading conditions a function of the number of metallic plates of the rotating mode stirrer

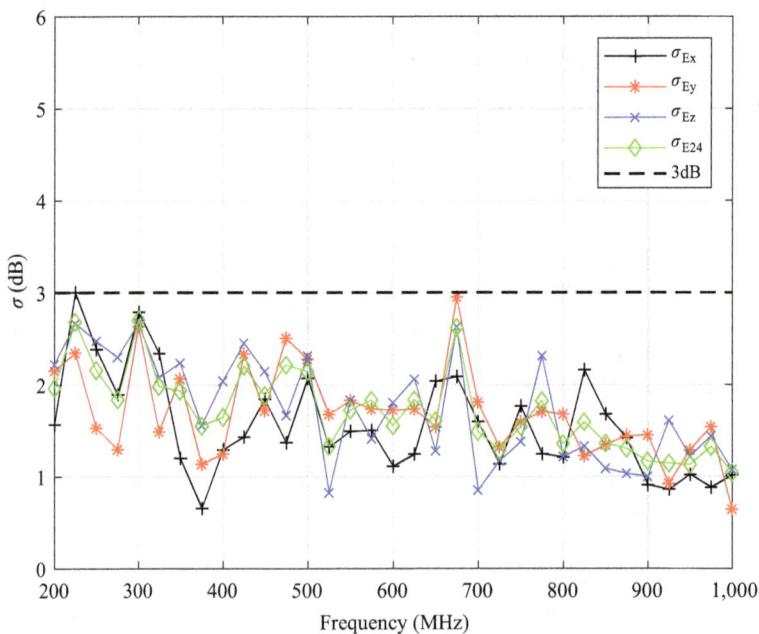

Figure 1.28   Electric field standard deviation obtained in the unloaded RC using the IEC 61000-4-21 procedure [1]

can be inferred from the method outputs. Indeed, above f_1, we know that the ideal EM field distributions are met. This means that a sufficient modal overlapping [26] is obtained within the RC, which implies itself a satisfying field uniformity. In other terms, a satisfying field uniformity is an automatic consequence of having the ideal EM field distributions matched within the RC. The approach followed by the method is the inverse with respect to the IEC procedure which considers the field uniformity independently of the EM field distributions.

The second one deals with the relevance of the concept of "working volume." In the IEC procedure, a satisfying field uniformity has to be obtained in such volume (from measurements performed at the corners). Above the LUF, the measurements can be made in such volume and only in this. There is here an important loss of available space within the RC. Indeed, we know from [20] that the distributions within an RC are the same in any point of the RC located sufficiently far from any object. There is therefore no reason to limit the usable volume within an RC to this so-called working volume. This is also why the results obtained with the well-stirred condition method are similar whatever the position of the antenna in the examples shown in this chapter.

The third important point to be evocated here concerns the assessment of the EM field strength (for instance the average total electric field strength $\langle |E_T| \rangle$) within the RC in order to be able to define the power to inject to the emitting antenna to get a given field strength. This is a logical objective of the RC calibration process, especially in the objective to perform a radiation susceptibility test after.

This is achieved in [1] through the electric field measurements required by the calibration process and the knowledge of the chamber loading factor (related to the losses involved by the EUT itself), a metric not provided by the well-stirred condition method.

However, we present hereafter a method only based on S_{11} measurements in order to predict $\langle |E_T| \rangle$ within an RC.

We first know that $\langle |E_T| \rangle$ is linked to the Q-factor

$$Q = \frac{\omega \varepsilon V \langle |E_T|^2 \rangle}{P_t}, \tag{1.11}$$

where ω is the angular frequency, ε the permittivity of the propagating medium, V the RC volume and P_t the transmitted power to the antenna. As shown for instance in [30], Q can be determined from S_{11} (even if the number of antennas in the RC is greater than 1):

$$Q = \langle |S_{11} - \langle S_{11} \rangle|^2 \rangle \frac{Z_0 \omega \varepsilon V}{\left(\lambda^2 / (4\pi) \right) \left(1 - |\langle S_{11} \rangle|^2 \right)^2 \eta^2}, \tag{1.12}$$

where Z_0 is the wave impedance, η the efficiency of the antenna and λ the wavelength. Combining (1.11) and (1.12), $\langle |E_T| \rangle_N$ (i.e. as a function of N) can then be estimated as follows:

$$\langle |E_T| \rangle_N = \sqrt{\frac{QP_t}{\omega \varepsilon V}} \frac{15}{16} \sqrt{\frac{\pi}{3}} \frac{\Gamma(3N)\sqrt{3N}}{\Gamma(3N + 1/2)}, \tag{1.13}$$

with Γ the mathematical function.

In (1.13), it should be emphasized that

- the term $15/16\sqrt{\pi/3}$ is related to the inequality $\sqrt{\langle|E_T|^2\rangle} \neq \langle|E_T|\rangle$;
- the N-dependent conversion factor $\left(\Gamma\left(3N\right)\sqrt{3N}\right)/\Gamma\left(3N+1/2\right)$ [31] (close to 1 for N values generally considered in RC) takes into account the difference between the expected value (for an infinite N) and the obtained value of $\langle|E_T|\rangle_N$ for N independent realizations.

Both correction terms can be easily retrieved through the multiple generation of random samples.

As an example, $\langle|E_T|\rangle$ is computed using (1.13) from S_{11} measurements performed for three different positions of the antenna within our unloaded RC. These results are compared in Figure 1.29 with direct measurements made with a triaxial electric field probe at three different locations, the injected power P_t being monitored in this case by a power meter and the power losses in the cables being taken into account.

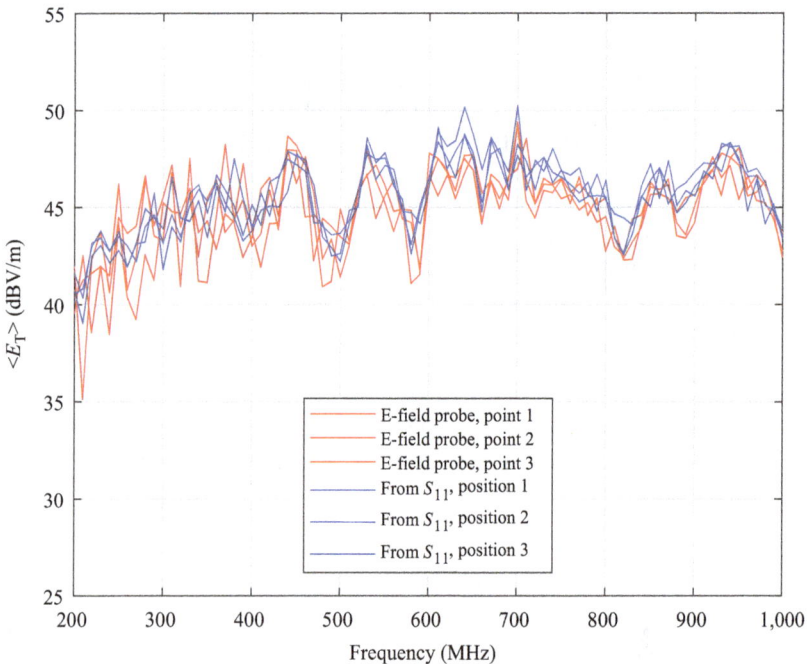

Figure 1.29   *Comparison of $\langle|E_T|\rangle$ (in dBV/m) measured in the unloaded RC with a triaxial electric field probe in three different positions and computed from three different sets (i.e. three positions of the antenna) of S_{11} measurements using (1.13). The efficiency η of the log-periodic antenna is assumed to be 0.95*

1.5　Conclusion

In this chapter, the well-stirred condition method is presented in the case of a mechanically stirred RC. This method is able to determine the frequency f_{wsc} where the well-stirred condition of an RC is achieved. The method focusing on the ability of the RC to produce stirred contributions compatible with the well-stirred condition presents the advantage to be independent of the type, location and orientation of the emitting antenna if the antenna is sufficiently matched over the whole bandwidth.

The method based on S_{11} measurements of an antenna is simple to set up (only a VNA is required) and fast (a few minutes). The method relies on a clear definition of the well-stirred condition and uses threshold values related to theory. Finally, the method is usable in the presence of an EUT (even a heavily lossy one) provided that the antenna is located sufficiently far from it. This has the double advantage to give an overview of the RC performance in the exact conditions of the subsequent test and also allows the field strength generated in the RC for a given injected power to be predicted. The inherent uncertainty of the method outputs combining the overall uncertainty of the RC itself and of the PCF process is estimated to be lower than 10%.

Considering all these advantages, the method constitutes an alternative RC calibration process proposed to the community of RC users. The quickness of the method gives also the possibility to perform parametric studies in a reasonable amount of time, for instance, in order to define its optimal amount of absorbers to insert in and, therefore, to optimize its performance, in particular, on the lowest frequency range.

In the next chapter of this book, the method is extended to the most common stirring techniques able to be implemented in our RC.

References

[1]　Reverberation Chamber Test Methods. Int Electrotech Commiss Standard IEC 61000-4-21:2011. 2011 January.

[2]　Serra R. Reverberation chambers through the magnifying glass: an overview and classification of performance indicators. IEEE Electromagnetic Compatibility Magazine. 2017 Second;6(2):76–88.

[3]　Arnaut LR, Serra R, and West PD. Statistical anisotropy in imperfect electromagnetic reverberation. IEEE Transactions on Electromagnetic Compatibility. 2017;59(1):3–13.

[4]　Lemoine C, Besnier P, and Drissi M. Investigation of reverberation chamber measurements through high-power goodness-of-fit tests. IEEE Transactions on Electromagnetic Compatibility. 2007;49(4):745–755.

[5]　Lundén O and Bäckström M. Stirrer efficiency in FOA reverberation chambers. Evaluation of correlation coefficients and chi-squared tests. In: IEEE International Symposium on Electromagnetic Compatibility. Symposium Record (Cat. No.00CH37016). IEEE, Washington DC, USA. vol. 1; 2000. p. 11–16.

[6]　Holloway CL, Hill DA, Ladbury JM, *et al*. On the use of reverberation chambers to simulate a Rician radio environment for the testing of wireless devices. IEEE Transactions on Antennas and Propagation. 2006;54(11):3167–3177.

[7]   Lemoine C, Amador E, and Besnier P. On the K-factor estimation for Rician channel simulated in reverberation chamber. IEEE Transactions on Antennas and Propagation. 2011;59(3):1003–1012.

[8]   Ladbury J and Hill DA. Enhanced backscatter in a reverberation chamber: inside every complex problem is a simple solution struggling to get out. In: 2007 IEEE International Symposium on Electromagnetic Compatibility. IEEE, Honolulu, USA.; 2007. p. 1–5.

[9]   Holloway CL, Shah HA, Pirkl RJ, *et al.* Reverberation chamber techniques for determining the radiation and total efficiency of antennas. IEEE Transactions on Antennas and Propagation. 2012;60(4):1758–1770.

[10] Dunlap CR, Holloway CL, Pirkl R, *et al.* Characterizing reverberation chambers by measurements of the enhanced backscatter coefficient. In: 2012 IEEE International Symposium on Electromagnetic Compatibility. IEEE, Pittsburgh, USA.; 2012. p. 210–215.

[11] Genender E, Holloway CL, Remley KA, *et al.* Simulating the multipath channel with a reverberation chamber: application to bit error rate measurements. IEEE Transactions on Electromagnetic Compatibility. 2010;52(4):766–777.

[12] Holloway CL, Shah HA, Pirkl RJ, *et al.* Early time behavior in reverberation chambers and its effect on the relationships between coherence bandwidth, chamber decay time, RMS delay spread, and the chamber buildup time. IEEE Transactions on Electromagnetic Compatibility. 2012;54(4):714–725.

[13] Andrieu G, Ticaud N, Lescoat F, *et al.* Fast and accurate assessment of the "well stirred condition" of a reverberation chamber from S_{11} measurements. IEEE Transactions on Electromagnetic Compatibility. 2019;1–9.

[14] Cozza A. The role of losses in the definition of the overmoded condition for reverberation chambers and their statistics. IEEE Transactions on Electromagnetic Compatibility. 2011;53(2):296–307.

[15] Kostas JG and Boverie B. Statistical model for a mode-stirred chamber. IEEE Transactions on Electromagnetic Compatibility. 1991;33(4):366–370.

[16] Lehman TH. A statistical theory of electromagnetic fields in complex cavities. EMP Interaction Note 494. 1993.

[17] Corona P, Ferrara G, and Migliaccio M. Reverberating chambers as sources of stochastic electromagnetic fields. IEEE Transactions on Electromagnetic Compatibility. 1996;38(3):348–356.

[18] Hill DA. Plane wave integral representation for fields in reverberation chambers. IEEE Transactions on Electromagnetic Compatibility. 1998;40(3): 209–217.

[19] Hill DA and Ladbury JM. Spatial-correlation functions of fields and energy density in a reverberation chamber. IEEE Transactions on Electromagnetic Compatibility. 2002;44(1):95–101.

[20] Hill DA. Boundary fields in reverberation chambers. IEEE Transactions on Electromagnetic Compatibility. 2005;47(2):281–290.

[21] Hill DA. Electronic mode stirring for reverberation chambers. IEEE Transactions on Electromagnetic Compatibility. 1994;36(4):294–299.

[22] Lemoine C, Amador E, Besnier P, *et al.* Antenna directivity measurement in reverberation chamber from Rician K-factor estimation. IEEE Transactions on Antennas and Propagation. 2013;61(10):5307–5310.

[23] Stephens MA. EDF statistics for goodness of fit and some comparisons. Journal of the American Statistical Association. 1974;69(347):730–737.

[24] Lemoine C, Besnier P, and Drissi M. Estimating the effective sample size to select independent measurements in a reverberation chamber. IEEE Transactions on Electromagnetic Compatibility. 2008;50(2):227–236.

[25] Adardour A, Andrieu G, and Reineix A. On the low-frequency optimization of reverberation chambers. IEEE Transactions on Electromagnetic Compatibility. 2014;56(2):266–275.

[26] Monsef F and Cozza A. Average number of significant modes excited in a mode-stirred reverberation chamber. IEEE Transactions on Electromagnetic Compatibility. 2014;56(2):259–265.

[27] Senic D, Cavaliere D, North MV, *et al.* Isotropy study for over-the-Air measurements in a loaded reverberation chamber. In: 2017 IEEE International Symposium on Electromagnetic Compatibility Signal/Power Integrity (EMCSI). IEEE, Washington DC, USA.; 2017. p. 124–129.

[28] Andrieu G, Tristant F, and Reineix A. Investigations about the use of aeronautical metallic halls containing apertures as mode-stirred reverberation chambers. IEEE Transactions on Electromagnetic Compatibility. 2013;55(1):13–20.

[29] Fall AK, Besnier P, Lemoine C, *et al.* Determining the lowest usable frequency of a frequency-stirred reverberation chamber using modal density. In: 2014 International Symposium on Electromagnetic Compatibility. IEEE, Gothenburg, Sweden.; 2014. p. 263–268.

[30] Besnier P, Lemoine C, and Sol J. Various estimations of composite Q-factor with antennas in a reverberation chamber. In: 2015 IEEE International Symposium on Electromagnetic Compatibility (EMC). IEEE, Dresden, Germany.; 2015. p. 1223–1227.

[31] Ladbury JM and Koepke GH. Reverberation chamber relationships: corrections and improvements or three wrongs can (almost) make a right. In: 1999 IEEE International Symposium on Electromagnetic Compatibility. Symposium Record (Cat. No.99CH36261). IEEE, Seattle, USA. vol. 1; 1999. p. 1–6.

Chapter 2

Review of the stirring techniques proposed in the literature – performance assessment of the most common ones using the well-stirred condition method

Guillaume Andrieu[1]

Since the appearance of reverberation chambers (RCs) some decades ago, numerous kinds of stirring techniques have already been proposed in the literature [1].

At its most basic level, an RC is composed of a metallic enclosure and an antenna exciting resonant modes according to its orientation, polarization and position within the enclosure. Without anything changing the boundary conditions of the system made of the enclosure and of the antenna, the electromagnetic (EM) field in the enclosure is spatially dependent and depends on the combination of all the excited modes. At this stage, in the absence of any mode stirring process, the enclosure is unable to be a facility ensuring reliable and reproducible measurements.

A mode stirring process refers to any mean able to modify the boundary conditions of the entire system to obtain on average a uniform and isotropic EM field in any point of the enclosure (with the exception of all the points close to any object [2]). When a satisfying mode stirring process is implemented, the RC can be used as a statistical tool (i.e., by considering the results obtained over different stirring configurations) with a reasonable and quantifiable level of uncertainty.

Most of the stirring techniques proposed in the literature can be sorted in three different categories depending on whether the stirring process is

- ensured by an additional moving metallic device called "mode stirrer" (generally in rotation) within the enclosure (method referred to as "mechanical stirring" techniques in this chapter);
- related to the source signal feeding the emitting antenna ("source signal stirring" techniques);
- due to the movement of the emitting antenna ("source position stirring" techniques).

This chapter presents a "critical" review of the stirring techniques already proposed in the literature based on some fundamental criteria; among them are the

[1]University of Limoges, Limoges, France

complexity of the required setup, the available space within the RC and the ease to assess the efficiency of the technique.

In addition, performance of the most common techniques are assessed by taking advantage of the "well-stirred condition method" presented in details in the first chapter of this book. In reality, the method is extended in this chapter to all the techniques able to be implemented quite easily in the RC of the XLIM Laboratory. The results presented in the first chapter with the RC mechanically stirred, which is traditionally the most used stirring technique, are considered as the reference results and allow the performance of all the tested stirring techniques to be compared. The interest in such analysis is

- to perform a fair comparison of the performance of each technique without introducing any bias related to the use of cavities of different sizes;
- to identify the key parameters playing a role on the efficiency of each stirring technique;
- to analyze the effect of a variation of the composite Q-factor (modified by the amount of absorber inserted in the RC) for each stirring technique.

To compare the performance of the different tested stirring techniques, the most important metric is the frequency f_{wsc} (with also f_1 and f_2, see the first chapter to get more information about these metrics) obtained with the well-stirred condition method for a given number N (with $N = 50$ in this chapter) of stirring configurations. It is assumed that the lower the f_{wsc}, the better the efficiency of the stirring technique.

Logically, vibrating intrinsic RCs (the so-called VIRCs) [3,4], being based on a specific stirring process using a time-varying modification of the enclosure shape, are discussed separately in the next chapter of this book.

2.1 Mechanical stirring techniques

2.1.1 Overview

As discussed extensively in the first chapter of this book, mechanical stirring techniques are the most commonly used by the worldwide community of RC users. This corresponds to the techniques where the stirring process is ensured by the movement of a metallic object called a mode stirrer.

Most of the time, the mode stirrer rotates on itself. Different kinds of rotating mode stirrers have been proposed, including Z-fold, bent-plates stirrer or Carousel stirrers [5] as mentioned in [1].

Mode stirrers having a different kind of movement have also been proposed. Some authors [6,7] have studied the possibility to move the walls of an enclosure instead of inserting a mode stirrer. The so-called "oscillating mode stirrer" [8] featuring an oscillating movement of a large metallic piece within the RC has also been proposed recently [8].

The well-stirred condition method can be applied on any mechanical mode stirrers in order to assess the stirring efficiency. As discussed in Chapter 1, their performance can be different from one shape to another. In the case of the mode stirrer having a translation movement, the method can be used without any adjustment by considering each fixed position of the stirrer as a stirring configuration.

2.1.2 Dual-mode stirrer

A solution widely used in the range of mechanically stirred RC consists in implementing a dual-mode stirring process as depicted in Figure 2.1. This is traditionally done through the use of two independent mode stirrers, each one being driven independently of the other through a step-by-step motor. It should first be mentioned that in reason of the impossibility to insert a second mode stirrer in the XLIM RC, no experimental results showing the interest to insert a second mode stirrer are shown in this chapter.

Two different stirring scenarios are possible when an RC is equipped of a dual-mode stirrer.

- If both the mode stirrers are moved simultaneously at N different fixed positions, there is no fundamental difference with a stirring process using only one mode stirrer. The addition of the second mode stirrer is expected of course to improve the overall performance of the stirring solution. Indeed, the presence of the second mode stirrer acts as if the dimensions of the first mode stirrer would be increased, contributing to the increase of its stirred volume, defined as the overall volume encompassing the displacement of the object during its movement. The well-stirred condition can be applied in these conditions as described in the first chapter of this book.
- If both the mode stirrers are moved asynchronously, the stirring technique leads to $N \times M$-measured samples if N and M are, respectively, the number of positions considered for each mode stirrer. This is here the main interest of the method

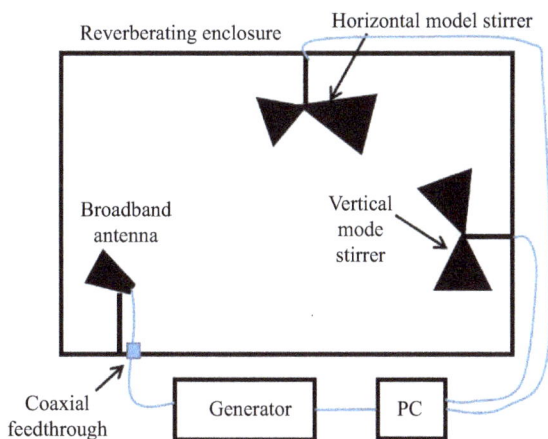

Figure 2.1 Typical experimental setup required with a dual rotating mode stirrer

which allows a larger number of independent samples to be obtained during the stirring process.

To assess the efficiency of the stirring process when both mode stirrers are moved asynchronously, it seems relevant to apply the method on each mode stirrer taken separately. Two different values of f_{wsc} are obtained in this case, one for each mode stirrer (i.e., f_{wsc}^1 for the first mode stirrer and f_{wsc}^2 for the second one). Above each f_{wsc} value, all the positions of the mode stirrer under investigations provide independent samples. Therefore, above the maximum value between f_{wsc}^1 and f_{wsc}^2, it is reasonable to assume that the dual-mode stirrer provides $N \times M$ independent samples.

Below this value, the number of independent samples is lower. Even if it is possible to assess the overall number of independent samples among a greater number of samples, it may be difficult in practice to identify a scenario dealing only with independent samples. This topic is discussed in detail in Chapters 5 and 7 of this book.

2.1.3 Reference results

The results presented in the first chapter of this book using a rotating mode stirrer in the RC of the XLIM Laboratory are the reference results of this chapter (it is suggested here to read this chapter to get more details about the mode stirrer geometry and the measurements performed).

The obtained results in this case using the well-stirred condition method are reported in Table 2.1 as a function of the amount of absorbers inserted, the data being averaged over four different positions of the antenna.

To compare the obtained results to the reference ones, the gain G (in %) is introduced

$$G_i = 100\frac{\left(f_i^{\text{ref}} - f_i\right)}{f_i^{\text{ref}}},\tag{2.1}$$

where i can be either "1," "2" or "wsc." f_i^{ref} is the value of f_1, f_2 or f_{wsc} obtained for the reference configuration (i.e. the RC mechanically stirred) and f_i the value of the same

Table 2.1   *Results (averaged over four positions of the antenna) obtained in the RC of the XLIM Laboratory mechanically stirred as a function of loading (from Chapter 1)*

	Unloaded	9 PA	One block	Four blocks
$V_{\text{abs}}/V_{\text{RC}}$ (%)	N/A	0.04	0.13	0.53
f_1 (MHz)	559	346	280	<200
f_2 (MHz)	289	339	397	663
f_{wsc} (MHz)	559	346	397	663
f_{wsc}/f_0	7.6	4.7	5.4	9

parameter obtained for the stirring technique under investigation, both values being obtained for the same loading condition. Thus, a positive value of G_i means that the mode stirring process shows better performance than the rotating mode stirrer.

It is worth noting here that G_i is a metric comparing two particular measurement configurations and is therefore subject to an inherent uncertainty. Logically, the lowest meaningful value of G_{wsc} equals the method uncertainty that is assumed to be 5% (a value demonstrated in the first chapter to be lower than 10%). In other terms, a value of G_i lower (in absolute value) than 5% is considered as not significant.

2.2 Source signal stirring techniques

The techniques where the stirring process is ensured by the source signal feeding a fixed antenna are investigated in this section. Often referred to as "electronic stirring" as the stirring process is achieved by electronic means, the terminology "source signal stirring techniques" is used in this chapter. The aim is to better dissociate these techniques with the stirring techniques investigated in the next section where the antenna is moved (literally or fictitiously through the use of a network of fixed antennas) in order to ensure the stirring process.

The great benefit of these techniques is to avoid to move any object within the RC. This leads to the simplest experimental setup, as sketched in Figure 2.2, ensuring an increase of the available space within the RC and a simplified maintenance in reason of the absence of a motor. The duration of the measurements is also expected to be shortened in reason of the cancellation of the time required to move and stabilize the stirrer (or the moving object) when it is moved from one position to the following one.

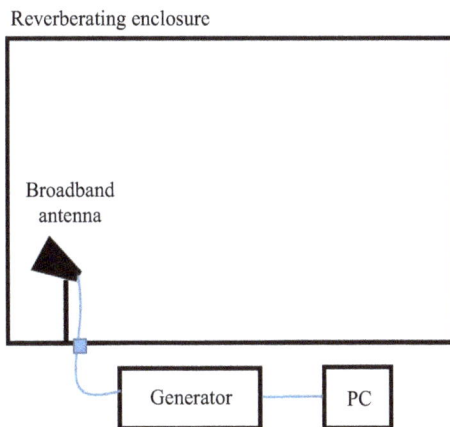

Figure 2.2 Typical experimental setup required for the implementation of a source signal stirring technique

Three kinds of "source signal stirring techniques" have at our knowledge already been proposed. They are presented successively in this section with a special focus on the frequency stepping technique for reasons developed hereafter.

2.2.1   The frequency stepping stirring technique

2.2.1.1   Principle

The basic principle of the frequency stepping technique, sketched in Figure 2.3, is straightforward and consists in exciting successively the metallic enclosure at different frequencies (i.e., with constant wave (CW) or sinusoidal signals) shifted from each other with a frequency step δ_f around the working frequency f_0.

Indeed, in a lossless enclosure, each mode occurs at a particular frequency with an infinite amplitude. In a lossy enclosure, each mode appears on a limited frequency range related to the average composite quality factor Q, or to the average mode bandwidth Δ_f defined as $\Delta_f = f/Q(f)$. This means that the modal density increases with the decrease of Q, the modal density being defined for instance as the number of significant modes occurring at one given frequency [9].

The frequency shift δ_f of the CW signal feeding the antenna leads therefore to excite differently each mode at each frequency of excitation. The total EM field that is the sum of the contribution of all the excited modes is expected to have different properties for each excited frequency. One understands here that the choice of the frequency shift δ_f is of fundamental importance. Intuitively, it makes sense to consider a value of δ_f greater than Δ_f in order to obtain independent realizations, an assumption discussed later in this chapter.

Unlike for instance the mechanical stirring technique, the measurements claimed to be performed at the working frequency f_0 are in reality performed on a total

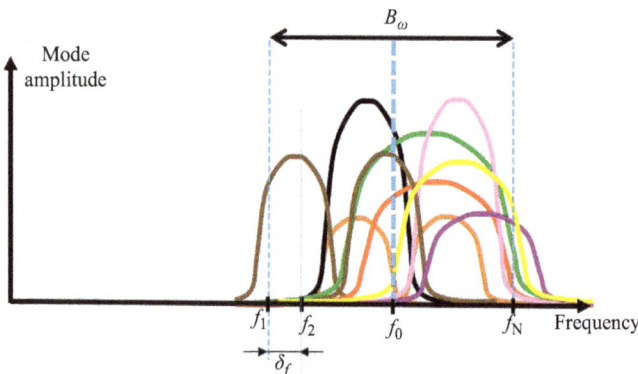

Figure 2.3   Schematic illustration of the frequency stepping principle. N frequencies shifted with a step δ_f are excited successively around the working frequency f_0 on a total bandwidth B_ω. It is shown that the weight of each mode is different at each frequency

bandwidth $B_\omega = (N-1)\delta_f$ with N the number of samples considered (here the number of shifted frequencies around f_0). Consequently, the characteristics of the equipment under test (EUT) is assumed to be constant on B_ω. We can cite for instance the immunity level of an EUT during a radiated susceptibility test (see Chapter 4 of this book) or the efficiency of an antenna (see Chapter 7) to be characterized. To be confident on this assumption considering as a constant the behavior of the EUT around each frequency over B_ω, it is reasonable to limit the ratio B_ω/f to a few percent [10], a limit of 10% being considered in this chapter.

2.2.1.2 Determination of the minimum step Δ_f

The assessment of Q (or Δ_f) for an RC is generally made in the frequency domain through scattering parameter measurements [11] (S_{11} or S_{21} for instance) obtained over N stirring configurations, generally N mode stirrer positions. In the case of the frequency stepping technique when only one antenna is inserted in the enclosure (as shown in Figure 2.2), only one S_{ii} reflection coefficient is measurable. Another method is therefore required to proceed to the assessment of the Q-factor of the metallic enclosure.

A common method [12,13] to reach this objective is based on the time-domain analysis of S_{11}. Indeed, Q is related to the chamber decay time τ_{RC}

$$Q = \omega \cdot \tau_{RC}, \tag{2.2}$$

where τ_{RC} is expressed in seconds and ω is the angular frequency in radians. To compute τ_{RC}, the first step consists in computing the power delay profile (PDP) of the enclosure from the impulse response of the chamber $h(t)$

$$PDP(t) = |h(t)|^2, \tag{2.3}$$

where $h(t)$ is the inverse Fourier transform (IFT) of S_{11}

$$h(t) = IFT(S_{11}). \tag{2.4}$$

$PDP(t)$ that decreases exponentially can be written on the following form

$$PDP(t) = \alpha \cdot e^{-t/\tau_{RC}}, \tag{2.5}$$

with α an arbitrary constant. τ_{RC} is therefore related to the slope of $ln[PDP(t)]$ which equals $-1/\tau_{RC}$. τ_{RC} can be easily computed from a linear regression of the natural logarithm $ln[PDP(t)]$ (provided for instance by MATLAB® and GNU Octave software packages). To get a good accuracy, the early-time (or build-up time) of the enclosure [14] has to be removed, this time being the time required by the enclosure to reach a reverberant regime. Following this process, the average mode bandwidth Δ_f is easily derived

$$\Delta_f = \frac{1}{(2 \cdot \pi \cdot \tau_{RC})}. \tag{2.6}$$

This method has been used in our RC on the four loading conditions considered in the first chapter of this book and reminded in Table 2.1 (see Section 1.2.1 of the first chapter for a complete description of the experimental setup). Figure 2.4 shows the plots of $ln[PDP(t)]$ obtained for one fixed position of the mode stirrer. In the figure, one curve shows the entire $ln[PDP(t)]$ plot, one is truncated with the early-time removed and the last one shows the linear regression of the truncated one, this last one leads to τ_{RC}, Q and Δ_f.

An important point concerns the accuracy of this method. To get some reliable elements, the results obtained for each mode stirrer position are compared. Table 2.2 contains the minimum, the maximum and the average value of Δ_f obtained for all

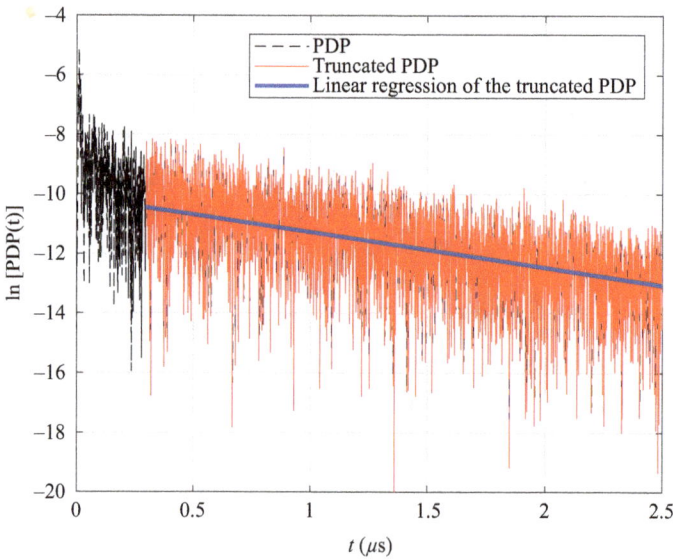

Figure 2.4   *PDP obtained in the unloaded RC using S_{11} measured for one-mode stirrer position. Plots of the entire $ln[PDP(t)]$, the truncated one with the early-time removed and the linear regression of the truncated signal are shown*

Table 2.2   *Minimum, maximum and average value of Δ_f (in MHz) of our RC using the time-domain method for different amount of absorbers inserted in*

	Unloaded	9 PA	One block	Four blocks
Δ_f^{min}	0.183	0.295	0.442	0.84
Δ_f^{max}	0.197	0.32	0.497	1.401
$\langle\Delta_f\rangle$	0.191	0.307	0.47	1.13
$|\Delta_f - \langle\Delta_f\rangle|_{max}$ (%)	4.2	4.2	5.9	25.7

the mode stirrer positions. The maximum deviation of the results with respect to the average value (i.e., $\left|\Delta_f - \langle\Delta_f\rangle\right|_{max}$) is lower than 6% for all the loading conditions except when four blocks of absorber are inserted in the enclosure. This is considered as an acceptable level of uncertainty. In the loading condition using four blocks of absorbers, this deviation reaches 25% as the $PDP(t)$ noise floor is obtained more rapidly (see for instance Fig. 2 of [13]). Therefore, the limited number of usable points in the $ln[PDP(t)]$ reduces the accuracy of the linear regression.

As mentioned before, no mode stirrer is needed with the frequency stepping technique. Figure 2.5 illustrates once again that only one S_{11} measurement is sufficient to get a satisfying accuracy for Δ_f. Indeed, despite the greater discrepancy between the points of $ln[PDP(t)]$ when only one S_{11} measurement is considered, the slope of both curves is assumed to be the same. This demonstrates that Δ_f can be determined accurately using the time-domain method even if the enclosure is not equipped with any mode stirrer. This is possible thanks to a single measurement of S_{11} performed on a large frequency range, a measurement requiring just a few seconds. It will be also shown hereafter that a high accuracy on Δ_f is not necessarily required.

2.2.1.3 Adaptation of the method

The extension of the well-stirred condition method for the frequency stepping stirring technique has to deal with a unique $S_{11}(f)$ measurement collected within the enclosure on the whole frequency range of interest. An example of measurement, taking only a few seconds of measurement, is shown in Figures 2.6 and 2.7 for two

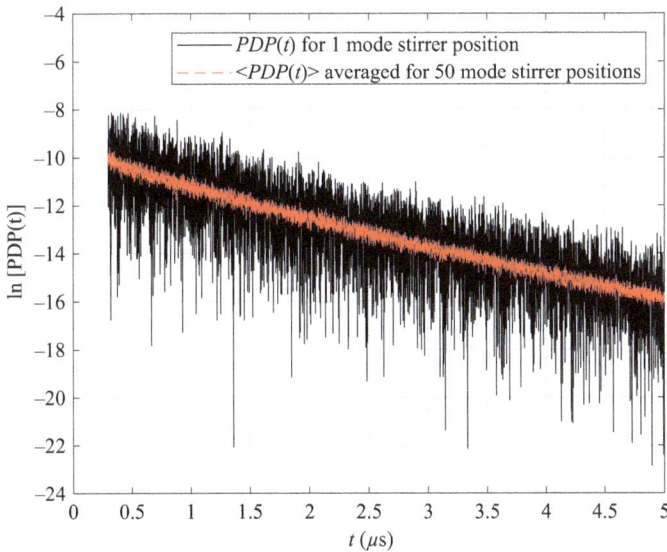

Figure 2.5   *Comparison of ln [PDP(t)] (early-time removed) obtained from S_{11} measurement in the unloaded RC made in 1 mode stirrer position and its average value over 50 mode stirrer positions*

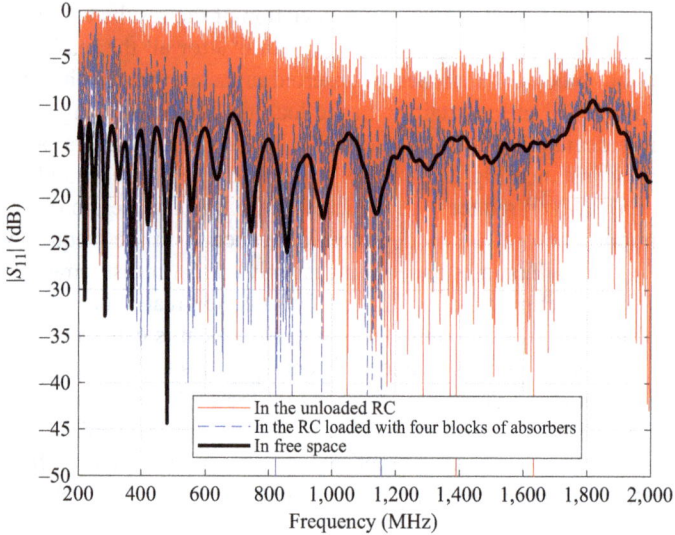

Figure 2.6   Magnitude of S_{11} measured in the RC for two loading conditions for one particular mode stirrer position. This is the unique measurement required to use the well-stirred condition method in the case of the frequency stepping technique. Comparison with the same measurement made in free space

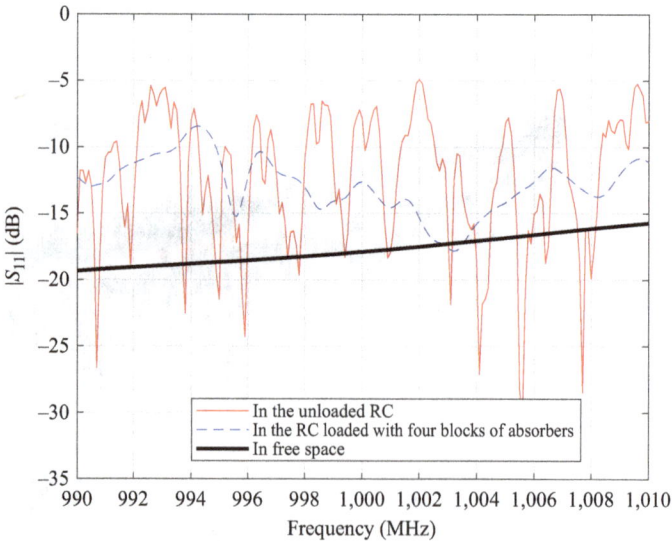

Figure 2.7   Magnitude of S_{11} measured in the RC for two loading conditions (for one particular mode stirrer position) on a 20-MHz bandwidth around 1 GHz. Comparison with the same measurement made in free space

loading conditions of the RC and also in free space. In reality, we are using here the results measured for one fixed position of our mechanical mode stirrer. The range of fluctuations of $S_{11}(f)$ at each frequency is clearly reduced when the RC is loaded.

Figure 2.8 presents in the complex plane the real and imaginary parts of S_{11} on a 25-MHz bandwidth measured in free space and in the unloaded RC around the working frequency $f_0 = 250$ MHz. Both results show a strong correlation between two successive frequencies. The free-space result shows a satisfying impedance mismatch on the entire bandwidth, which is not true for most of the measured data in the RC.

From this single $S_{11}(f)$ measurement, the first task is to remove the S_{11}^{fs} parameter of the antenna when measured in free space at each measured frequency

$$S_{11}^{\text{cor}}(f) = S_{11}(f) - S_{11}^{\text{fs}}(f).$$ (2.7)

Indeed, it is fundamental to discard the variation of $S_{11}(f)$ due to the variation of the antenna behavior as a function of the frequency. The aim is to focus the analysis on the variation of $S_{11}(f)$ due to the RC behavior itself (the modal combination of modes being different between two close frequencies as shown in Figure 2.3).

The second step consists in gathering together N corrected samples S_{11}^{cor} shifted with a step δ_f around each frequency of interest f_0

$$\left[S_{11}^{\text{cor}}(f_1) \ S_{11}^{\text{cor}}(f_2) \ ... \ S_{11}^{\text{cor}}(f_N) \right],$$ (2.8)

with $f(i + 1) - f(i) = \delta_f$.

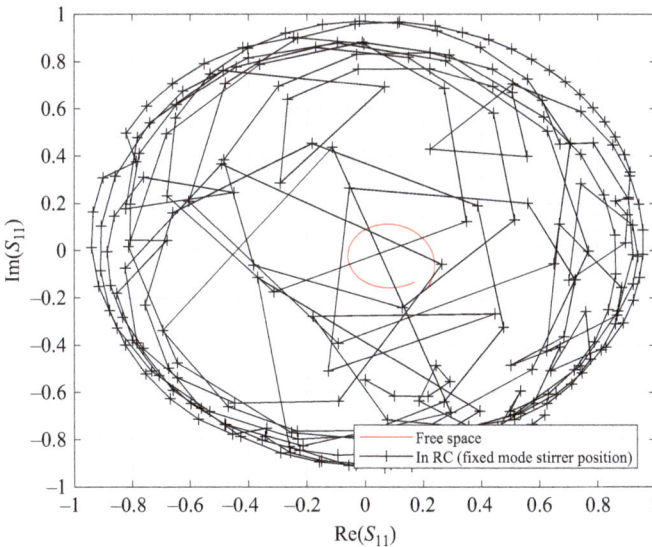

Figure 2.8   *Real and imaginary parts of S_{11} measured in free space and in the unloaded RC (for one particular mode stirrer position) on a 25-MHz bandwidth around the working frequency $f_0 = 250$ MHz*

Then, in the third and last step, the stirred contribution of the N samples, i.e., S_{11}^{sti}, is calculated by removing their average value

$$S_{11}^{sti}(f_1) = S_{11}^{cor}(f_1) - \langle S_{11}^{cor}\rangle. \tag{2.9}$$

The performance of this stirring process can also be assessed using electric field measurements collected with a triaxial electric field probe. In this case, the application of the method requires some minor adjustments described in [15]. It is worth noting that the conclusions drawn from S_{11} or electric field measurements are similar.

2.2.1.4   Example of results

In this subsection, we analyze the measurements made in the RC mechanically stirred, the results obtained for each position of the mode stirrer being treated separately. This allows the process described in Section 2.2.1.3 to be repeated for each position. Then, it is possible to analyze the variation of the method outputs (if any) as a function of the mode stirrer position.

At each frequency of interest f_0, both metrics of the method can be calculated using the N S_{11}^{sti} samples.

From the EM field distributions point of view, the fitted modified AD statistic A_m^2 calculated for the unloaded RC using a frequency step δ_f of 500 kHz is shown in Figure 2.9. As for mechanical stirring, the decrease of this metric with the frequency is clear. Moreover, a remarkable similarity of the curves obtained for each mode stirrer position is observed, leading to a small dispersion of the values of f_1 obtained.

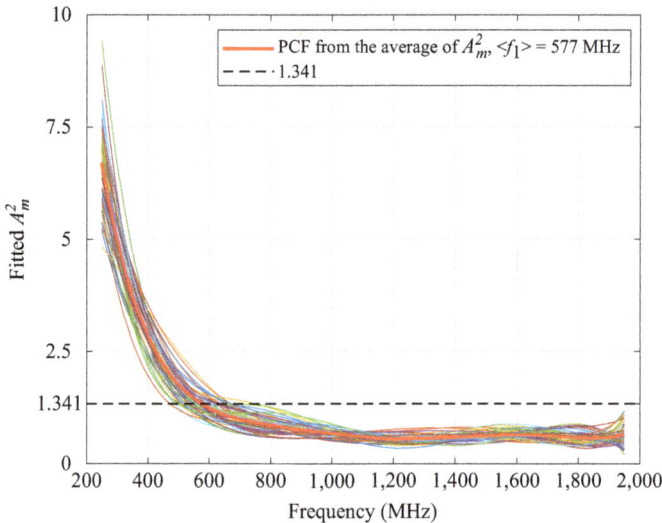

Figure 2.9   Fitted modified AD statistic A_m^2 obtained in the unloaded RC using a frequency step δ_f of 500 kHz (each curve corresponds to a fixed mode stirrer position)

To go further, the same result plotted in Figure 2.10 as a function of the step δ_f shows a small effect of δ_f (if ≥ 200 kHz) on the field distributions. For $\delta_f = 100$ kHz, f_1 increases in reason of the greater correlation of the S_{11}^{sti} samples.

From the sample correlation point of view, the first-order autocorrelation coefficient $r(1)$, shown in Figure 2.11 for the unloaded RC in the case of a frequency step δ_f of 500 kHz, is remarkably different to the results obtained in the case of mechanical stirring. Indeed, $r(1)$ is in this case independent of the frequency. In reason of the flatness of $r(1)$ with the frequency, the PCF process does not present any interest in this case. Instead, it is more relevant to compute the average value of $r(1)$, i.e., $\langle r(1) \rangle$, obtained on the whole frequency range and to compare this value to the threshold value defined in the first chapter of this book (i.e., $t_2 = 0.28$ for $N = 50$ samples). These N samples are considered as uncorrelated on the whole frequency range if $\langle r(1) \rangle < 0.28$. The frequency f_2 is not defined for this particular stirring process. Here the results are binary: the samples are correlated or not depending on the value of $\langle r(1) \rangle$ with respect to t_2. For this stirring technique, the well-stirred condition is considered to be reached above f_1 when $\langle r(1) \rangle < 0.28$.

Figure 2.12 shows, through the analysis of $r(1)$, that the sample correlation decreases on the whole frequency range if the frequency step δ_f increases. Indeed, the sample correlation is driven on the whole frequency range by the relationship between δ_f and the average mode bandwidth Δ_f. The optimal frequency step δ_f^{opt}, defined as the minimum frequency step leading to have $\langle r(1) \rangle$ lower than 0.28, equals 300 kHz, which roughly corresponds to 1.5 Δ_f.

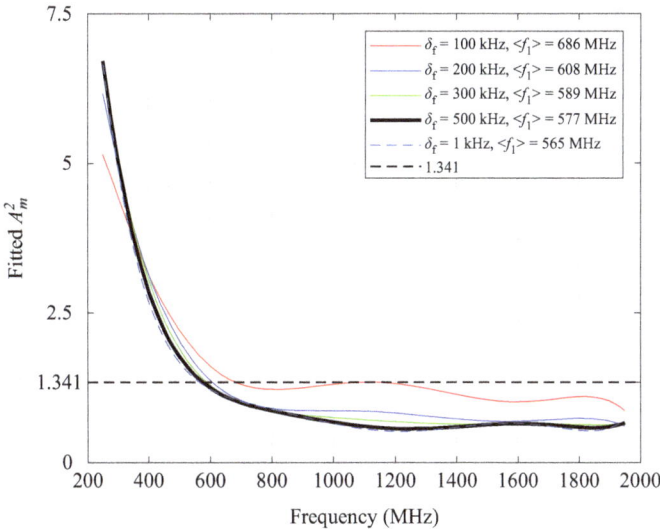

Figure 2.10   Fitted modified AD statistic A_m^2 obtained in the unloaded RC as a function of the frequency step δ_f (for one fixed mode stirrer position)

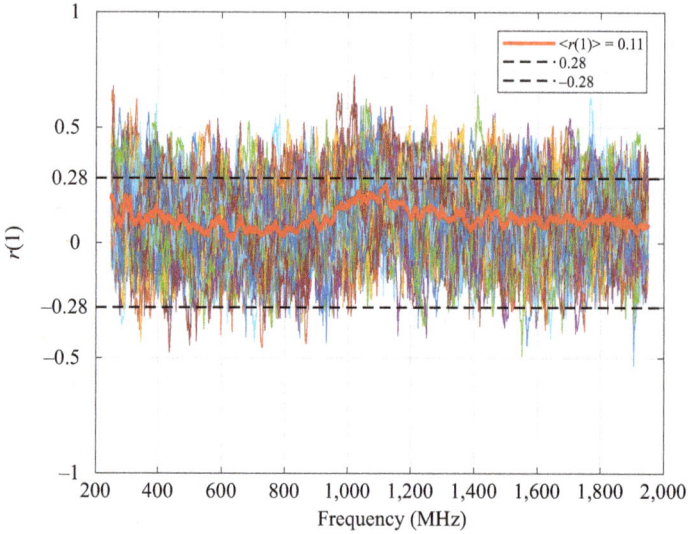

*Figure 2.11   First-order autocorrelation coefficient r(1) in the unloaded RC for a
frequency step δ_f of 500 kHz (each curve corresponds to a fixed mode
stirrer position). The thick-line curve is the average at each frequency
over the 50 mode stirrer positions*

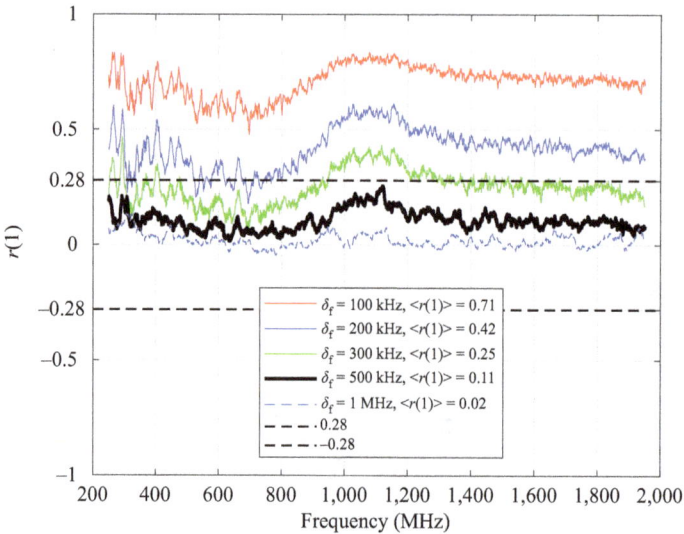

*Figure 2.12   First-order autocorrelation coefficient r(1) (average over the 50 mode
stirrer positions) in the unloaded RC as a function of the frequency
step δ_f (for one fixed mode stirrer position)*

The decrease of correlation when the frequency step δ_f increases is clearly shown in Figure 2.13 where the N S_{11}^{sti} samples collected in the unloaded RC are presented in the complex plane. Indeed, the samples are more randomly distributed for the larger step δ_f of 500 kHz.

The influence of δ_f in the case of the unloaded RC is summarized in the results shown in Table 2.3. It is confirmed that the optimal frequency step equals 300 kHz in this case.

Table 2.4 compiles the results obtained for the four loading conditions of the RC, each time for the optimal frequency step δ_f^{opt}. It is shown that δ_f^{opt} increases with the amount of loading inserted in the enclosure. However, it is particularly interesting to see that the ratio δ_f^{opt}/Δ_f is roughly equal to 2 for all the configurations. This is therefore the recommended ratio when implementing the frequency stepping

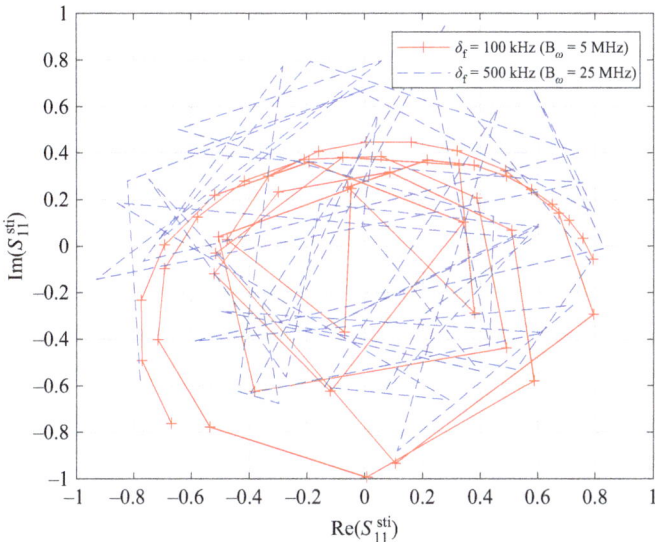

Figure 2.13   *Real and imaginary parts of $N = 50$ S_{11}^{sti} samples obtained in the*
unloaded RC (for one particular mode stirrer position) for two
different frequency steps δ_f around the working frequency
$f_0 = 250$ MHz

Table 2.3   *Results of the frequency stepping for the unloaded RC as a function of*
the frequency step δ_f

δ_f (kHz)	100	200	300	400	5,000	1,000
$\langle f_1 \rangle$ (MHz)	686	608	589	583	577	565
$\langle r(1) \rangle$	0.71	0.42	0.25	0.16	0.11	0.02

Table 2.4   *Results obtained with the frequency stepping technique as a function of the loading configuration for the optimal frequency step δ_f^{opt}*

	Unloaded	9 PA	One block	Four blocks
Δ_f (MHz)	≈ 0.2	≈ 0.3	≈ 0.5	≈ 1.1
δ_f^{opt} (MHz)	0.3	0.6	1	2.2
δ_f^{opt}/Δ_f	≈ 1.5	≈ 2	≈ 2	≈ 2
$\langle f_1 \rangle$ (MHz)	583	411	335	< 250
$\langle r(1) \rangle$	0.25	0.21	0.17	0.2
B_ω (MHz)	14.7	29.4	49	107.8
f_{wsc} (MHz)	583	411	335	< 250
B_ω/f_{wsc} (%)	2.5	7.1	14.6	43.1
f_{min} (MHz)	583	411	490	1,078

technique. It has been reported in [15] that a larger shift degrades the adequacy of the EM field distributions with respect to the ideal ones when the frequency step becomes unreasonable with respect to Δ_f.

From the EM field distribution point of view, f_1 decreases with the amount of absorbers inserted in the RC as for mechanical stirring techniques. Indeed, the decrease of Q involves a better modal overlapping at any frequency. From the sample correlation point of view, the sample can be uncorrelated at 200 MHz.

The lowest f_{wsc} value is obtained for $\delta_f = 2.2$ MHz with four blocks of absorbers inserted in the RC. However, this result is misleading and has to be interpreted carefully. Indeed, this configuration involves a totally unreasonable ratio B_ω/f_{wsc} of almost 43%, larger than the maximum tolerated value of 10%.

When analyzing the results from the point of view of the B_ω/f_{wsc} ratio, only two loading conditions (unloaded and loaded with only 9 PAs) of the RC allow a B_ω/f ratio lower than 10% at f_{wsc}. A new frequency f_{min} defined as the minimum usable frequency for each configuration is introduced. If $B_\omega/f < 10\%$ at f_{wsc}, f_{min} equals f_{wsc}. Otherwise, f_{min} is the frequency when the ratio $B_\omega/f = 10\%$.

The configuration leading to the minimum value of f_{min} is obtained for a frequency shift δ_f of 600 kHz when the enclosure is loaded with a 9 PA absorber. It is worth noting that, as for mechanical stirring, this is the loading condition leading to the minimum value of f_{min} (or f_{wsc} for mechanical stirring). In these conditions, f_{min} is only 411 MHz with a ratio B_ω/f_{wsc} of 7%.

To go further, Table 2.5 presents a comparison between f_{min} obtained with the frequency stepping technique and f_{wsc} obtained with the reference results using the rotating mode stirrer. It can be concluded that the frequency stepping technique gives lower performance than mechanical stirring techniques in our measurement conditions, the lowest absolute value of G_{wsc} being 4.1% for the unloaded configuration.

To conclude, the frequency stepping stirring technique is a relevant technique requiring a simpler experimental setup than mechanical stirring (no movement of any object being required within the enclosure) which simplifies the maintenance and optimizes the available space within the enclosure. The duration of the subsequent test is expected to be shortened (as well as the time required to calibrate the RC with the well-stirred condition method), the time required to move and stabilize the stirrer being no longer necessary.

This technique is recommended if the fact to work on a bandwidth B_ω around the working frequency is tolerated during the subsequent tests or measurements. The frequency stepping technique is often combined with other stirring techniques in order to increase the number of independent samples as discussed in particular in Chapters 5 and 7 of this book.

2.2.2 Variants

Two other source signal stirring techniques have also been proposed in the literature. Instead of exciting the enclosure at different frequencies close to each other by a CW signal, their principle is to excite the enclosure at different frequencies at the same time, from a narrowband white Gaussian noise (NWGN) [16–18] or from a frequency-modulated signal [19] as sketched in Figure 2.14.

Considering the principle of these techniques, it seems difficult to assess their performance with the well-stirred condition method, which requires samples collected at a given frequency to be analyzed. Therefore, these methods have not been implemented in our RC as they also involve a non-negligible increase of the experimental setup complexity, through the use of a more complex generator.

Table 2.5 *Comparison of results obtained with the mechanical and the frequency stepping stirring techniques*

		Unloaded	9 PA	One block	Four blocks
Mechanical (f_{wsc} (MHz))		559	346	397	663
Freq. stepping	δ_f^{opt} (MHz)	0.3	0.6	1	2.2
	f_{min} (MHz)	583	411	490	1,078
G_{min} (%)		−4.3	−18.8	−23.4	−62.6

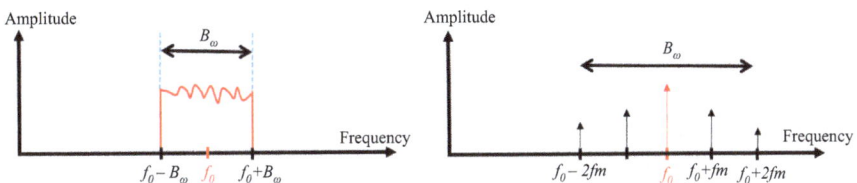

Figure 2.14 *Sketch of the basic principle of the NWGN technique (on the left) and of the frequency-modulated technique (on the right) presenting the power spectrum of the excitation signal in both cases*

The NWGN has been proposed first. At any instant, the NWGN signal excites the modes of the enclosure according to the amplitude of each frequency contained in the input signal. Due to the time-varying nature of the source, the amplitude of each frequency contained in the NWGN spectrum varies over the time. Each enclosure mode is therefore supposed to be excited differently at any moment resulting in an efficient stirring process.

If each frequency of the NWGN spectrum is considered separately, the same enclosure modes are excited by a CW signal of time-varying amplitude. Considering the linearity of the problem, there is no stirring process at this level and the total field is perfectly correlated with the power level of the source at this frequency. The stirring process is achieved because the field produced at any point of the enclosure is the summation of all the excited modes by all the frequencies contained in the NWGN signal. If two different frequencies (separated by a frequency shift δ_f) excited by the NWGN signal are now considered, the study looks similar to the case of the frequency stepping technique and depends on the δ_f / Δ_f ratio. The key point is to define the bandwidth B_ω (with respect to Δ_f) of the NWGN signal ensuring a satisfying stirring process. It makes sense to consider that the bandwidth B_{NWGN} of the NWGN signal has to be greater than the average mode bandwidth Δ_f of the RC. Therefore, the extension described previously of the well-stirred condition method to the frequency stepping technique is assumed to provide all the required information in order to apply successfully the NWGN stirring technique after.

This technique exhibits a great interest for subsequent EMC measurements such as radiated susceptibility tests or shielding effectiveness measurements [20]. Indeed, the time-varying excitation of each frequency contained in the NWGN bandwidth is expected to drastically shorten the measurement time with respect to the frequency stepping technique where each frequency around the working frequency is excited separately. Indeed, when considering for instance a radiated susceptibility test, the EUT exposition time required in order to obtain a large number of independent states of the EM field (for instance $N = 50$) is supposed to be extremely short (i.e., a few seconds), similarly to what is discussed in the third chapter of this book devoted to the VIRCs. Moreover, at any instant, the richness of the spectral content of the incoming waves arriving on the EUT gives a great confidence in the completeness of the test.

Based on the same principle than the NWGN method presented in the previous subsection, a variant presented some years ago [19] consists in exciting the emitting antenna with a frequency-modulated signal. Indeed, if a low-frequency sinusoidal signal f_m frequency modulated around an RF frequency f_0 is considered (but other kinds of modulated signals would be usable), the power spectrum related to this frequency-modulated signal contains different frequencies $f_0 \pm k \cdot f_m$ (with k being an integer), each having an amplitude linked to the amplitude (i.e., the modulation index) of the f_m signal. It is then possible to excite the enclosure with different closed frequencies at the same time while controlling the magnitude of each excited frequency. Then, a slight modification of either f_0 or f_m leads to shift all the frequencies of the spectrum at the same time and to excite the enclosure in a different manner. In the opinion of the author, this technique is less promising than the NWGN technique as the richness of the excitation signal spectrum is lower.

2.3   Source position stirring techniques

The techniques where the stirring process is ensured by the movement of the emitting antenna itself are analyzed in this section. Often referred to as "source stirring" techniques, the terminology "source position stirring" is used in this chapter in order to better emphasize the differences with "source signal stirring" techniques described in the previous section.

In this framework, the basic technique consists in moving the antenna without moving anything else in the RC. The same principle can also be applied by inserting a network of different fixed antennas. Even if the experimental setup is more complex, this method avoids the movement of any object in the RC.

These stirring techniques are often coupled (see Chapters 5 and 7 of this book) with a mechanical stirring technique (i.e., one-mode or two-mode stirrers) in order to increase the number of independent samples and therefore reduce the uncertainty level [21,22] of the measurements.

2.3.1   Moving antenna stirring technique

The idea of moving the emitting antenna in an RC has been first proposed in [23] and investigated theoretically and numerically in [24,25], among other papers. This technique, sketched in Figure 2.15, requires moving the antenna in different positions. It represents an interesting alternative to the construction and the installation of a rotating mode stirrer.

The physical principle of such technique is described as follows. In an enclosure, each mode is excited differently according to the source position. If a point source is considered, the excitation of each mode is related to its relative amplitude in this particular position. In other terms, if the cartography of the mode presents a maximum in this position, the excitation of the mode is also maximum. Then, a stirring process

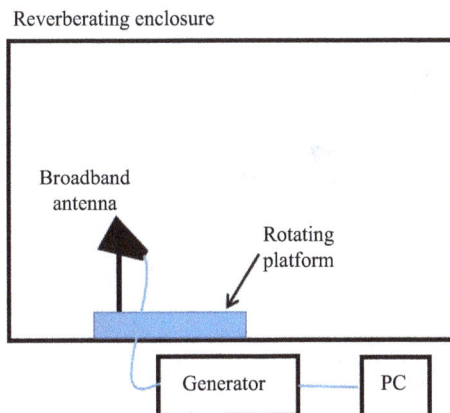

Figure 2.15   Typical experimental setup required for the moving antenna technique using a rotating platform (i.e., platform stirring)

can be achieved if the position of the source (i.e., the antenna) is moved within the enclosure, each mode being excited differently for each position of the source. It is worth noting that this technique could also be considered as a mechanical stirring technique as the antenna (a metallic object of non-negligible size) is moved within the enclosure. A classical implementation of this technique concerns the installation of the antenna upon a rotating platform (or a turntable), a method referred in the literature as "platform stirring" [26–28].

The key point is the distance between two successive positions of the antenna which has to be sufficient to generate independent samples of the EM field in any position of the enclosure. A commonly accepted criterion is to consider a $\lambda/2$ minimum distance between two successive positions of the antenna [29]. If this criterion is not matched during the rotation of the antenna, for instance if the radius (or the rotation angle) of the circular movement of the antenna is too small, a correlation of the EM field obtained for two successive positions of the antenna is expected.

The experimental setup implemented in the XLIM RC is depicted in Figure 2.16. Instead of using a turntable, the log-periodic antenna has been attached to the metallic mast linked to the step-by-step motor thanks to a dielectric bar, the rotating mode stirrer being therefore removed during these experiments. Different radii r (15, 30 and 45 cm) of displacement have been considered, r being the distance between the metallic mast and the point where the antenna is attached to the dielectric support. It is important to mention that $r = 45$ cm is the maximum achievable value in the RC according to the dimensions of both the enclosure and the antenna and to the relative position of the metallic mast within the enclosure. The measurement time (10 min) is the same as the measurement time required for the mechanical stirring technique.

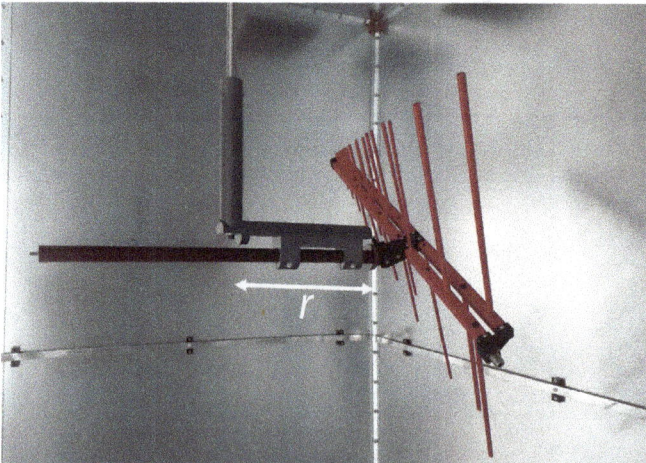

Figure 2.16   Experimental setup related to the rotating antenna stirring technique installed in our RC (with a definition of the radius r of the circular movement)

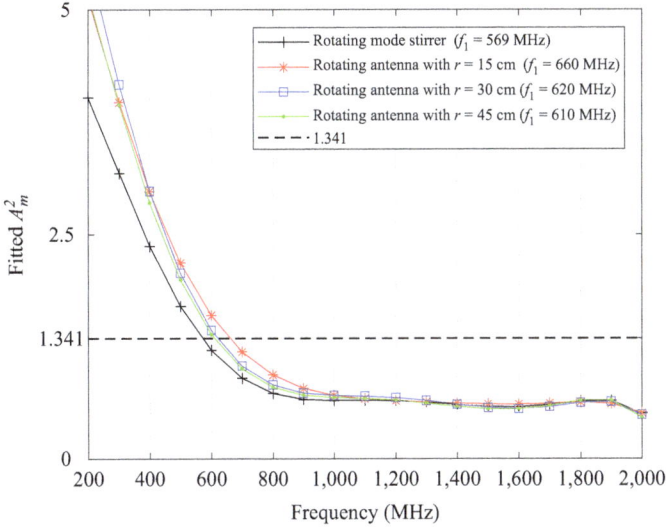

Figure 2.17   Fitted modified AD statistic A_m^2 in the unloaded RC for a rotating antenna with different radii of displacement r. Comparison with the reference results obtained with the rotating mode stirrer for the same loading condition

As for the other techniques, the aim is to assess its performance as a function of both the amount of absorber inserted in (i.e., Q or Δ_f) and the radius r. $N = 50$ samples are collected at each frequency. This corresponds to a rotation of the antenna of $7.2°$ between two successive positions. Figures 2.17 and 2.18 present the results obtained for the unloaded RC as a function of r also compared with the reference configuration. In addition, complete results obtained for each loading configuration are collected in Table 2.6.

It should first be noticed that the decreasing trends of both metrics with the frequency are similar to the results obtained with a rotating mode stirrer. Moreover, as expected, the performance of this technique illustrated by f_1, f_2 and f_{wsc} are rapidly improved with larger values of r because of the increase of the stirred volume and of the average angular velocity of the antenna [30].

Finally, these results also show, in these experimental conditions, that this technique has lower performance compared to mechanical stirring. The performance of the method from the EM field distribution (i.e., f_1) point of view is not so far from the results obtained in the reference configuration for all the loading conditions. For instance, G_1 equals only -9% for the unloaded RC. However, the decrease of performance is more sensitive on the sample correlation (i.e., f_2) as G_2 (in absolute value) has a minimum value of 53% when four blocks of absorbers are inserted in the RC and is greater than 100% for the other configurations. The sample correlation being related to the movement of the antenna between two successive stirring configurations, it is clearly confirmed that this movement has to be as large as possible.

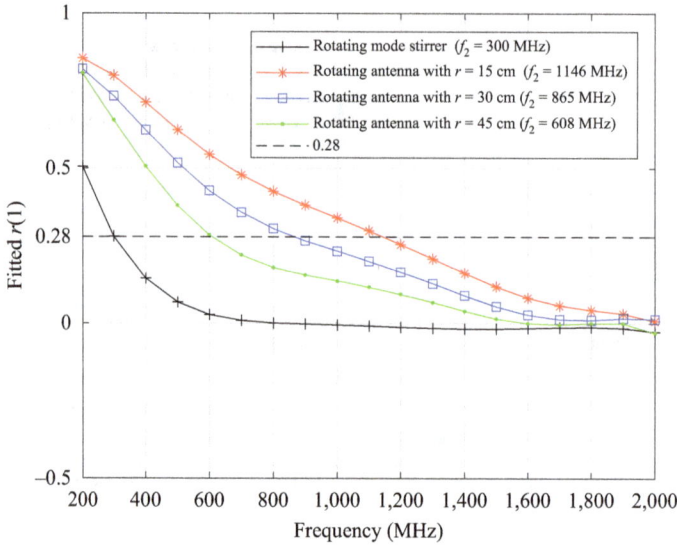

*Figure 2.18   Fitted first-order autocorrelation coefficient r(1) in the unloaded RC
for a rotating antenna with different radii of displacement r.
Comparison with the reference results obtained with the rotating
mode stirrer for the same loading condition*

*Table 2.6   Results obtained for the moving antenna stirring technique for different
amount of absorbers inserted in and different displacement radii r (G$_1$,
G$_2$ and G$_{wsc}$ are calculated for the r = 45 cm configuration)*

		Unloaded	9 PA	One block	Four blocks
f_1 (MHz)	$r = 15$ cm	659	511	479	443
	$r = 30$ cm	620	486	450	390
	$r = 45$ cm	610	461	408	328
	G_1 (%)	−9.1	−33	−46	N/A
f_2 (MHz)	$r = 15$ cm	1,146	1,253	1,301	1,384
	$r = 30$ cm	865	1,052	1,111	1,219
	$r = 45$ cm	608	789	860	1,012
	G_2 (%)	−110	−133	−117	−53
f_{wsc} (MHz)	$r = 15$ cm	1,146	1,253	1,301	1,384
	$r = 30$ cm	865	1,052	1,111	1,219
	$r = 45$ cm	610	789	860	1,012
	G_{wsc} (%)	−9.1	−128	−117	−53

The large values of f_2 obtained with the moving antenna technique means that
f_{wsc} is equal to f_2 in most of the configurations. Therefore, f_{wsc} increases with the
amount of absorbers inserted in the RC. However, an in-depth analysis of f_1 and f_2 as
a function of the amount of loading shows similar influence on both metrics, i.e., f_1

decreases while f_2 increases with the insertion of absorbers for the reasons described in Chapter 1.

It should be emphasized that a larger radius r would probably lead to similar results (i.e., f_{wsc}) with respect to mechanical stirring but this was impossible to test in our RC as already mentioned. It is also important to mention that the movement of the emitting antenna does not reduce the setup complexity (as the use of a motor is still required) and does not increase the available space within the RC, particularly for large r values.

2.3.2 Antenna network stirring technique

An alternative to the movement of the source consists in inserting different fixed antennas sequentially fed. An example of (simplified) experimental setup is shown in Figure 2.19 with only six antennas (for the sake of readability of the figure). The stirring process is here ensured by the different coupling obtained between each antenna and the enclosure modes. The different antennas are being located in any position of the RC (with a $\lambda/2$ minimum distance between two antennas [29]); this leads rationally to decrease the correlation of the EM field in comparison to the antenna moving stirring technique.

This technique requires the use of a broadband "reflective" switch in order to excite alternatively each antenna, the term reflective meaning that each antenna is connected on an open-circuit impedance if the antenna is not excited. This allows a sufficient Q to be maintained, only one antenna being active for each stirring configuration. For N values larger than 10, the use of different switches put in cascade seems to be required. It should be noticed that the performance of this technique can be studied (experimentally or numerically) by moving one antenna in different positions of the cavity assuming that the effect of the other non-excited antennas is quite negligible. Some examples of studies can be found in the literature. In [31], a network of 20 conical wire antennas located on a metallic cube of 1 m side size has been proposed with a minimum distance of 20 cm between two antennas. In [32], a

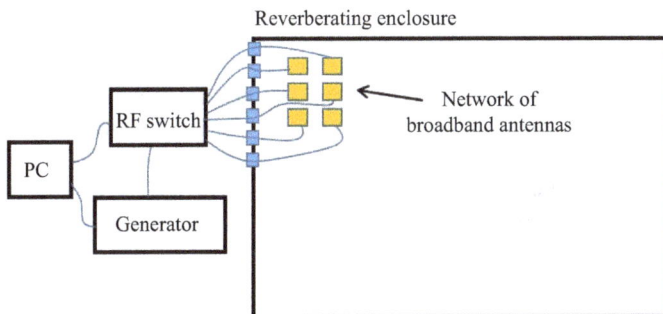

Figure 2.19 Typical experimental setup required for the network antenna technique using six broadband antennas

network of 154 patches located on a wall of an enclosure has been studied numerically with the CST software.

Unlike the antenna moving technique, this technique does not require a movement of any object within the enclosure during the stirring process. However, this technique is difficult to implement in real conditions in reason of the complexity and of the cost of the experimental setup. Indeed, N broadband antennas (with also N coaxial cables and connectors feedthrough on the RC walls) are required if N stirring configurations are desired. This leads for sure to highly reduced the available space within the enclosure with respect to traditional stirring techniques.

In reason of all these drawbacks, this stirring technique has not been implemented in our RC. However, a hybrid approach combining the benefits of using multiple antenna and of the frequency stepping technique is described in the next subsection.

2.3.3   Hybrid stirring technique combining multiple antennas and frequency stepping

As mentioned earlier, a hybrid approach combining the benefits of using multiple antennas and of the frequency stepping technique has been implemented. The aim is to use a reasonable number N_a of broadband antennas within the enclosure and to excite each of them successively at N_f shifted frequencies (with a δ_f step) in order to get N samples (with $N = N_a N_f$). The aim is to reduce the bandwidth B_ω around each working frequency f with respect to the frequency stepping technique in order to maintain the B_ω/f ratio below 10%. Indeed, the bandwidth B_ω equals $(N_f - 1)\,\delta_f$ in this case and is therefore reduced by a ratio almost equal to the number of antennas N_a with respect to the frequency stepping technique.

In the experiments carried out in our RC, $N_a = 5$ broadband antennas shown in Figure 2.20 and $N_f = 10$ shifted frequencies around each working frequency have been chosen in order to work once again with $N = 50$ samples. Two log-periodic antennas and three horn antennas (respectively, matched in the frequency range 0.2–2 and 0.5–2.8 GHz) are used (as well as a broadband "reflective" switch) because five identical broadband antennas were not available in the XLIM EMC Laboratory.

The free-space reflection coefficient S_{11}^{fs} of each model of antenna is plotted in Figure 2.21. A minimum distance of at least $\lambda/2$ at the minimum frequency of interest [29] has been respected when positioning the different antennas within the RC in order to ensure a low sample correlation.

Because of the bandwidth of the antennas, the measurements have been performed only in the frequency range between 0.5 and 2 GHz over 15,001 working frequencies linearly spaced with a shift of 100 kHz. The application of the well-stirred condition method requires the measurement of the S_{11} parameter of each antenna within the RC as depicted in Figure 2.22. In these measurement conditions, the measurement time is lower than a minute.

As described in Section 2.2.1.3 for the frequency stepping technique, the post-treatment process requires

- to remove for each set of data the free-space reflection coefficient S_{11}^{fs} of the corresponding antenna;

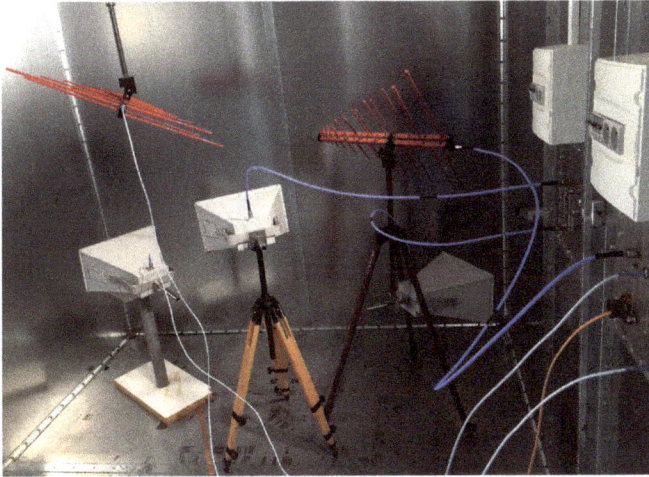

Figure 2.20 Experimental setup of the hybrid stirring technique showing two log-periodic antennas (red color) and three horn antennas (gray color) inserted in the RC

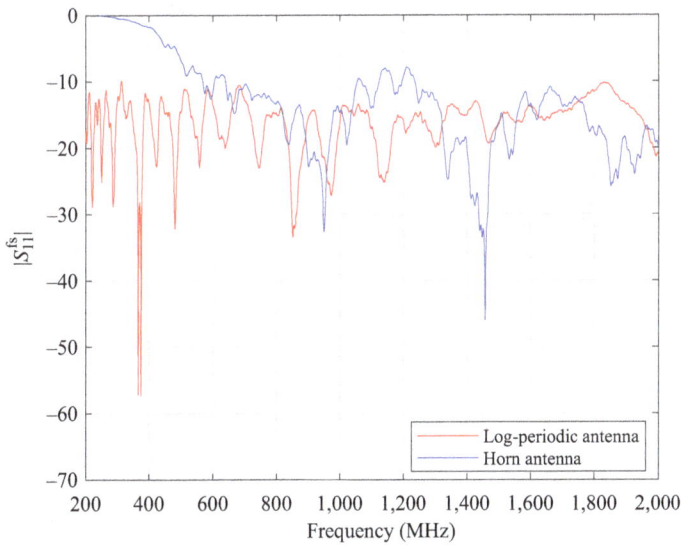

Figure 2.21 Free-space reflection coefficient S_{11}^{fs} of both kinds of antennas used

- to gather together N samples around each frequency of interest f_0;
- to remove the average of these samples in order to get S_{11}^{sti}.

Figure 2.23 shows the data obtained after the first step of the procedure. It is particularly interesting to see that it is not possible anymore to determine what is the antenna related to each curve.

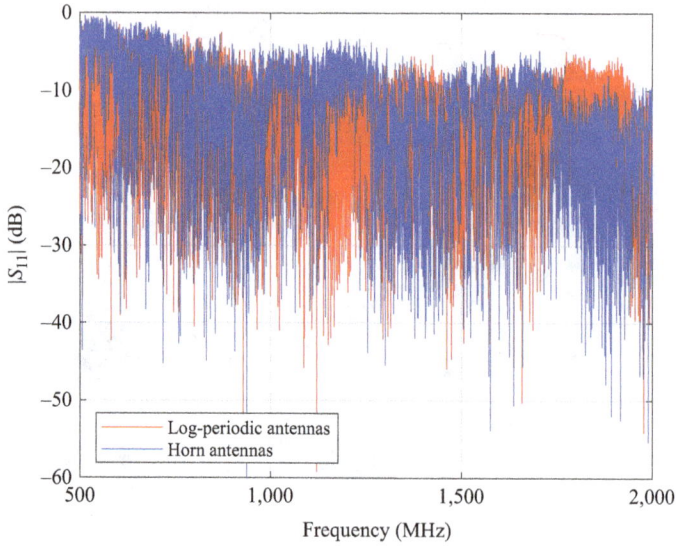

Figure 2.22   Magnitude of S_{11} measured in the unloaded RC for each of the five antennas. It is easy to distinguish the plots related to each kind of antennas in reason of the difference of S_{11}^{fs}

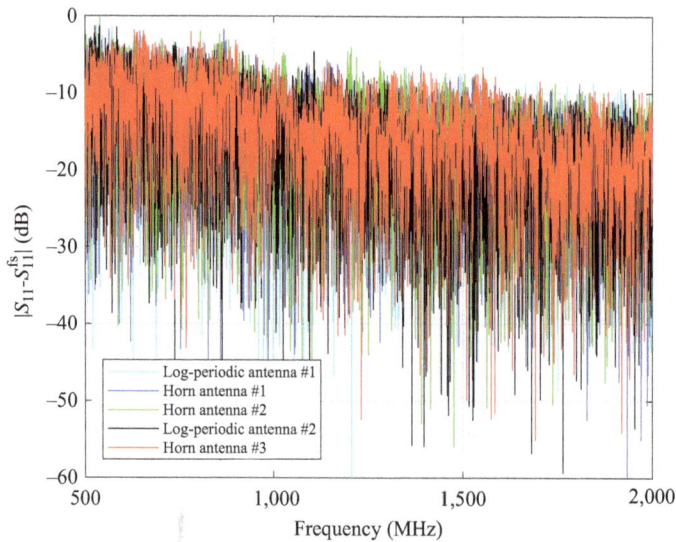

Figure 2.23   Magnitude of $S_{11} - S_{11}^{fs}$ measured in the unloaded RC. All the curves have now the same shape whatever the antenna used during the measurements

At each frequency of interest, the AD GOF test is performed over the N samples collected at each working frequency as shown in Figure 2.24 in the case of the unloaded RC for different frequency steps δ_f. The Rayleigh distribution is most of the time accepted over 500 MHz with no major differences observed between the different values of δ_f considered (with the exception of the lowest frequency step of 100 kHz in reason of the sample correlation occurring in this case). This conclusion is similar to the case of the frequency stepping technique. Below 500 MHz, it is assumed that this distribution would be massively rejected at 200 MHz and progressively obtained with the increase of the frequency.

Instead of computing the PCF process, it is more relevant in this case to compute the average value of the modified AD statistic A_m^2, i.e., $\langle A_m^2 \rangle$ (indicated in the legend of the figure) in the whole frequency range and to compare it with the corresponding critical value of the AD GOF test for such number of samples N (1.341 for $N = 50$). It is shown that the frequency step δ_f of 300 kHz allows a value of $\langle A_m^2 \rangle$ significantly lower than the critical value to be obtained.

A major difference with other stirring techniques has to be highlighted concerning the sample correlation assessment. Thus, in traditional stirring techniques, data are collected in such a way that the first-order $r(1)$ of the autocorrelation coefficient (when data are shifted of one rank) corresponds to the order leading to the maximum of the correlation. With the hybrid stirring technique, data are gathered differently. The way we have proceeded (but other solutions exist) is described as follows

$$\left[S_{11}^{A_1, f_1} \; S_{11}^{A_2, f_1} \; \cdots \; S_{11}^{A_5, f_1} \; S_{11}^{A_1, f_2} \; \cdots \; S_{11}^{A_5, f_{10}} \right], \tag{2.10}$$

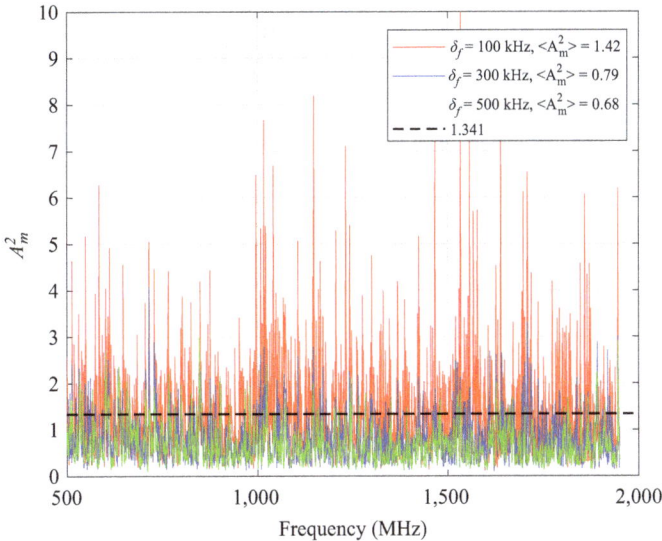

Figure 2.24   *Modified AD statistic A_m^2 obtained with the hybrid stirring technique in the unloaded RC as a function of the frequency step δ_f*

with $S_{11}^{A_1, f_1}$ corresponds to the measurements of S_{11} when the first antenna is excited at the first frequency around f_0.

When the data are arranged like in (2.10), two orders of the autocorrelation coefficient $r(n)$ are particularly interesting

- The first order, i.e., $r(1)$, quantifies the correlation obtained when two different antennas are excited at the same frequency. Considering that the distance between each antenna is at least $\lambda/2$, $r(1)$ is supposed to be low [29];
- The fifth order, i.e., $r(5)$, quantifies the degree of correlation obtained when the same antenna is excited at two successive frequencies. This is the critical order to be checked.

Figure 2.25 presents the autocorrelation coefficient r obtained in the unloaded RC for the orders 1, 5 and 10 in the case of a small frequency step δ_f of 100 kHz. The figure shows first that there is no frequency dependence of these coefficients as the results are relatively flat with the frequency. This is the same phenomenon already observed for the frequency stepping technique.

The critical order is, as expected, the order 5, i.e., $r(5)$. The order 10 represents the correlation of samples obtained when the same antenna is excited at two frequencies shifted by a step of $2\delta_f$. The absence of correlation at the order 1, i.e., $r(1)$, is clearly emphasized. It is worth noting that the results obtained at all the other orders (except $r(15)$, not shown in the figure for the sake of clarity), exhibit a clear absence of correlation of the samples.

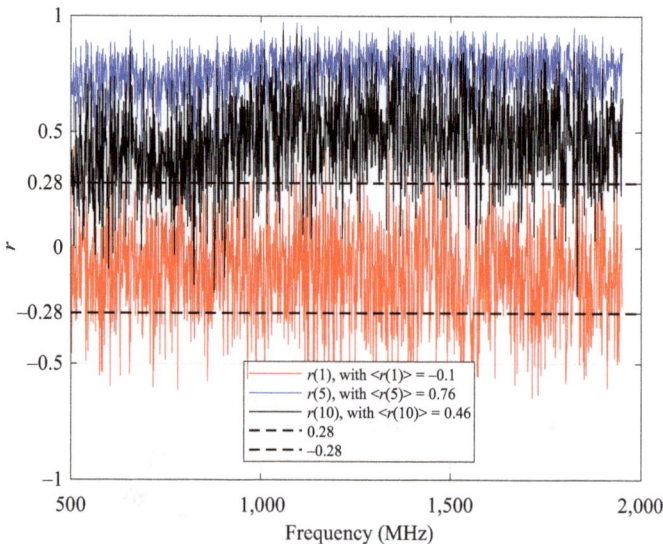

Figure 2.25   Autocorrelation coefficient r(n) obtained with the hybrid stirring technique in the unloaded RC as a function of the order n for a frequency step δ_f of 100 kHz

In reason of the absence of frequency dependence of these results, it is therefore interesting to compute the average of the autocorrelation coefficient at each order on the whole frequency range, i.e., $\langle r(n) \rangle$, and to compare this value to the threshold value t_2, i.e., 0.28 for $N = 50$. In Figure 2.26 showing the decrease of $\langle r(5) \rangle$ when δ_f grows, the minimum value of δ_f leading to $\langle r(5) \rangle < 0.28$ equals 500 kHz.

Table 2.7 presents the results obtained for the unloaded RC as a function of the frequency step δ_f. It is confirmed that the minimum frequency step ensuring that the well-stirred condition is reached (i.e., $\langle A_m^2 \rangle < 1.341$ and $\langle r(5) \rangle < 0.28$) equals 500 kHz.

Table 2.8 presents the results obtained for the optimal frequency step δ_f^{opt} for each loading configuration, while Table 2.9 allows the results to be compared with the reference results. It should be emphasized here that no frequency step, even large, has permitted to reach the well-stirred condition for the four blocks of absorber condition.

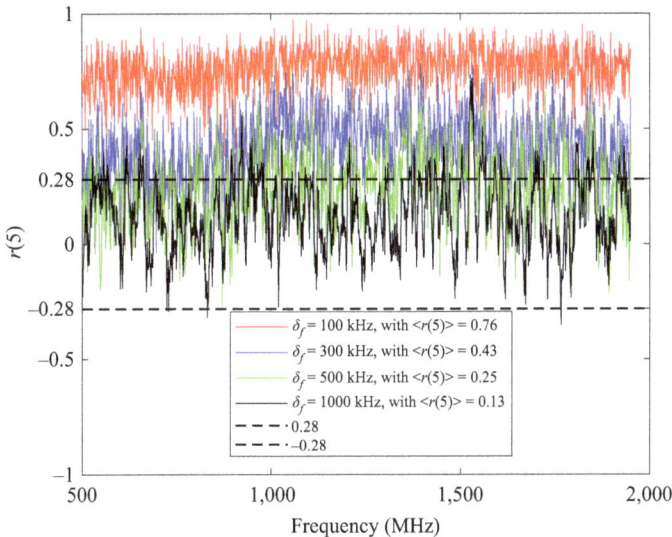

Figure 2.26 Autocorrelation coefficient r(5) obtained with the hybrid stirring technique in the unloaded RC as a function of the frequency step δ_f

Table 2.7 Results of the hybrid stirring technique for the unloaded RC as a function of the frequency step δ_f

δ_f (MHz)	0.1	0.2	0.3	0.4	0.5	0.6	1
$\langle A_m^2 \rangle$	1.42	0.95	0.79	0.72	0.68	0.66	0.68
$\langle r(5) \rangle$	0.76	0.58	0.43	0.32	0.25	0.2	0.13

Table 2.8   Results obtained with the hybrid stirring technique in the RC as a function of the loading configuration for the optimal frequency step δ_f

	Unloaded	9 PA	One block
Δ_f (MHz)	≈ 0.2	≈ 0.3	≈ 0.5
δ_f^{opt} (MHz)	0.5	0.8	1.2
δ_f^{opt}/Δ_f	≈ 2.5	≈ 2.7	≈ 2.4
B_ω (MHz)	4.5	7.2	10.8
$\langle A_m^2 \rangle$	0.68	0.69	0.69
$\langle r(5) \rangle$	0.25	0.27	0.28
f_{wsc} (MHz)	<400	<400	<400
B_ω/f_{wsc}^* (%)	1.1	1.8	2.7
f_{min} (MHz)	<400	<400	<400

*$f_{wsc} = 400$ MHz is considered here.

Table 2.9   Comparison of results obtained with the mechanical and the hybrid stirring techniques in the RC

		Unloaded	9 PA	One block
Mechanical (f_{wsc} (MHz))		559	346	397
Hybrid stirring	δ_f^{opt} [MHz]	0.5	0.8	1.2
	f_{min} [MHz]	<400	<400	<400
G_{min}^* (%)		28.4	−15.6	−0.8

*$f_{min} = 400$ MHz is considered here.

As a conclusion on these results, the hybrid stirring technique exhibits similar performance with respect to the frequency stepping technique. However, the major interest of this technique is the reduction of the bandwidth B_ω of a factor almost equal to the number of antennas N_a used.

However, this bandwidth reduction is achieved at the cost of a strong increase of the experimental setup complexity requiring multiple antennas, numerous coaxial cables inside and outside the enclosure and a switching control unit. Therefore, this stirring technique seems adapted only when (1) it is impossible to move any object (no motor available) within the enclosure, (2) when a reduction of the available space within the enclosure is not essential and (3) when the reduction of B_ω is of prime importance.

During the subsequent tests or measurements performed after this calibration procedure, it is also important to make sure that each antenna is sufficiently matched (in free space) over the whole bandwidth. Indeed, this ensures that the power radiated by each antenna is similar, which is fundamental. Otherwise, the power radiated by each antenna will be really different which would contribute to create important differences in the EM field strength produced in the enclosure by each antenna.

2.4 Other mode stirring techniques

Two other stirring techniques (not implemented in our RC in reason of the complexity of the required setup) deserve to be evocated in this chapter reviewing the performance of stirring techniques already proposed for RCs.

2.4.1 Reactively loaded antenna stirring technique

This method proposed in [33] consists in installing several antennas in the RC, one being used as the emitting antenna and the others being used as passive antennas receiving and reemitting the energy in the RC. Each passive antenna is loaded by a controllable reactance circuit, the real part of the loading impedance being null in order to avoid to severely decrease the average composite quality factor Q of the enclosure. The stirring process is ensured in this technique by the variation of the reactance circuit connected to each passive antenna.

This approach is shown in [33] to modify the scattering parameters between a set of two antennas, the autocorrelation coefficient obtained at a frequency (2 GHz) highly greater than the fundamental resonance of the enclosure being also almost null. Despite these encouraging results, this stirring technique quite comparable in the principle with the hybrid stirring technique presented before constitutes an incremental step from the experimental setup complexity in reason of the use of a controllable reactance connected to each passive antenna. Moreover, interrogations remain on the bandwidth of operation of such controllable reactances, which have to be broadband.

We encourage RC users to investigate in the future the performance of this stirring technique, for instance using the well-stirred condition method, in order to compare with the performance (i.e., f_{wsc}) obtained with traditional stirring techniques and in particular mechanical stirring. Indeed, it is still an open question to prove that the modifications of the impedance of passive antennas is sufficient to obtain a satisfying stirring process, this on a large frequency range.

2.4.2 Switched stirred reverberation chamber

The so-called switched stirred RC is a recently patented stirring technique [34] not yet tested experimentally at our knowledge. The method consists in installing fixed metallic plates in the enclosure. By means of different switches, these plates are connected or disconnected to the enclosure walls. This process ensuring a modification of the boundary conditions of the whole system constitutes the stirring process. It is therefore possible to obtain 2^N different stirring configurations if N metallic plates are installed within the enclosure. As mentioned in [1], this technique seems interesting in the context of large structures such as aeronautic halls [35] where the building and installation of a rotating mode stirrer is difficult. It would be interesting in the near future to prove that the modifications of the boundary conditions involved by the different connections and disconnections of the metallic plates are sufficient to ensure a satisfying stirring process, especially around the frequency f_{wsc} which would be obtained by traditional stirring techniques.

2.5   Conclusion

This chapter presents a critical review of the stirring techniques already proposed in the literature as well as a fair comparison of the performance of the most popular stirring techniques implemented in the same enclosure (i.e., the XLIM RC) using the well-stirred condition method. Using the same enclosure, the observed differences can only be attributed to the performance of the stirring process itself. Even if the relative comparison between each stirring technique is obviously dependent of our experimental conditions, the purpose of this study is to identify the main parameters playing a role on the efficiency of each stirring technique. These rules are assumed to be extendable to an RC of any shape. Once these parameters identified, the optimization of each stirring technique becomes feasible.

In the experimental conditions we have described, each stirring technique has shown comparable performance with respect to the reference results obtained with a mechanical stirrer. It is important to mention here that in the case of an enclosure stirred successively by a "small" mode stirrer or by a rotating antenna having a large displacement radius, the relative differences between stirring techniques would be for sure different in these new experimental conditions while staying in agreement with the main conclusions summarized as follows for each stirring technique.

The frequency stepping stirring technique requires a simpler experimental setup and optimizes the available space within the enclosure. This technique is an interesting alternative if it is tolerated to perform the measurements on a narrowband around the working frequency. However, it seems reasonable to limit the ratio between the bandwidth B_ω of the excitation signal and of the working frequency f, for instance using a 10% criterion.

The fundamental point for the moving antenna stirring technique is to increase the stirred volume of the antenna during its movement. An interesting variant consists in inserting a reasonable number of fixed antennas and use the frequency stepping stirring technique on each of them. This hybrid technique allows the bandwidth B_ω to be reduced of a factor almost equal to the number of antennas used.

This chapter is written with the intention to be helpful for any RC user having to choose the best stirring technique to implement in an RC under construction. Indeed, this chapter brings elements to make the better choice according to the constraints of the user, the main ones being the enclosure dimensions, the experimental setup complexity required by the stirring process, the available space within the enclosure and the measurement time required during both the calibration phase and the subsequent test.

As already mentioned in this chapter, it should be emphasized that two or more stirring techniques are often combined together in order to increase the number of available independent samples and therefore reduce the uncertainty of the measurements. This is extensively discussed in this book, in particular, in Chapters 5 and 7.

References

[1]   Serra R, Marvin AC, Moglie F, *et al.* Reverberation chambers a la carte: an overview of the different mode-stirring techniques. IEEE Electromagnetic Compatibility Magazine. 2017 First;6(1):63–78.

[2]   Hill DA. Boundary fields in reverberation chambers. IEEE Transactions on Electromagnetic Compatibility. 2005;47(2):281–290.

[3]   Leferink F, Boudenot JC, and van Etten W. Experimental results obtained in the vibrating intrinsic reverberation chamber. In: IEEE International Symposium on Electromagnetic Compatibility. Symposium Record (Cat. No.00CH37016). vol. 2; 2000. p. 639–644.

[4]   Leferink F. Patent NL1010745, Test Chamber, 7-12-1998.

[5]   Fedeli D, Iualè M, Primiani VM, *et al.* Experimental and numerical analysis of a carousel stirrer for reverberation chambers. In: 2012 IEEE International Symposium on Electromagnetic Compatibility; 2012. p. 228–233.

[6]   Cappetta L, Feo M, Fiumara V, *et al.* Electromagnetic chaos in mode-stirred reverberation enclosures. IEEE Transactions on Electromagnetic Compatibility. 1998;40(3):185–192.

[7]   Kouveliotis NK, Trakadas PT, and Capsalis CN. Theoretical investigation of the field conditions in a vibrating reverberation chamber with an unstirred component. IEEE Transactions on Electromagnetic Compatibility. 2003;45(1):77–81.

[8]   Barakos D and Serra R. Performance characterization of the oscillating wall stirrer. In: 2017 International Symposium on Electromagnetic Compatibility - EMC EUROPE; 2017. p. 1–4.

[9]   Monsef F and Cozza A. Average number of significant modes excited in a mode-stirred reverberation chamber. IEEE Transactions on Electromagnetic Compatibility. 2014;56(2):259–265.

[10] Fall AK, Besnier P, Lemoine C, *et al.* Design and experimental validation of a mode-stirred reverberation chamber at millimeter waves. IEEE Transactions on Electromagnetic Compatibility. 2015;57(1):12–21.

[11] Besnier P, Lemoine C, and Sol J. Various estimations of composite Q-factor with antennas in a reverberation chamber. In: 2015 IEEE International Symposium on Electromagnetic Compatibility (EMC); 2015. p. 1223–1227.

[12] Holloway CL, Shah HA, Pirkl RJ, *et al.* Reverberation chamber techniques for determining the radiation and total efficiency of antennas. IEEE Transactions on Antennas and Propagation. 2012;60(4):1758–1770.

[13] Genender E, Holloway CL, Remley KA, *et al.* Simulating the multipath channel with a reverberation chamber: application to bit error rate measurements. IEEE Transactions on Electromagnetic Compatibility. 2010;52(4):766–777.

[14] Holloway CL, Shah HA, Pirkl RJ, *et al.* Early time behavior in reverberation chambers and its effect on the relationships between coherence bandwidth,

chamber decay time, RMS delay spread, and the chamber buildup time. IEEE Transactions on Electromagnetic Compatibility. 2012;54(4):714–725.

[15]   Andrieu G and Ticaud N. Performance comparison and critical examination of the most popular stirring techniques in reverberation chambers using the "well-stirred" condition method. IEEE Transactions on Electromagnetic Compatibility. 2019;1–13.

[16]   Loughry TA. Frequency stirring: an alternate approach to mechanical mode-stirring for the conduct of electromagnetic susceptibility testing. In: Phillips Lab., Tech. Rep. PL-TR-91-1036; 1991.

[17]   Crawford ML, Loughry TA, Hatfield MO, *et al.* Bandlimited, White Gaussian Noise Excitation for Reverberation Chambers and Applications to Radiated Susceptibility Testing. National Institute of Standards and Technology (NIST), Technical Note 1375. 1996.

[18]   Hill DA. Electronic mode stirring for reverberation chambers. IEEE Transactions on Electromagnetic Compatibility. 1994;36(4):294–299.

[19]   Maarleveld M, Hirsch H, Obholz M, *et al.* Experimental investigation on electronic mode stirring in small reverberation chambers by frequency modulated signals. In: 10th International Symposium on Electromagnetic Compatibility; 2011. p. 143–147.

[20]   Hatfield MO. Shielding effectiveness measurements using mode-stirred chambers: a comparison of two approaches. IEEE Transactions on Electromagnetic Compatibility. 1988;30(3):229–238.

[21]   Holloway CL, Hill DA, Ladbury JM, *et al.* On the use of reverberation chambers to simulate a Rician radio environment for the testing of wireless devices. IEEE Transactions on Antennas and Propagation. 2006;54(11):3167–3177.

[22]   Lemoine C, Besnier P, and Drissi M. Estimating the effective sample size to select independent measurements in a reverberation chamber. IEEE Transactions on Electromagnetic Compatibility. 2008;50(2):227–236.

[23]   Huang Y and Edwards DJ. A novel reverberating chamber: the source stirred chamber. In: Proc. Int. Conf. Electromagn.Compat., Edinburgh, U.K.; 1992. p. 120–124.

[24]   Cerri G, Primiani VM, Pennesi S, *et al.* Source stirring mode for reverberation chambers. IEEE Transactions on Electromagnetic Compatibility. 2005;47(4):815–823.

[25]   Cerri G, Primiani VM, Monteverde C, *et al.* A theoretical feasibility study of a source stirring reverberation chamber. IEEE Transactions on Electromagnetic Compatibility. 2009;51(1):3–11.

[26]   Kildal PS, Chen X, Orlenius C, *et al.* Characterization of reverberation chambers for OTA measurements of wireless devices: physical formulations of channel matrix and new uncertainty formula. IEEE Transactions on Antennas and Propagation. 2012;60(8):3875–3891.

[27]   Rosengren K, Kildal PS, Carlsson C, *et al.* Characterization of antennas for mobile and wireless terminals by using reverberation chambers: improved accuracy by platform stirring. In: IEEE Antennas and Propagation Society International Symposium. 2001 Digest. Held in Conjunction With:

USNC/URSI National Radio Science Meeting (Cat. No.01CH37229). vol. 3; 2001. p. 350–353.

[28] Kildal PS and Carlsson C. Detection of a polarization imbalance in reverberation chambers and how to remove it by polarization stirring when measuring antenna efficiencies. Microwave and Optical Technology Letters. 2002;34:145–149.

[29] Hill DA and Ladbury JM. Spatial-correlation functions of fields and energy density in a reverberation chamber. IEEE Transactions on Electromagnetic Compatibility. 2002;44(1):95–101.

[30] Soltane A, Andrieu G, and Reineix A. Doppler spectrum analysis for the prediction of rotating mode stirrer performances in reverberation chamber. IEEE Transactions on Electromagnetic Compatibility. 2018; p. 1–6.

[31] Primiani VM, Russo P, and Cerri G. Experimental characterization of a reverberation chamber excited by the source stirring technique. In: International Symposium on Electromagnetic Compatibility – EMC EUROPE; 2012. p. 1–6.

[32] Tian Z, Huang Y, and Xu Q. Stirring effectiveness characterization based on Doppler spread in a reverberation chamber. In: 2016 10th European Conference on Antennas and Propagation (EuCAP); 2016. p. 1–3.

[33] Voges E and Eisenburger T. Electrical mode stirring in reverberating chambers by reactively loaded antennas. IEEE Transactions on Electromagnetic Compatibility. 2007;49(4):756–761.

[34] Klingler M. Dispositif et procédé de brassage électromagnétique dans une chambre réverbérante brassage de modes – Patent FR 2 887 337, 17/06/2005.

[35] Andrieu G, Tristant F, and Reineix A. Investigations about the use of aeronautical metallic halls containing apertures as mode-stirred reverberation chambers. IEEE Transactions on Electromagnetic Compatibility. 2013;55(1):13–20.

Chapter 3

Frequency and time-domain performance assessment of vibrating intrinsic reverberation chambers

Guillaume Andrieu[1]

Vibrating intrinsic reverberation chambers (VIRCs) proposed at the beginning of the 2000s [1,2] are a particular kind of reverberation chamber (RC) based on the time-varying modification of the cavity shape. They are generally made of a flexible tent of conducting material mounted in a rigid structure. An example of VIRC is depicted in Figure 3.1 which shows the VIRC recently installed in Safran Electrical & Power. This is the facility used in the experimental campaign described in this chapter.

VIRCs have many great advantages with respect to "classical" RCs, for instance parallelepipedic ones. First, VIRCs have a lower cost with respect to a classical RC of similar volume. Moreover, as described later in this chapter, the stirring process being induced by the movement of the metallic tent can be made in a simple manner, and it is no longer required to use a mechanical mode stirrer

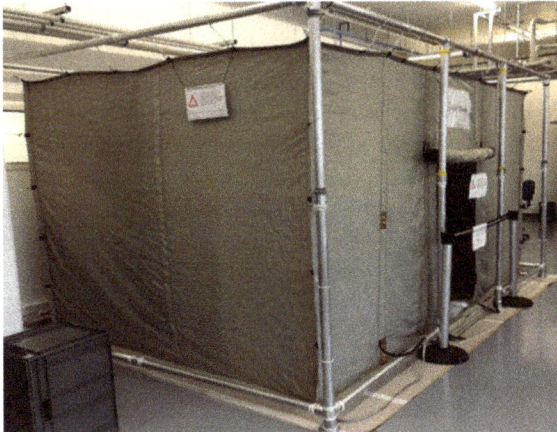

Figure 3.1   Picture of the VIRC installed in Safran Electrical & Power (Toulouse, France)

[1]XLIM Laboratory, UMR7252, University of Limoges, Limoges, France

which greatly simplifies the setup complexity (and therefore the total cost). Second, a VIRC can be installed and uninstalled easily in a limited amount of time (i.e., a few hours) to realize in situ tests for instance. This induces a change in the classical way to perform such tests as the testing facility can move toward the system to test, which is the inverse of what happens generally.

Devoted at the beginning to the purpose of performing in situ tests, their principle has been extended for shielding effectiveness measurements with the use of a "dual-VIRC" system [3,4], the sample to characterize being inserted at the interface between two VIRCs.

As for any other kind of RCs, it is of fundamental interest to characterize the VIRC performance (and in particular of the associated stirring process) from the point of view of the quality of the electromagnetic field generated inside before performing any test or measurement. However, maybe in reason of the continuous movement of the metallized tent, the standard procedure [5] (which probably would require some adjustments in this case) does not mention how to calibrate a VIRC.

This chapter based on a recent work [6] first shows how the "well-stirred" condition method can be extended to the case of VIRCs in order to characterize their performance in the frequency-domain. This is achieved by collecting the measurements made using a succession of vector network analyzer (VNA) sweeps. However, the VIRC case is significantly different to the case of "classical" RCs studied in the first two chapters of this book in reason of the continuous movement of the metallic tent. This implies to add a time-domain analysis and justifies why a separated chapter is devoted to this particular kind of RCs. Therefore, a method also proposed recently in [6] and based on the autocorrelation of S_{11} is shown in order to determine the "decorrelation time" t_d of the VIRC. The double objective of these time-domain measurements is first to determine the minimum time to wait before collecting two independent samples at the same point, and second to detect eventual periodical pattern of the stirring process. These results have a strong impact in terms of the number of collectable independent samples (as a function of time) which is strongly related to the measurement uncertainty [7–9]. It should be emphasized that both these frequency and time-domain methods are assumed to be usable in any VIRC independently of the stirring process implemented, and without loss of generality.

Before showing how the VIRC performance can be assessed in a reliable manner both in frequency and time-domains, this chapter starts first by presenting the characteristics of the VIRC we have used and how it is possible to proceed in order to implement an efficient stirring process.

3.1   Description of the VIRC used

3.1.1   General characteristics

The VIRC used in this chapter (shown in Figure 3.1) has been recently installed in Safran Electrical & Power in Toulouse, France. The dimensions of the VIRC are approximately 3.66 m long, 2.44 m wide, 2.2 m high, which corresponds roughly to

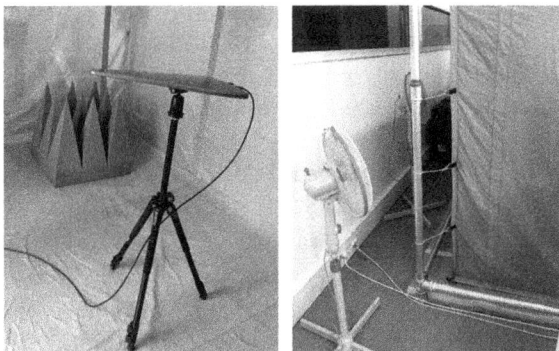

Figure 3.2   Picture of the log-periodic antenna and the block of absorber (on the left). Picture of a fan oriented toward a corner of the VIRC (on the right)

a volume of 19.6 m^3. According to these dimensions, the fundamental resonance of the cavity f_0 can be estimated around 74 MHz. As shown in the picture, the VIRC lies on the ground. Therefore, the tent floor is not moving under the action of the stirring process we have implemented (discussed later in this chapter) unlike all the five other sides of the tent.

The antenna used in the experiments is a log-periodic antenna (0.38–10 GHz) shown in Figure 3.2 (on the left). Two loading configurations of the VIRC have been considered. In the "unloaded" configuration, the VIRC is left empty, while in the "loaded" configuration, a block of nine pyramidal absorbers shown in the left of Figure 3.2 (having a total squared ground surface of 61 by 61 cm^2 with a height of 51 cm) has been inserted within the VIRC. The volume of the block of absorbers equals 0.06 m^3, which represents about 0.03% of the VIRC volume.

3.1.2 Stirring strategy

To the knowledge of the chapter contributor, two main kinds of stirring processes have been proposed in the literature for such chambers. The main approach is to attach to the metallized tent some strings driven by one or several independent motors, potentially working with different rotational speeds [10].

A second approach consists in using one or several fans installed outside the VIRC but oriented toward it in order to induce a tent movement [11]. As the latter solution is extremely simple to set up, low cost and does not risk to tear the metallic tent, this is the solution that has been implemented in the VIRC previously described.

This stirring strategy has to be optimized in order to determine for instance the number of fans to use and their relative positions with respect to the VIRC. This can be done according to a visual inspection of the tent, in particular, by looking at several points of the tent in order to check if the local movements of the tent (even between two close points) seem random. This is made possible by the extreme lightness of the metallic tent, which allows very strong contortions of each moving side of the metallic tent. To illustrate in concrete terms what we are expecting, it is more like the water surface in open sea than a boat sail stretched by strong winds.

As shown in Figure 3.3 (on the left), it has been observed that when a fan is located toward one particular side of the VIRC, the movement of the tent may be limited as this side of the tent is stretched. This can also be observed when the speed of the fans is too strong, generating a too important air flux on the VIRC tent.

According to this visual inspection, it becomes rapidly clear that a satisfying solution (but probably not the optimized one) consists in locating the fan toward a VIRC corner in order to produce smooth variations of the two closest adjacent sides. However, this solution implies a movement on the two closest sides only. Therefore, the final solution we have implemented is to install two fans oriented toward both opposite corners of the VIRC as illustrated in Figure 3.3 (on the right). The possibility offered by many fans to have an azimuthal rotation (which is periodic) around its mast has not been used.

It is difficult to give clear guidelines about the optimal parameters (position, optimal distance from the VIRC, orientation, air flux, etc.) related to each fan (as well as the optimal number of fans) required to obtain a satisfying stirring process. We can only propose a "trial-and-error" approach leading to choose all these parameters from a visual inspection of the metallic tent movements and then to analyze using the experimental methods presented hereafter if the stirring process is satisfying. This is facilitated as these two methods imply fast measurements of only a few minutes.

One understands that the volume of the VIRC may be continuously changing as a function of time in reason of the tent movement. Numerous approaches in the field

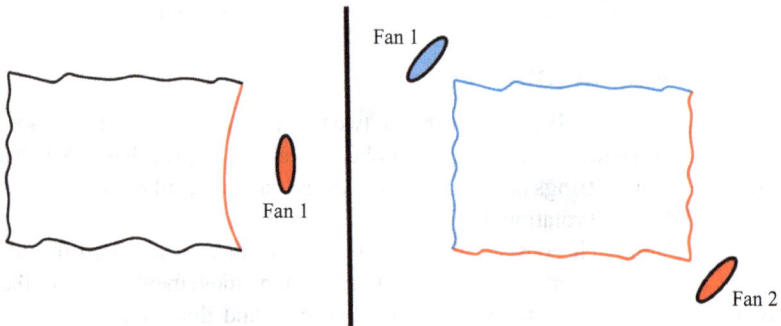

Figure 3.3   Schematic description of two stirring strategies for our VIRC (seen from the top). On the right, it is shown that a fan oriented toward a side of the VIRC may stretch the tent and then limit its movement. On the right, a satisfying stirring strategy we have implemented using two fans located toward both opposite corners of the VIRC. In the latter configuration, each fan has a stirring action on both adjacent sides of the tent

of RCs are based on the knowledge of the cavity volume (always known with a given uncertainty). This occurs for instance when measuring the antenna efficiency using the three-antenna approach described in [12] and in Chapter 7 of this book. In the case of a VIRC, we can consider as a first approach that the volume is continuously changing. Obviously, this could lead to increase the measurement uncertainty in all the approaches where the volume needs to be accurately known. However, a rapid calculation allows some doubts about this potential problem to be dissipated. If we consider that each point of the VIRC is able to move inward or outward over a distance of 5 mm (which is probably greater than the reality) under the action of the air flux produced by the fans, the VIRC volume would be increased by 0.5% if all the points were moved outward (and of −0.5% in the other direction) under these conditions. In reality, in reason of the "sweet" stirring process involved by the fans, the volume increase related to a point moving outward can reasonably be assumed to be compensated by another point moving in the opposite direction. Therefore, the VIRC volume variation as a function of time is assumed to be negligible, in particular, using the stirring process proposed in this chapter.

3.1.3 Q-factor and average mode bandwidth Δ_f

Before analyzing VIRC performance in both frequency and time-domains, it is important to determine its Q-factor. This is a study of prime importance to perform before any measurement within an RC. Indeed, as developed in several chapters of this book, the behavior of an RC is closely related to its ability to store the energy [13,14].

Similarly to the well-stirred condition method that requires to measure only the S_{11} parameter of a unique antenna, we have used this parameter collected from $N = 50$ successive VNA sweeps (i.e., at N different instants for each frequency) to compute the Q-factor of the VIRC following (3.1) [15]:

$$Q = \langle |S_{11} - \langle S_{11} \rangle|^2 \rangle \frac{Z_0 \omega \varepsilon V}{\left(\lambda^2/(4\pi)\right)\left(1 - |\langle S_{11} \rangle|^2\right)^2 \eta^2}. \tag{3.1}$$

To deepen the analysis, we compare in Figure 3.4 the Q-factor for both loading configurations with results obtained with the classical RC of the XLIM Laboratory unloaded and loaded with one block of absorbers (more details about the XLIM characteristics and this absorber can be found in the first chapter of this book). It should be noticed that the volumes of both facilities are of the same order of magnitude.

We can see that the Q-factor for the unloaded and the loaded configurations is quite similar for both facilities on the whole bandwidth. This can dissipate any doubt on the ability of the metallized tent (made from woven metal fibers) to store the energy when comparing to a "classical" Faraday cage. The VIRC Q-factor is around 5,000 and 700 at a frequency of 1 GHz for the unloaded and the loaded configurations, respectively. The knowledge of Q leads also to the average mode

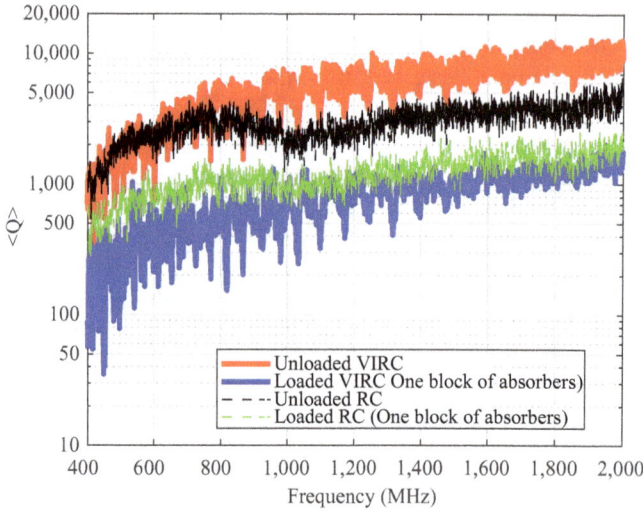

Figure 3.4   Measured averaged Q-factor of the VIRC for both loading configurations. Comparison with the results obtained in the RC of the XLIM Laboratory. Note that the block of absorbers used in both facilities is not the same

bandwidth Δ_f defined in the first chapter of this book (Δ_f being respectively close to 0.2 and 1.4 MHz for the unloaded and the loaded configurations).

3.2   Frequency-domain characterization

The "well-stirred condition" method presented in the first chapter of this book and applied in the second chapter on the most common stirring techniques is extended to the VIRC case in this third chapter. As mentioned before, the main challenge in this case is to deal with the continuous movement of the metallized tent.

This issue is solved in a simple manner by collecting S_{11} on the desired frequency range from a succession of N VNA sweeps ($N = 50$ in the results shown in this chapter) of duration τ done successively and shifted with a time step $\delta_t^{freq} > \tau$ (the superscript "freq" meaning here that the measurements are collected in the frequency-domain). On a mathematical point of view, the first two samples of S_{11} collected at the frequency f_0 can be, respectively, written $S_{11}(f_0, t = 0)$ and $S_{11}\left(f_0, t = \delta_t^{freq}\right)$, while in a mechanically stirred RC, this variable can be expressed as a function of the angle θ (i.e., $S_{11}(f_0, \theta)$), θ being the mode stirrer angle. Finally, it should be emphasized that, for the "well-stirred condition" method, the continuous modification of the VIRC geometry during an entire VNA sweep is not a problem as this is the variation of the N S_{11}^{sti} samples collected from each sweep at a given frequency which is of interest.

As an example, the frequency-domain results presented here are obtained from the S_{11} measurements of the emitting antenna collected using $N = 50$ successive VNA sweeps on 20,001 frequencies linearly spaced between 0.2 and 2 GHz (using also a resolution bandwidth of 10 kHz). This data corresponds to a frequency resolution δ_f of 90 kHz. Two samples collected at the same frequency are consequently measured at an interval of time t_{sweep} equal to the VNA sweep, i.e., $\tau = 2.4$ s in these measurement conditions. These measurements are repeated for two positions of the antenna within the VIRC and for both loading configurations.

Figures 3.5 and 3.6 present the magnitude of S_{11} measured for each VNA sweep in the unloaded and the loaded VIRC, respectively, as well as the same parameter measured in free space (i.e., S_{11}^{fs}). It is clear that $|S_{11}|$ measured in the VIRC exhibits large fluctuations around $|S_{11}^{fs}|$, especially for the unloaded VIRC. It is assumed that all the information required to assess the RC performance are contained in these fluctuations.

To have an idea about the respective amplitude of the stirred and of the unstirred contributions (calculated using the equations given in the first chapter of this book), Figure 3.7 presents these contributions on the S_{11} parameter in the unloaded VIRC. It is shown that, in our VIRC, the stirred contributions have a larger magnitude on the

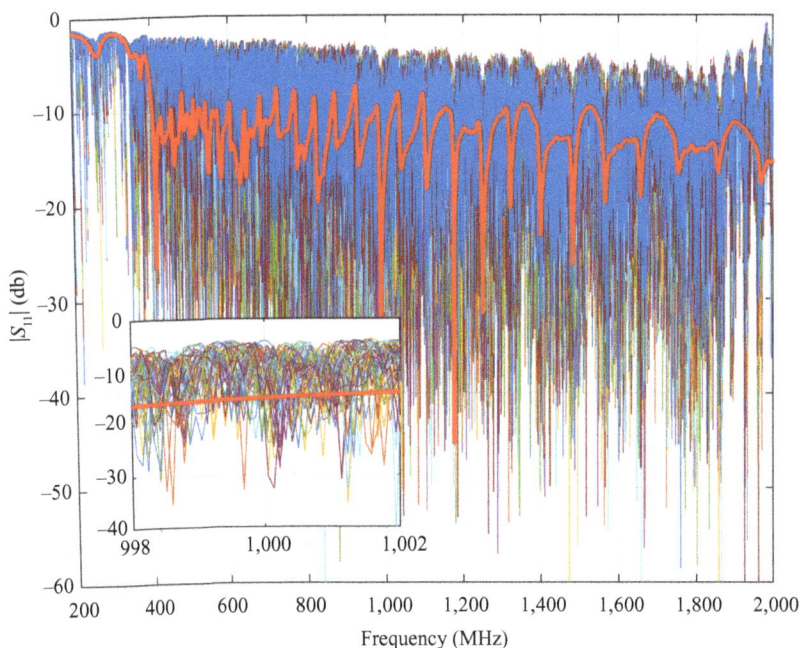

Figure 3.5   *Magnitude of S_{11} measured from $N = 50$ successive VNA sweeps in the unloaded VIRC. Comparison with the same result in free space (thick curve). On the bottom subplot, zoom over a 4-MHz bandwidth around 1 GHz*

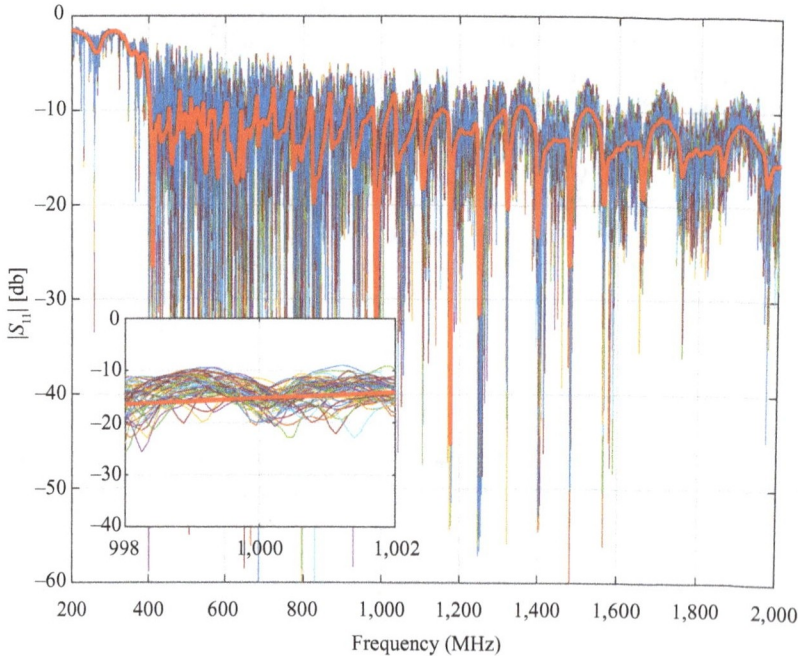

Figure 3.6   Magnitude of S_{11} measured from N = 50 successive VNA sweeps in the loaded VIRC. Comparison with the same result in free space (thick curve). On the bottom subplot, zoom over a 4-MHz bandwidth around 1 GHz

whole frequency range, except below 500 MHz where both contributions are similar. This is a first good indication of the efficiency of the implemented stirring process that is supposed to increase the power related to the stirred paths with respect to the unstirred one (or in other terms to decrease the K-factor). It is therefore a good practice to plot these metrics before proceeding to the analysis presented hereafter.

After this preliminary analysis, the "well-stirred condition" method can be directly applied on the measured data. Figure 3.8 shows the modified Anderson-Darling (AD) statistic (and the polynomial curve fitting at the fifth order) obtained for both loading configurations of the VIRC. The results are similar to the ones obtained in a classical RC. Indeed, a decreasing trend with the frequency is observed as well as better results for the unloaded configuration. As mentioned in the previous chapters, this is due to a better modal overlapping (or in other terms to an increase of the number of significant modes) when the Q-factor is decreased by the absorbers inserted in [14].

From the sample correlation point of view, it is first important to emphasize that two different kinds of correlation are possible.

First, a correlation related to the insufficient modification of the tent geometry during the interval t_{sweep} could be observed. Considering the large value of t_{sweep} used

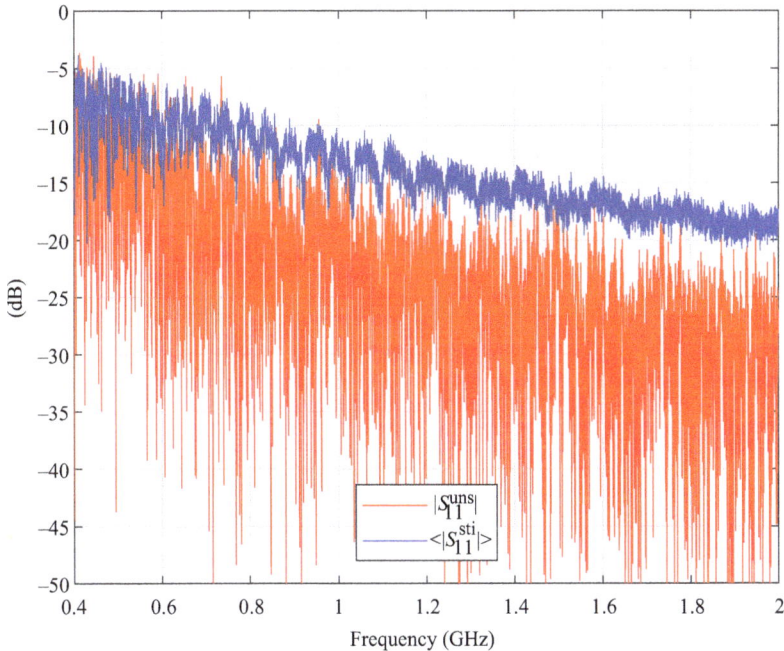

Figure 3.7 *Magnitude of the unstirred (i.e., $|S_{11}^{uns}|$) contribution and of the average stirred contributions (i.e., $\langle |S_{11}^{sti}| \rangle$) measured from $N = 50$ successive VNA sweeps in the unloaded VIRC*

in these frequency-domain measurements, this first kind of correlation is assumed to be null. This is a hypothesis confirmed from time-domain measurements, as described in the next section of this chapter.

Second, a correlation related to the insufficient modal density at a given frequency could be observed as seen previously for other kind of stirring techniques. This is particularly the case for low Q-factor, i.e., when each mode has a significant impact on a large bandwidth. This is the type of correlation that is investigated here. Results shown in Figure 3.9 show an absence of this kind of correlation on the whole frequency range for the unloaded configuration but a slight one (on the lowest frequency range) for the unloaded configuration. These results match with the ones obtained using a mechanically stirred RC (as shown in the first chapter) where the sample correlation has been proved to increase with the amount of absorbers inserted in [14–16]. This sample correlation problem is solved, as for mechanically stirred RC, when the frequency increases. Further measurements have shown that the insertion of an additional time interval (of a few seconds) between the different VNA sweeps has no effect on the results, confirming that the sample correlation observed here for the loaded case is not due to an insufficient value of t_{sweep}.

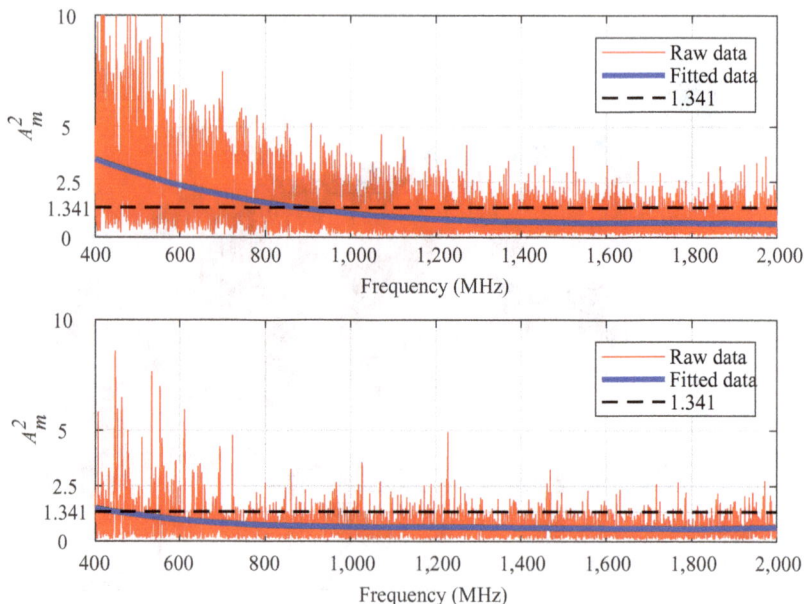

Figure 3.8   *Modified AD statistic A_m^2 (with the corresponding fifth-order PCF) calculated from N = 50 successive frequency sweeps of the VNA for two different loading conditions (unloaded on the top subplot, loaded on the bottom) of the VIRC*

A comparison of the results (f_1, f_2 and f_{wsc}) obtained in the VIRC and the mechanically stirred parallelepipedic RC of the XLIM Laboratory for different loading configurations is made in Table 3.1. There is no clear difference of performance between the VIRC and the XLIM RC (both facilities having a similar volume) in the frequency-domain. Indeed, f_{wsc} is, respectively, lower than 900 and 500 MHz for the unloaded and the loaded configurations, these values being slightly higher with respect to the XLIM RC (having a slightly larger volume). It is worth noting that the optimal quantity of absorber, i.e., leading to the lowest f_{wsc} of the VIRC, is probably not one of both configurations used during the experiments.

It is important to mention here that the accuracy on f_2 for the loaded VIRC is not optimal, probably because the antenna is satisfyingly matched only above 400 MHz. This involves a non-negligible inaccuracy of the polynomial curve fitting (PCF) process in this case and therefore on the f_2 value obtained.

As a conclusion on this frequency-domain analysis, the "well-stirred" condition method is extendable to the case of VIRCs, using S_{11} measurements collected on N successive VNA sweeps, whatever the stirring process implemented.

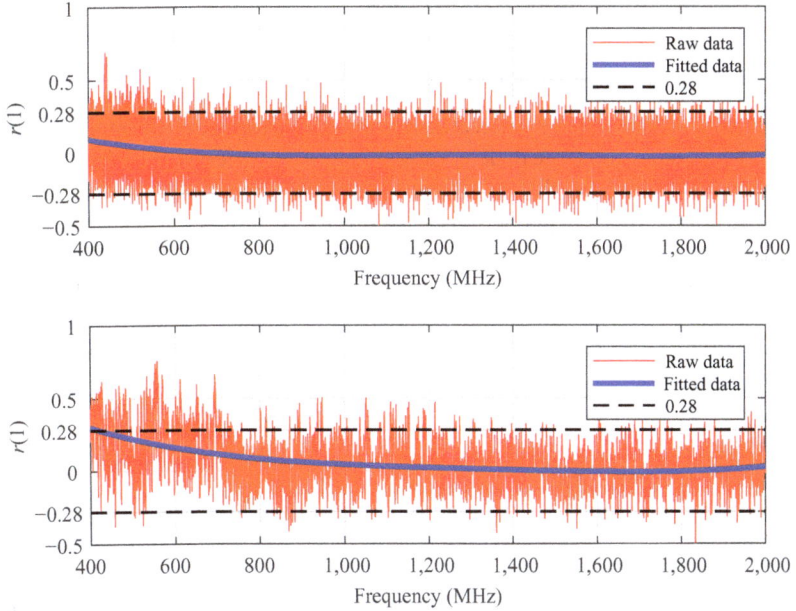

Figure 3.9   First-order autocorrelation coefficient r(1) (with the corresponding fourth-order PCF) calculated from N = 50 successive frequency sweeps of the VNA for two different loading conditions (unloaded on the top subplot, loaded on the bottom) of the VIRC

Table 3.1   Comparison of results obtained in the VIRC and in the XLIM RC

VIRC		f_1 (MHz)	f_2 (MHz)	f_{wsc} (MHz)
Unloaded	Position 1	880	<400	880
	Position 2	862	<400	862
Loaded	Position 1	434	424	434
	Position 2	476	491	491
RC		f_1 (MHz)	f_2 (MHz)	f_{wsc} (MHz)
Unloaded		559	289	559
Loaded		280	397	397

"<400" means that f_2 is lower than 400 MHz; RC results are averaged over four positions of the antenna.

3.3   Time-domain characterization

3.3.1   *Decorrelation time and detection of periodicity*

The frequency-domain analysis described in the previous section shows that t_{sweep} is, in most of the cases, too large to detect any sample correlation in the VIRC. However, it makes sense to consider that it exists at a given frequency a minimum time to wait in order to collect two independent (i.e., uncorrelated) samples at the same location. Indeed, for an interval of time lower than t_d defined as the "decorrelation time," we can consider that the overall shape of the metallized tent has not been sufficiently modified to provide a new independent sample. In addition, it is also important to be able to detect an eventual periodicity of the stirring process, even if this cycle could be larger than a few seconds.

Following these two objectives, it is of interest to perform time-domain measure-ments (using again S_{11}) at a fixed frequency with a VNA working in constant wave (CW) mode.

The metric of interest is again the autocorrelation of the measured signal. How-ever, unlike the samples collected for instance at different positions of a rotating mode stirrer, the data collected here as a function of time are not circular. In this case, the autocorrelation $r(k)$ (k being the order) of a time-domain signal $x(t)$ (in this case, the real or the imaginary part of S_{11}) collected at N different instants is computed as follows:

$$r(k) = \frac{\sum_{i=k+1}^{N} [x(i) - \langle x \rangle] \cdot [x(i-k) - \langle x \rangle]}{\sigma_x^2}, \tag{3.2}$$

with σ_x^2 being the standard deviation of $x(t)$. The order $k = 1$ corresponds therefore to the time interval δ_t^{time} (with the superscript "time" meaning that the measurements are made in the time-domain) between each measured sample. It is therefore easy to express the autocorrelation r as a function of time, i.e., $r(t)$. When examining (3.2), it is also clear that r is not dependent of the signal average value. The formula can therefore be directly applied on S_{11}.

There is a need for a threshold value in order to define the so-called decorrelation time t_d on a rigorous basis. This is done by defining t_d as the instant when the autocorrelation becomes lower than 0.5, similarly as it is done for instance for the coherence bandwidth [17]. Another threshold (0.3 or $1/\sqrt{2}$ for instance) could be used and would lead to (slightly) different results.

As an example, the S_{11} parameter related to the emitting antenna has been col-lected during 60 s at 10,001 instants (i.e., $\delta_t^{\text{time}} = 6$ ms) and at 8 frequencies linearly spaced between 0.5 and 4 GHz ($\delta f = 500$ MHz). These measurements have been repeated for two different positions of the antenna and for both loading configurations of the VIRC.

The variation as a function of time (over 60 seconds) of the real and the imaginary parts of S_{11} at $f = 1$ GHz are presented in Figure 3.10. It is clearly emphasized that the amplitude of the fluctuations of S_{11} are lower when the VIRC is loaded. Their related autocorrelations are plotted in Figures 3.11 and 3.12.

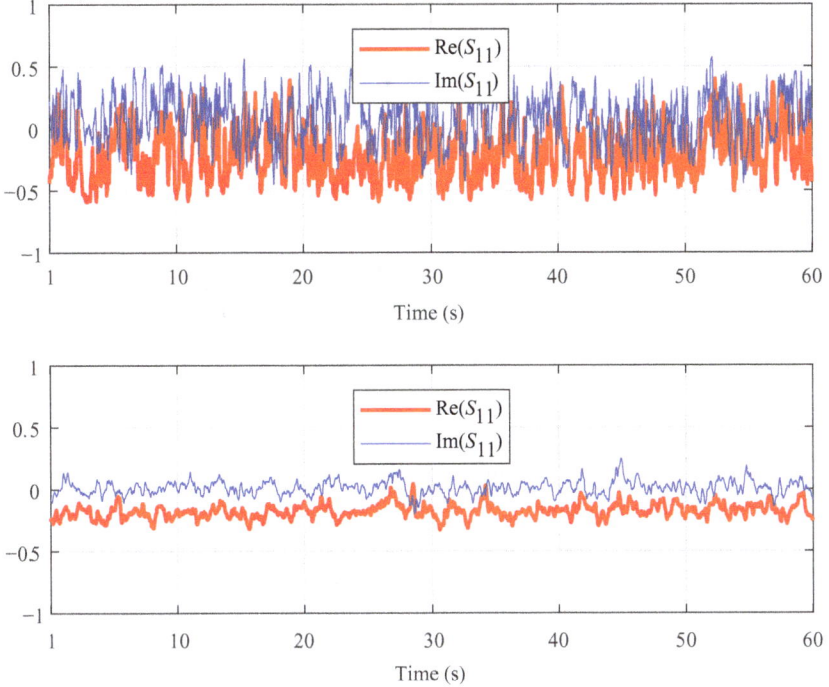

Figure 3.10 Time-domain variation over 60 seconds of the real and imaginary parts of S_{11} at $f = 1$ GHz for two different loading conditions (unloaded on the top subplot, loaded on the bottom) of the VIRC

From these figures, it is clear that, as expected, the correlation decreases rapidly with the time. Whatever the threshold value and the loading condition considered, the decorrelation time t_d is clearly lower than 1 second at $f = 1$ GHz as shown in both insets, confirming the hypothesis made in the previous section of this chapter. The analysis of the results obtained in the first second shows also that the correlation is a stable metric giving similar results whatever the position of the antenna and if the real or the imaginary part of S_{11} is considered.

The absence of clear correlation above 1 second leads to reject the hypothesis of a clear periodical repetition of the stirring process induced by the implemented stirring process using two fans. The decrease of $r(t)$ for values close to 60 seconds is due to the decrease of the number of samples considered in (3.2) for these extreme values. Finally, the decorrelation time t_d is larger for the "loaded" VIRC which is similar to what is observed for a CW excitation (i.e., in the frequency-domain) of an RC.

To deepen the analysis, the decorrelation time t_d obtained as a function of the frequency is depicted in Figure 3.13 for both loading conditions, each result being plotted for two antenna positions. The decorrelation time t_d decreases with the frequency and

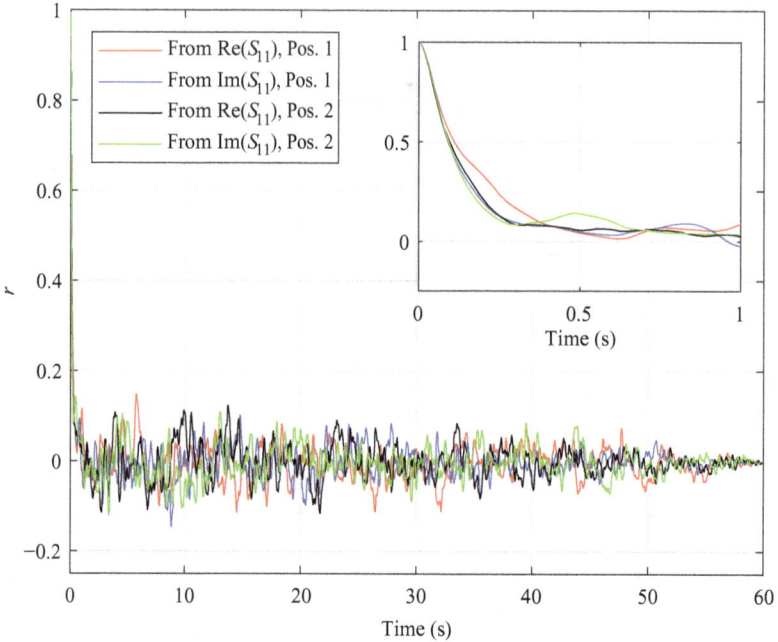

Figure 3.11   Autocorrelation r(t) of the real and the imaginary parts of S_{11} calculated over 60 seconds at $f = 1$ GHz for two positions of the emitting antenna when the VIRC is unloaded. In the inset, a zoom on the first second

is remarkably independent of the antenna position within the VIRC. Moreover, t_d increases for the loaded VIRC as previously shown at 1 GHz.

Maximum values of t_d (obtained at $f = 0.5$ GHz) are, respectively, 0.16 and 0.29 second for the unloaded and the loaded VIRCs. Assuming the absence of repetition of the stirring process, this would lead to consider that 375 and 207 independent samples (when using a time step equal to t_d) could respectively be collected in the VIRC in 1 minute at 0.5 GHz (and more at higher frequencies). One understands here that in the absence of any periodicity of the stirring process, the number of collectable independent samples would be proportional to the measurement time. Similarly, the number of independent states applied to the device under test (DUT) during a radiated susceptibility tests (see the next chapter) would also be proportional to the exposition time.

3.3.2   Rayleigh distribution in the time-domain

Once t_d is known, it is interesting to confirm that the ideal distribution (i.e., Rayleigh for $|S_{11}^{sti}|$) is also matched in the time-domain when considering samples collected with a time step equal to t_d. Therefore, at each frequency of interest, $N = 50$ samples

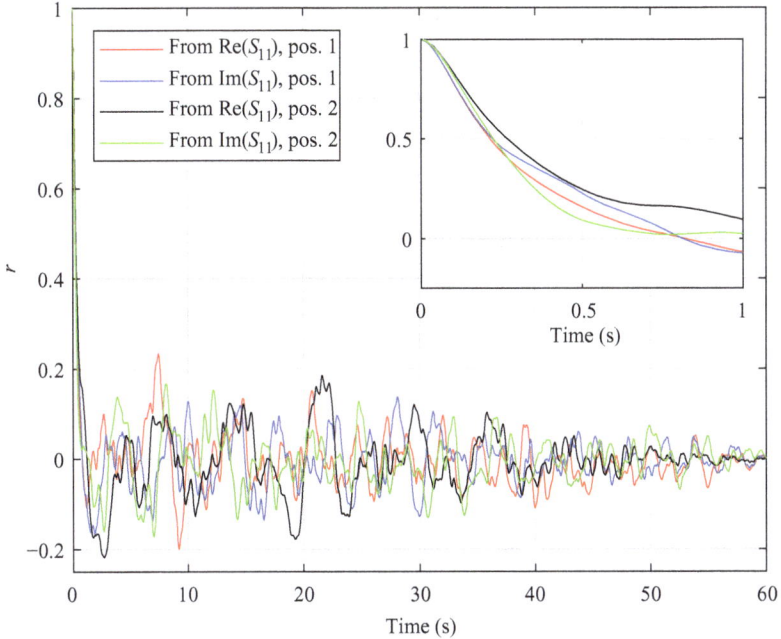

Figure 3.12   Autocorrelation r(t) of the real and the imaginary parts of S_{11}
calculated over 60 seconds at $f = 1$ GHz for two positions of the
emitting antenna when the VIRC is loaded. In the inset, a zoom on the
first second

of $|S_{11}^{sti}|$ shifted with a time step t_{step} slightly greater than the maximum value of t_d obtained for each loading condition (i.e., t_{step} equals, respectively, 0.2 and 0.3 s, respectively, for the "unloaded" and the "loaded" configurations) have been gathered together. As the first and the last samples of a serie of 50 samples are separated by a time interval of 9.8 and 14.7 s, respectively, six and four AD Goodness-of-fit (GOF) tests from the S_{11} samples collected during 60 seconds can be carried out.

The average value of the AD GOF statistic is presented in Figure 3.14 for both loading configurations as a function of frequency. It is shown that the samples collected at different instants shifted with a time step slightly larger than t_d can match the ideal distributions of a "well-stirred" RC, in this case the Rayleigh distribution. It makes sense to observe that the results are quite similar to the results plotted in Figure 3.8 from the frequency-domain analysis (with in this case a larger time step, i.e., t_{sweep}). In particular, it is shown that the AD GOF test is highly rejected for the unloaded VIRC at $f = 0.5$ GHz.

To go further, the same result is plotted in Figure 3.15 for the loaded VIRC for different t_{step} values, most of them being lower than t_d. As expected, the adequacy of the samples with respect to the Rayleigh distribution becomes poorer when t_{step} is chosen lower than t_d in reason of the correlation of the considered samples.

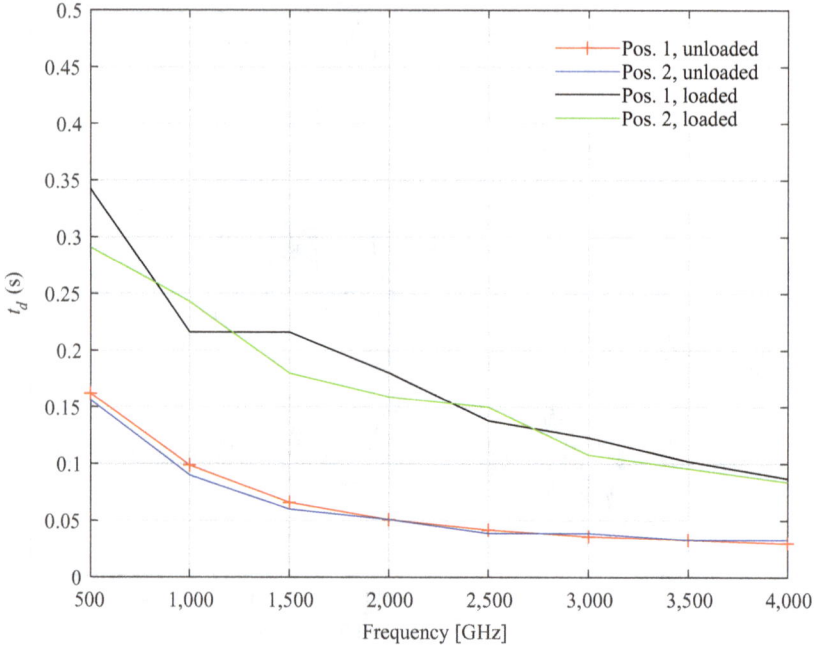

Figure 3.13   Decorrelation time t_d as a function of the frequency for two different loading conditions of the VIRC and two antenna positions

3.3.3   Discussion

The absence of detection of any periodical repetition of the stirring process shown in this time-domain analysis makes sense when considering that most of the geometry of the metallic tent is continuously moving. It is therefore unlikely that the whole tent geometry becomes similar (and therefore involving correlated measured samples) between two instants separated by a time step greater than t_d. We are here close to a perfectly chaotic RC in the sense that it is impossible to predict the path of any wave traveling inside. This result is of fundamental importance. Indeed, as mentioned previously, it would mean (if confirmed in the future by further analysis) that the number of collectable samples could be proportional to the measurement time.

There is a major difference with a classical mechanically stirred RC in continuous stirring (with the mode stirrer continuously rotating) where the number of independent samples collectable is finite and equals the ratio t_r/t_d with t_r being the revolution time of the stirrer. In such RC using one rotating mode stirrer, the unique way to increase the number of independent samples requires to open the RC, to modify the position of one object (antenna or DUT for instance) and to repeat the measurements for this new configuration. This is why when a large number of independent samples are required in order to limit the uncertainty of the results, this is generally achieved by combining

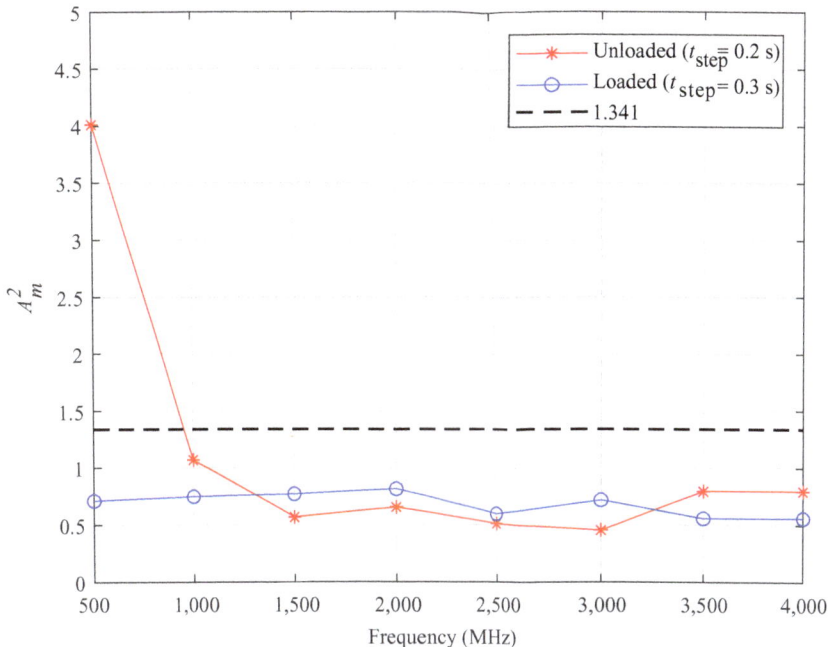

Figure 3.14   Average modified AD statistic A_m^2 as a function of the frequency for both loading configurations of the VIRC from the analysis of $N = 50$ samples of $|S_{11}^{sti}|$ shifted with a time step t_{step}. Each point is an average of the 6 or 4 modified AD statistic obtained for the unloaded and the loaded VIRCs, respectively. Comparison with the 1.341 threshold

different stirring processes, as described for instance in Chapter 5 of this book. For instance, in the case of antenna efficiency [12,18] measurements or over-the-air tests of wireless devices [19,20], several rotating mode stirrers (i.e., two most of the time) are generally combined with platform stirring [21,22]. This involves a more complex setup, a reduction of the available space within the RC and more difficulties to treat the data, in order to determine for instance, the number of independent samples provided by each stirring process [23]. Moreover, this (high) number of independent samples is finite at a given frequency.

In the case of the VIRC, the number of collectable independent samples seems to be directly proportional to the measurement time and is therefore potentially greater than what we can obtain in classical RCs. As a consequence, VIRCs seem promising for RC measurements requiring a low uncertainty budget as for instance the applications mentioned before. Moreover, this could be achieved with a reasonable measurement time, a low-cost reverberating enclosure and a basic stirring process.

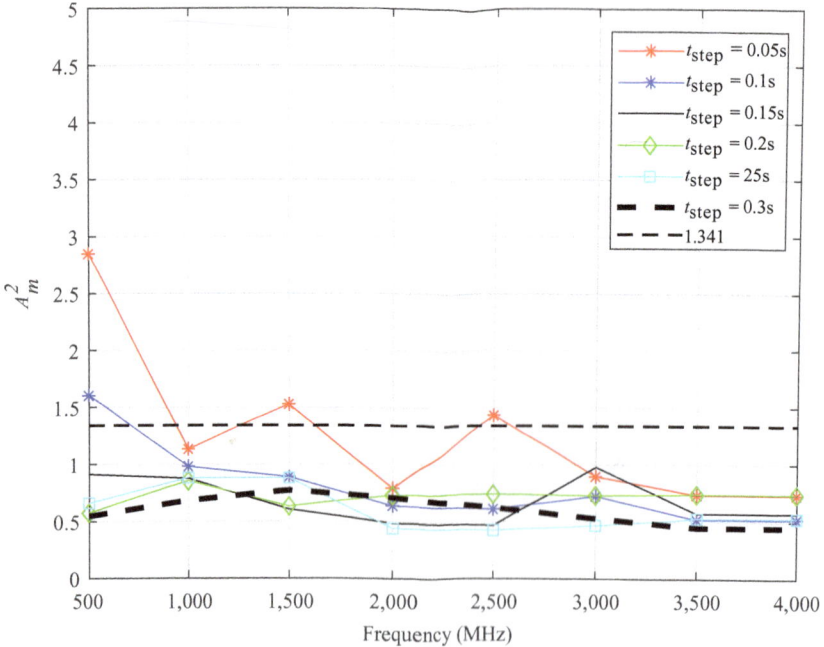

Figure 3.15 *Average modified AD statistic $A_m{}^2$ as a function of the frequency for the loaded VIRC for different t_{step} values. Each point is an average of the four modified AD statistic obtained for each AD GOF test. Comparison with the 1.341 threshold*

3.4 Conclusion

In this chapter, a complete framework able to characterize the performance of a VIRC and of its associated stirring process in both frequency and time-domains has been presented. This work tries to fill the gap in the literature, and particularly of the standard procedure [5], which does not mention how to calibrate such kind of RCs.

In the frequency-domain, the "well-stirred" condition method is extended to the case of VIRCs, using N successive VNA sweeps. The method highlights the fact that VIRC performance in the frequency-domain are similar to the ones obtained in a mechanically stirred RC of similar volume.

In the time-domain, the approach aims first at determining the decorrelation time t_d. In reason of the continuous movement of the metallized tent, it is shown that the decorrelation time t_d (defined as the minimum time interval to respect in order to collect two independent samples at the same location) can be lower than 1 second using the straightforward proposed stirring process. Second, no periodical repetition of the stirring process as a function of time has been observed. This fundamental result implying a proportionality between the number of collectable independent samples and the acquisition time could contribute to decrease the uncertainty budget [7–9] of

RC measurements (for instance antenna efficiency or over-the-air testing of wireless devices) with respect to measurements performed in classical parallelepipedic RCs. Moreover, there is probably, at a given frequency, an upper limit potentially linked to the modal density at this frequency. It is therefore suggested to the community of VIRC users to investigate in the near future this fundamental question.

References

[1] Leferink F, Boudenot JC, and van Etten W. Experimental results obtained in the vibrating intrinsic reverberation chamber. In: IEEE International Symposium on Electromagnetic Compatibility. Symposium Record (Cat. No.00CH37016). IEEE, Washington DC, USA. vol. 2; 2000. p. 639–644.

[2] Leferink F. Patent NL1010745, Test Chamber, 7-12-1998; 1998.

[3] Skrzypczynski J. Dual vibrating intrinsic reverberation chamber used for shielding effectiveness measurements. In: 10th International Symposium on Electromagnetic Compatibility. IEEE, Amsterdam, Netherlands; 2011. p. 133–136.

[4] van de Beek S, Vogt-Ardatjew R, Schipper H, *et al.* Vibrating intrinsic reverberation chambers for shielding effectiveness measurements. In: International Symposium on Electromagnetic Compatibility – EMC EUROPE. IEEE, Rome, Italy; 2012. p. 1–6.

[5] Reverberation Chamber Test Methods. Int Electrotech Commiss Standard IEC 61000-4-21:2011. 2011 January.

[6] Andrieu G, Meddeb N, Jullien C, *et al.* Complete Framework for Frequency and Time-Domain Performance Assessment of Vibrating Intrinsic Reverberation Chambers. IEEE Transactions on Electromagnetic Compatibility. 2020:1–10.

[7] Remley KA, Pirkl RJ, Shah HA, *et al.* Uncertainty From Choice of Mode-Stirring Technique in Reverberation-Chamber Measurements. IEEE Transactions on Electromagnetic Compatibility. 2013;55(6):1022–1030.

[8] Remley KA, Wang CJ, Williams DF, *et al.* A Significance Test for Reverberation-Chamber Measurement Uncertainty in Total Radiated Power of Wireless Devices. IEEE Transactions on Electromagnetic Compatibility. 2016;58(1):207–219.

[9] Remley KA, Pirkl RJ, Wang C, *et al.* Estimating and Correcting the Device-Under-Test Transfer Function in Loaded Reverberation Chambers for Over-the-Air Tests. IEEE Transactions on Electromagnetic Compatibility. 2017;59(6):1724–1734.

[10] Serra R and Rodríguez AC. The Ljung-Box test as a performance indicator for VIRCs. In: International Symposium on Electromagnetic Compatibility – EMC EUROPE. IEEE, Rome, Italy; 2012. p. 1–6.

[11] Serra R and Leferink F. Optimizing the stirring strategy for the vibrating intrinsic reverberation chamber. In: 9th International Symposium on EMC and 20th International Wroclaw Symposium on Electromagnetic Compatibility. IEEE, Wroclaw, Poland; 2010. p. 457–462.

[12] Holloway CL, Shah HA, Pirkl RJ, *et al.* Reverberation Chamber Techniques for Determining the Radiation and Total Efficiency of Antennas. IEEE Transactions on Antennas and Propagation. 2012;60(4):1758–1770.

[13] Cozza A. The Role of Losses in the Definition of the Overmoded Condition for Reverberation Chambers and Their Statistics. IEEE Transactions on Electromagnetic Compatibility. 2011;53(2):296–307.

[14] Adardour A, Andrieu G, and Reineix A. On the Low-Frequency Optimization of Reverberation Chambers. IEEE Transactions on Electromagnetic Compatibility. 2014;56(2):266–275.

[15] Besnier P, Lemoine C, and Sol J. Various estimations of composite Q-factor with antennas in a reverberation chamber. In: 2015 IEEE International Symposium on Electromagnetic Compatibility (EMC); 2015. p. 1223–1227.

[16] Andrieu G, Ticaud N, Lescoat F, *et al.* Fast and Accurate Assessment of the "Well Stirred Condition" of a Reverberation Chamber From S_{11} Measurements. IEEE Transactions on Electromagnetic Compatibility. 2019;61(4):974–982.

[17] Holloway CL, Shah HA, Pirkl RJ, *et al.* Early Time Behavior in Reverberation Chambers and Its Effect on the Relationships Between Coherence Bandwidth, Chamber Decay Time, RMS Delay Spread, and the Chamber Buildup Time. IEEE Transactions on Electromagnetic Compatibility. 2012;54(4):714–725.

[18] Chen X. On Statistics of the Measured Antenna Efficiency in a Reverberation Chamber. IEEE Transactions on Antennas and Propagation. 2013;61(11):5417–5424.

[19] Kildal PS, Chen X, Orlenius C, *et al.* Characterization of Reverberation Chambers for OTA Measurements of Wireless Devices: Physical Formulations of Channel Matrix and New Uncertainty Formula. IEEE Transactions on Antennas and Propagation. 2012;60(8):3875–3891.

[20] Genender E, Holloway CL, Remley KA, *et al.* Simulating the Multipath Channel With a Reverberation Chamber: Application to Bit Error Rate Measurements. IEEE Transactions on Electromagnetic Compatibility. 2010;52(4):766–777.

[21] Rosengren K, Kildal PS, Carlsson C, *et al.* Characterization of antennas for mobile and wireless terminals by using reverberation chambers: improved accuracy by platform stirring. In: IEEE Antennas and Propagation Society International Symposium. 2001 Digest. Held in conjunction with: USNC/URSI National Radio Science Meeting (Cat. No.01CH37229). IEEE, Boston, USA. vol. 3; 2001. p. 350–353.

[22] Kildal PS and Carlsson C. Detection of a Polarization Imbalance in Reverberation Chambers and How to Remove It by Polarization Stirring When Measuring Antenna Efficiencies. Microwave and Optical Technology Letters. 2002;34:145–149.

[23] Chen X. Experimental Investigation of the Number of Independent Samples and the Measurement Uncertainty in a Reverberation Chamber. IEEE Transactions on Electromagnetic Compatibility. 2013;55(5):816–824.

Chapter 4
Probabilistic model about the influence of the number of stirring conditions considered during a radiated susceptibility test
Guillaume Andrieu[1]

Reverberation chambers (RC) are used among other applications for radiated suscep-tibility (RS) tests in order to expose an equipment under test (EUT) to a homogeneous and isotropic electromagnetic (EM) field. The aim is to validate the electromagnetic compatibility (EMC) compliance of the EUT by verifying that the EUT is working normally when a given EM field strength is applied. In such RS tests, the number n of stirring configurations (for instance, the number of mode stirrer positions for a mechanically stirred RC, which is the most popular stirring technique) considered at each frequency of interest is of prime importance. Indeed, one understands that a lower n reduces the measurement time but in the meantime decreases the probability to observe a susceptibility of a faulty EUT for a given magnitude of the EM distur-bance. Therefore, the risk grows to consider erroneously the EUT as EMC compliant. The choice of n is, therefore, necessarily a trade-off between the RS test duration and the degree of confidence requested by the users.

This chapter details a probabilistic model recently published [1] quantifying the risk r to observe no susceptibility of a faulty EUT during an RS test as a function of n. The method is general as the risk r can be computed for any (fictitious) EUT susceptibility level by quantifying the probability of the maximum stress imposed onto the EUT [2] to be greater. This approach presenting the great advantage to be independent of the EUT size and directivity is, therefore, applicable on any EUT. The model presented is applicable in the case of an RS test done in constant wave mode when the EUT susceptibility is verified after the application of each stirring config-uration, or in other terms when all the successive configurations are not analyzed together.

The chapter is organized as follows. Section 4.1 introduces generalities about RS tests in RC with in particular a discussion on the best metric to deal with in order to assess the EM field strength applied on the EUT. The probabilistic model is described in Section 4.2 together with a presentation of some numerical results as well as exper-imental ones. The interest of overtesting the EUT during such tests is presented and discussed in Section 4.3. Finally, the possibility to extract the minimum susceptibility

[1]XLIM Laboratory, UMR7252, University of Limoges, Limoges, France

level of an EUT as a function of the testing conditions when no susceptibility is observed during the test is discussed in Section 4.4.

4.1　About RS tests in RC

4.1.1　*Principle of an RS test*

The principle of an RS test in an RC is sketched in Figure 4.1. The signal produced by an RF generator fed the antenna after amplification, the generator and the amplifier being located outside the RC. A power divider allows the input power to be measured while a (generally triaxial) electric field probe monitored the field within the RC. Therefore, the power of the generator can be modified in order to obtain the desired level of the EM field within the RC.

During the test, the EUT is exposed at each frequency of interest f_0 to an impinging EM field of desired strength for n successive stirring configurations assumed to be independent, each of them being applied during an exposure time t_{exp}, generally specific for each category of EUT. When the test is finished at a first frequency of interest f_0, the test is repeated at the next frequency of interest until the last one. The aim of an RS test is to make sure that the EUT does not exhibit any susceptibility due to the applied disturbance at all the frequencies considered successively and for each n.

The detection of an EUT failure is specific to each EUT. It can be done for instance with a video camera located in the RC (if the failure is expected to be visible), or with a computer linked to the EUT in order to detect its failure in real time or after the test during the post-processing.

Different kinds of susceptibility can be observed: (1) a temporary malfunction of the EUT which disappears when the EM disturbance is turned off and (2) a permanent malfunction of the EUT requiring for instance to reset the EUT or more rarely to replace one or more electronic component.

If no susceptibility is observed or if it is considered tolerable, the EUT is declared EMC compliant. Otherwise, the RS test can either be stopped (the EUT being

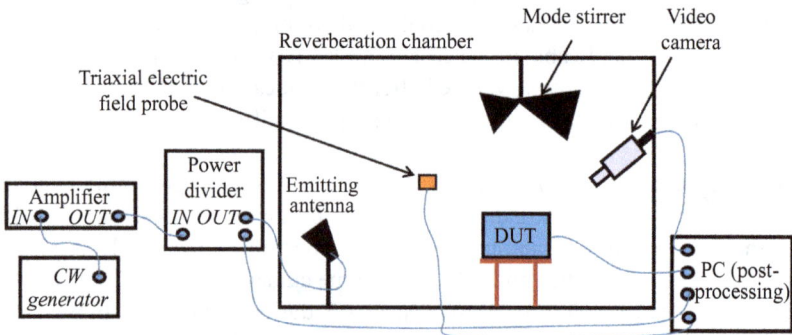

Figure 4.1　Example of setup used during a radiated susceptibility test in RC

considered as non-EMC compliant) or repeated for a lower EM field strength in order to define accurately its immunity level [3].

4.1.2 Values of n recommended by the standard

In the context of an RS test, the choice of n has a great incidence on the test duration and reliability. Table 4.1 indicates the value of n recommended by the IEC standard [4] for different frequency ranges defined with respect to the starting frequency f_s, f_s being generally a multiple (between 3 and 7) of the fundamental resonance (calculated theoretically) of the RC.

As stated in [4], the proposed values of n correspond to the number of samples required to determine the electric field strength with a 6-dB uncertainty within a 5% confidence interval. These values have been derived from the ideal distributions of the field according to the expected modal density [5,6]. Therefore, at low frequencies, i.e. at lower modal densities, the required value of n increases rapidly with this approach.

4.1.3 EM field distributions in a "well stirred" RC

The EM field distributions of a "well stirred" RC (see the first chapter of this book for a complete definition of this condition) are known for years [6–9]. For instance, the magnitude of each rectangular component $|E_R|$ of the electric field follows the Rayleigh distribution. Consequently, the total electric field $|E_T|$ follows the χ distribution with 6 degrees of freedom. The probability density function (PDF) of a random variable X following the latter distribution with the scale parameter σ is (for $x \geq 0$)

$$f_X(x) = \frac{x^5}{8\sigma^6} e^{-x^2/2\sigma^2}. \tag{4.1}$$

Therefore, one understands here that an EUT do not endure a constant strength of the impinging field for each n. This is illustrated in Figure 4.2 from measurements collected with a triaxial electric field probe in the RC of the XLIM Laboratory (the characteristics of this RC are given in details in the first chapter), the RC used in all the experiments shown in this chapter, at $f = 1$ GHz for $n = 50$ mode stirrer positions. Therefore, it is useful to investigate what metric is the best in order to characterize the magnitude of the field during an RS test.

Table 4.1   Value of n recommended in [4] during the RC calibration and a subsequent test according to the starting frequency

Frequency range	n
f_s to $3f_s$	50
$3f_s$ to $6f_s$	18
$6f_s$ to $10f_s$	12
Above $10f_s$	12

4.1.4   On the choice of the metric representing the EM field strength

Four different kinds of metrics listed hereafter can be used in order to assess the EM constraint applied on the EUT

- the maximum magnitude of one rectangular component E_R, i.e. $|E_R|_{max}$;
- the average magnitude of one rectangular component E_R, i.e. $\langle|E_R|\rangle$;
- the maximum of the total electric field E_T, i.e. $|E_T|_{max}$;
- the average of the total electric field E_T, i.e. $\langle|E_T|\rangle$.

All these metrics are generally assessed during the calibration of the RC using s electric field measurements performed at each frequency of interest with a triaxial field probe. For instance, in the IEC standard procedure [4], s equals 96 samples (i.e. $n = 12$ stirring configurations for 8 measurement points) above $6f_s$.

To assess the EM constraint applied on the EUT, most RS tests use peak values rather than average values in order to focus on the maximum stress imposed on the EUT. Indeed, peak values are considered to be the worst case for the RS test and then the most interesting configuration from an RS test point of view [10] or, in other terms, the configuration having more chances to exhibit an EUT failure. However, Höijer recommended in [2] to use average values considered as more stable metrics.

A numerical analysis presented hereafter gives some reliable elements about this important point. Indeed, in order to reproduce numerically the measurements of a triaxial electric field probe made in m different positions of a "well-stirred" RC (all positions being considered independent from the others), three random variables following the Rayleigh distribution have been defined, each one representing the magnitude of one rectangular component $|E_R|$ of the electric field (i.e. $|E_x|$, $|E_y|$,

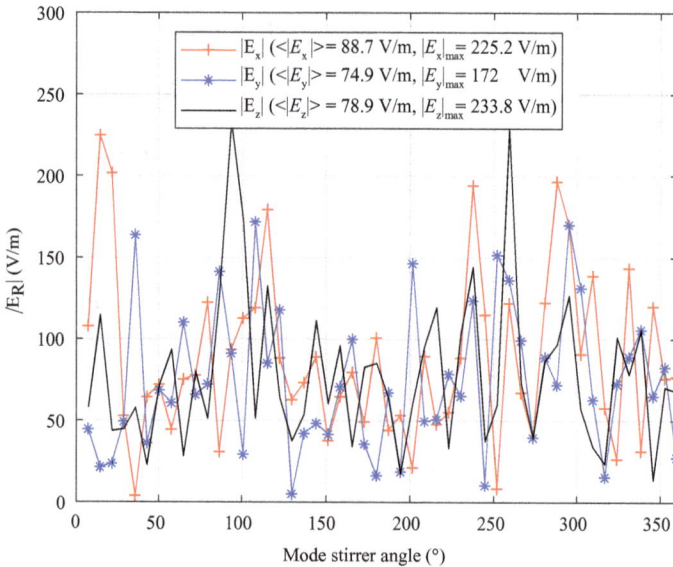

Figure 4.2   Measurement of $|E_R|$ in the "unloaded" (no absorber inserted in) RC at $f = 1$ GHz for $n = 50$ mode stirrer positions

$|E_z|$). For each random variable, $m = 10^4$ repeated runs (reproducing different field probe locations within this "ideal" RC) of $n = 50$ samples have been generated with the condition $\langle |E_R| \rangle = 50$ V/m. The magnitude of $|E_T|$ related to each sample n of each run m is then computed with $\sqrt{|E_x|^2 + |E_y|^2 + |E_z|^2}$.

Figure 4.3 presents the empirical PDF of $|E_x|_{\max}$, $|E_T|_{\max}$, $\langle |E_x| \rangle$ and $\langle |E_T| \rangle$. In addition, Figure 4.4 illustrates that the PDF of the average values fits the Gaussian

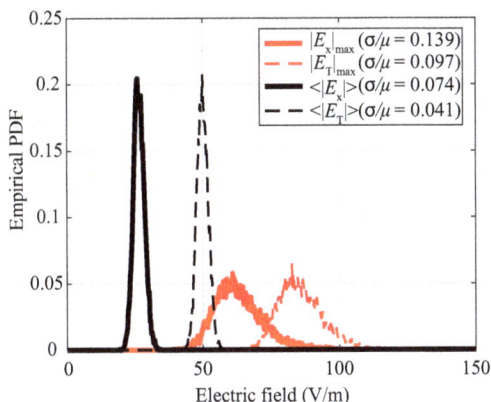

Figure 4.3 *Empirical PDF of peak and average values of $|E_x|$ and $|E_T|$ (for $m = 10^4$ repeated runs of $n = 50$ samples) with $\langle |E_R| \rangle = 50$ V/m*

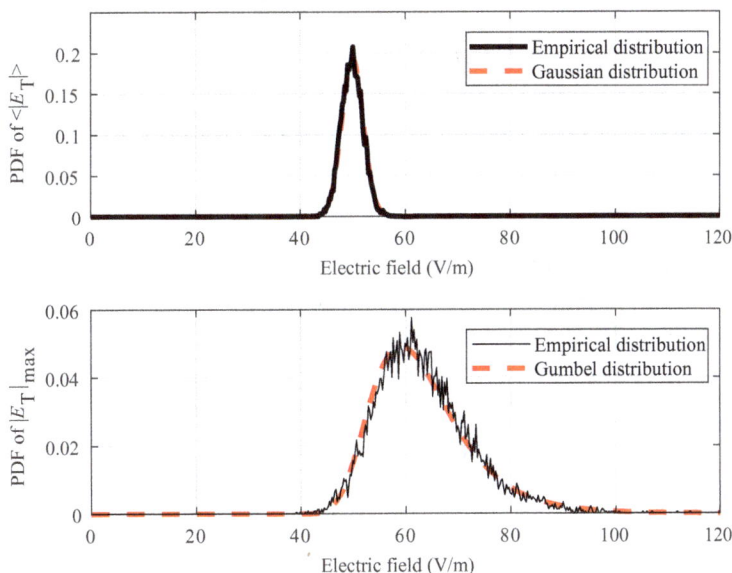

Figure 4.4 *Theoretical and empirical PDF of peak and average values of $|E_T|$ (from $m = 10^4$ repeated runs of $n = 50$ samples). Theoretical distributions fitting the empirical ones are respectively Gaussian (for average values) and Gumbel (for peak values) distributions*

distribution while the PDF of the peak values fits the Gumbel distribution [11,12], which is related to extreme value theory.

It is clearly emphasized on both figures that the spread of the PDF related to average values is narrower. To deepen the analysis, the legend of Figure 4.3 indicates that the lowest σ/μ ratio (with σ and μ, respectively, the standard deviation and the average of each metric) is obtained for $\langle|E_T|\rangle$. This fundamental result implies that the average value of $|E_T|$ measured in one or several points of the RC is statistically closer to the expected value if we compare with the three other metrics mentioned before. This is, therefore, the more stable one. Logically, the same study based on different values of n leads to the same conclusions.

According to these conclusions, it is recommended to assess the strength of the EM disturbance applied on the EUT during an RS test by the average value of the total electric field strength $\langle|E_T|\rangle$ measured within an RC. It is also what is done in the model described in this chapter.

After the RC calibration [4] (sometimes called also "validation"), the power to inject to the antenna in order to generate $\langle|E_T|\rangle$ is known. This is achieved through the electric field measurements collected during the calibration process and the knowledge of the chamber loading factor (related to the losses involved by the EUT itself). However, $\langle|E_T|\rangle$ can also be predicted directly from the average composite quality factor Q of the RC (calculated from S_{11} measurements for instance) in presence of the EUT as already shown in the first chapter of this book.

4.2 Description of the probabilistic model

4.2.1 Assumptions and definitions

The method is applicable for any RC satisfying the "well-stirred" condition. Therefore, the EM field distributions in the RC are considered matching the ideal ones while all the n stirring configurations are considered independent.

For the sake of clarity, some definitions about the strength of $\langle|E_T|\rangle$ are given hereafter.

The desired value of the average value of $|E_T|$, i.e. $\langle|E_T|\rangle$, during the RS test is denoted $\langle|E_T|\rangle_d$. However, this is not the real value generated in the RC in reason of the finite number s of electric field measurements collected during the RC calibration process.

The real value of $\langle|E_T|\rangle$ (expected to be close to $\langle|E_T|\rangle_d$) generated during the RS test is denoted $\langle|E_T|\rangle_r$. This would be the obtained value with an infinite number of samples. $\langle|E_T|\rangle_r$ is consequently unknown and can only be estimated.

The measured value of $\langle|E_T|\rangle$ for a given number of samples s is written $\langle|E_T|\rangle_m^s$. $\langle|E_T|\rangle_m^s$ is, therefore, an estimation of $\langle|E_T|\rangle_r$. Moreover, $\langle|E_T|\rangle_m^s$ is a random variable depending on the measurement conditions, for instance the position of the antenna or of the field probe.

4.2.2 Uncertainty on the estimation of $\langle|E_T|\rangle_m^s$

As mentioned previously, $\langle|E_T|\rangle_m^s$ is only an estimation of $\langle|E_T|\rangle_r$ depending on the finite number of samples s considered during the RC calibration process. It is therefore of interest to assess the uncertainty of $\langle|E_T|\rangle^s_m$ according to s and to integrate this level of uncertainty in the model.

A method using experimental (for a "real" RC) or numerical data (for an "ideal" RC) to assess this uncertainty is described hereafter. The uncertainty of $\langle|E_T|\rangle^s_m$ as a function of s has been estimated in our RC after measuring $|E_T|$ on 360 mode stirrer positions (rotation of 1 degree between two successive positions) for 10 different configurations. For each configuration, the position of both the triaxial field probe and the emitting antenna has been modified. The average value of these 3,600 samples, i.e. $\langle|E_T|\rangle^{s=3,600}_m$, is assumed here to be the expected value, i.e. $\langle|E_T|\rangle_r$.

$\langle|E_T|\rangle^s_m$ is then computed as a function of s from $m = 10^4$ subsets of s samples extracted randomly among these 3,600 samples, the samples being replaced in the pool of samples after the draw of each subset. The standard deviation σ_{dB} of $\langle|E_T|\rangle^s_m$ (for any value of s considered) obtained from the m subsets with respect to $\langle|E_T|\rangle_r$ is obtained as follows

$$\sigma_{dB} = 20 \log \left(\frac{\sigma + \langle|E_T|\rangle_r}{\langle|E_T|\rangle_r} \right), \tag{4.2}$$

with σ being the standard deviation of $\langle|E_T|\rangle^s_m$.

These experimental results have been obtained at different frequencies (0.5, 2 and 3 GHz) and for two different loading conditions of the RC: the "empty" configuration when no additional loading is inserted in the RC and when the RC is loaded with one block of 36 pyramidal absorbers (PAs), this loading condition being described in details in the first chapter of this book.

In addition, this process has been implemented numerically after generating samples following the χ distribution with 6 degrees of freedom (i.e. the ideal distribution of $|E_T|$), each subset being made of new samples in order to avoid re-sampling.

Results reported in Figure 4.5 logically show the decrease of σ_{dB} with the number of samples s. The trends of all the curves are similar with the exception of the results obtained from measurements performed at 0.5 GHz in reason of the residual sample correlation occurring at this frequency. For all the other experimental curves, the "well-stirred" condition is reached in the RC as discussed in the first chapter of this book. For all the other curves including the simulated one, the differences are lower than 0.15 dB, a reasonable value related to the intrinsic uncertainty of this method of determination of σ_{dB}. Moreover, the simulated curve being similar to the experimental ones (with the exception of the curve obtained at $f = 0.5$ GHz as discussed previously), it is worth noting that the effect of resampling during this Monte Carlo analysis does not introduce any artificial correlation on the experimental results.

As a conclusion, the value of σ_{dB} is independent (1) of the frequency provided that the "well-stirred" condition is valid and (2) of the RC quality factor. It only depends of s. The standard deviation σ_{dB} of $\langle|E_T|\rangle^s_m$ obtained numerically as a function of s (for instance the value of s used during the RC calibration process) can, therefore, be directly inserted in the probabilistic model (as shown later in this chapter) in order to take the uncertainty of $\langle|E_T|\rangle$ into account. Frequency-dependent values calculated from measurements can also be inserted in the model in order to take into account the lowest frequency range when the RC is not perfectly "well stirred."

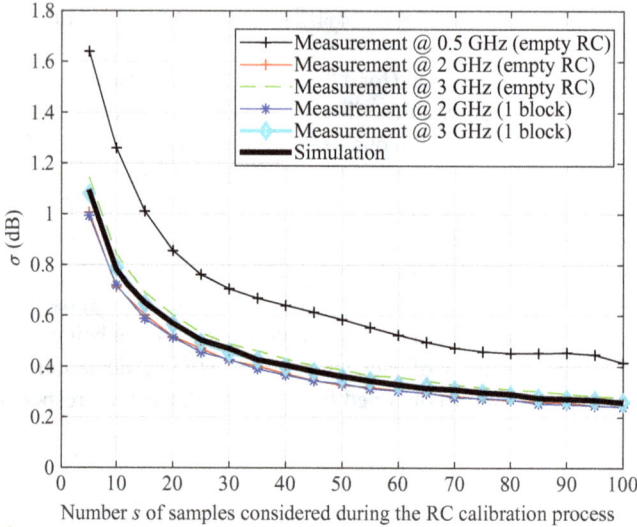

Figure 4.5   Measured and simulated standard deviation σ_{dB} of $\langle|E_T|\rangle_m^s$ with respect to the expected value $\langle|E_T|\rangle_r$ as a function of the number of samples s considered during the RC calibration process. "Empty" means that no absorber is inserted in the RC. "One block" means that one block of 36 PAs is inserted in the RC

In the results shown in this chapter, σ_{dB} obtained for $s = 100$ samples is considered (i.e. $\sigma_{dB} = 0.254$ dB), a close value with respect to the number of 96 samples used in [4] above $6f_s$ as mentioned before. Then, for $\langle|E_T|\rangle_d = 50$ V/m (or 34 dBV/m) and $\sigma_{dB} = 0.254$ dB, $\langle|E_T|\rangle_d - \sigma_{dB}$ and $\langle|E_T|\rangle_d + \sigma_{dB}$ equal, respectively, 48.6 and 51.5 V/m (or 33.75 and 34.25 dBV/m).

4.2.3   Demonstration of the method applicability for any EUT

The susceptibility level $|E_S|$ of an EUT tested in RC (which is unknown) is defined as the amplitude of $|E_T|$ leading to an EUT susceptibility for one particular stirring configuration, i.e. when $|E_T| > |E_S|$ [3]. The total electric field $|E_T|$ being χ-distributed with 6 degrees of freedom in a "well-stirred" RC, Hill demonstrates in [9] that the power P_{EUT} induced on an antenna of any size or receiving polarization pattern is as a consequence exponentially distributed.

From an RS test point of view, an EUT is no more than a lossy impedance-mismatched antenna and ideally the worst possible receiving one [2]. Therefore, the EUT exhibits a susceptibility if the power induced on the EUT (and, in particular, on its critical component) is higher than a threshold power P_T, which is also unknown.

In reason of the relationship discussed above between the distribution of the total electric field and of the power induced on an antenna, one understands here that the definition of the EUT susceptibility from the induced power on the EUT ($P_{\text{EUT}} > P_{\text{T}}$) or from the total electric field generated in the RC ($|E_{\text{T}}| > |E_{\text{S}}|$) is strictly equivalent, a reasoning already used in [13].

Therefore, the theory developed hereafter is applicable on any EUT independently of its size with respect to the wavelength (provided that the EUT volume is no more than a few percent of the RC volume) and its directivity if the RC matches the "well-stirred" condition.

4.2.4 Computation of the risk r

The definition of $|E_{\text{S}}|$ given above allows a probability of susceptibility p to be defined according to $\langle|E_{\text{T}}|\rangle_d$ and $|E_{\text{S}}|$ from (4.1) with $X = |E_{\text{T}}|$. For instance, an EUT having a susceptibility level $|E_{\text{S}}|$ of 87.2 V/m and tested under the condition $\langle|E_{\text{T}}|\rangle_d = 50$ V/m (the value used throughout this chapter in order to comment the presented results) has a probability of susceptibility p of 1%.

Table 4.2 contains different $|E_{\text{S}}|/\langle|E_{\text{T}}|\rangle_d$ ratios related to specific values of p, each value corresponding to a specific EUT susceptibility level. It is important to remind here that the susceptibility level of an EUT is unknown. However, the aim of the model is to investigate the influence of n for any value of p, so for any value of the EUT susceptibility level.

When reducing n, one understands intuitively that the probability to observe a susceptibility for a faulty EUT decreases, especially for large ratio $|E_{\text{S}}|/\langle|E_{\text{T}}|\rangle_d$ (or low p values). Consequently, the risk grows to consider EMC compliant a faulty EUT.

Thanks to a Monte Carlo approach, the model aims at quantifying the risk r to observe during an RS test an absence of susceptibility of a faulty EUT whatever its susceptibility level (or its probability of susceptibility p), and this as a function of n.

To reach this objective, $m = 10^4$ repeated runs of n samples following the χ distribution with 6 degrees of freedom are generated randomly in order to obtain the desired $\langle|E_{\text{T}}|\rangle_d$ value. For each run (representing an RS test performed with n stirring configurations), the aim is to control if $|E_{\text{T}}|$ is greater than $|E_{\text{S}}|$ at least once, this case corresponding to an EUT susceptibility. For instance, for $p = 1\%$, the model checks if $|E_{\text{T}}|$ is greater than 87.2 V/m (with $\langle|E_{\text{T}}|\rangle_d = 50$ V/m) at least once for each run. When considering all the runs together, the risk r is derived for any couple of values of p and m, r being the ratio between the number of runs when $|E_{\text{T}}|$ is greater than $|E_{\text{S}}|$ at least once and the total number of runs m. The risk r is, therefore, the probability

Table 4.2 Relationship between $|E_S|$ and $\langle|E_T|\rangle_d$ for different probabilities of susceptibility p. Example for $\langle|E_T|\rangle_d = 50$ V/m.

p	1%	5%	10%	20%
$|E_S|/\langle|E_T|\rangle_d$	1.74	1.51	1.39	1.24
$|E_S|$ (V/m) for $\langle|E_T|\rangle_d = 50$ V/m	87.2	75.5	69.4	62.2

of the maximum stress imposed onto the EUT to be greater than $|E_S|$ for each (n,p) couple of values. The flowchart of the model is sketched in Figure 4.6.

As discussed in Section 4.2.2, the uncertainty of $\langle|E_T|\rangle_m^s$ as a function of n can be introduced in the model. Therefore, for each run, the process described before is also repeated two times (for $\langle|E_T|\rangle_d \pm \sigma_{dB}$, $\langle|E_T|\rangle_d$ being expressed here in dBV/m) in order to deal with the $\langle|E_T|\rangle_m^s$ uncertainty. The value of $\sigma_{dB} = 0.254$ dB is used in this chapter in order to work with the standard deviation of $\langle|E_T|\rangle_m^{s=100}$ previously obtained numerically.

Figure 4.7 presents the risk r (decreasing with n) to consider EMC compliant a faulty EUT for different p values (equal or lower than 20%) as a function of n, the

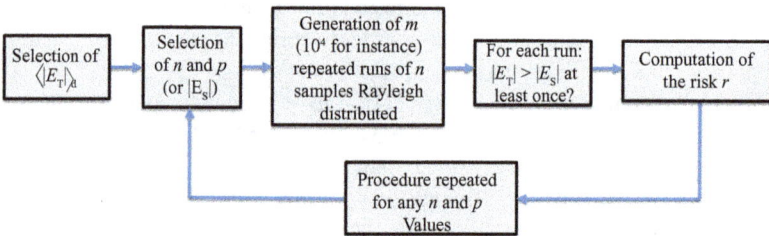

Figure 4.6   Flowchart of the probabilistic model

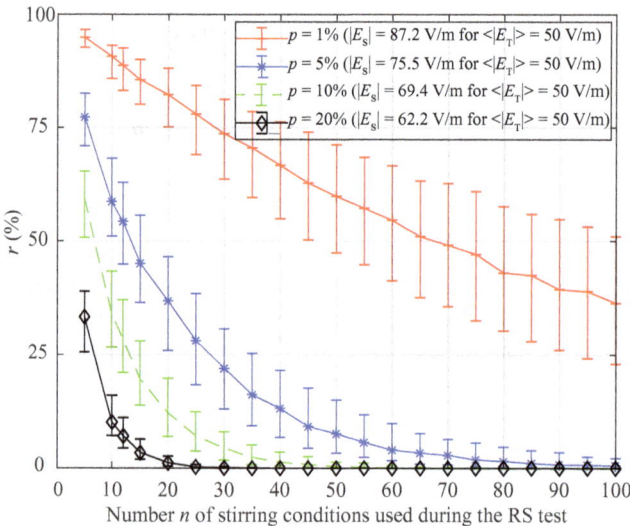

Figure 4.7   Risk r to observe no susceptibility of faulty EUTs of different probability of susceptibility p as a function of n (for m = 10⁴ repeated runs). Uncertainty bars correspond to the uncertainty $\sigma_{dB} = 0.254$ dB on $\langle|E_T|\rangle_m^{s=100}$ obtained numerically in Section 4.2.2

uncertainty bars corresponding to the uncertainty of $\langle |E_T| \rangle_s$. The upper bound of r is related to the case when the EUT is exposed to $\langle |E_T| \rangle_d - \sigma_{dB}$ and inversely the lower bound corresponds to $\langle |E_T| \rangle_d + \sigma_{dB}$. It is worth noting that these results are general (with the exception of the level of uncertainty of $\langle |E_T| \rangle_d$ which can be modified), as they are only dependent of p and n.

To explain the results provided by the model, it is useful to distinguish different cases.

- For $p \geq 20\%$, i.e. for "highly susceptible EUTs," the risk r vanishes if n equals at least 25, which is a reasonable number. RS tests performed for such p values reveal the EUT susceptibility in these conditions. In concrete terms, if the susceptibility level of the EUT equals 62.2 V/m or less, the EUT failure is for sure observed if the RS test is done using the condition $\langle |E_T| \rangle_d = 50$ V/m with $n = 25$ (or more).
- For $10\% \leq p \leq 20\%$, the upper bound of r tends toward 0% for a reasonable n, i.e. $r = 1.6\%$ with $n = 50$ for $p = 10\%$ for instance. Therefore, the RS test would reveal an EUT susceptibility for such p values. In other terms, if no EUT susceptibility is observed for $n = 50$, it is possible to conclude with an extremely good confidence that the EUT susceptibility level $|E_S|$ is greater than 69.4 V/m (for $\langle |E_T| \rangle_d = 50$ V/m) according to Table 4.2.
- For $p \leq 10\%$, i.e. for "weakly susceptible EUTs," it is impossible to draw strong conclusions from the RS test, even for large values of n. In other terms, the absence of EUT susceptibility does not lead to conclude definitely about the EMC compliance of the EUT for such susceptibility levels. As an example, when considering an EUT having a susceptibility level $|E_S|$ of 87.2 V/m exposed to $\langle |E_T| \rangle_d = 50$ V/m (i.e. $p = 1\%$), the risk r to observe no susceptibility for $n = 100$ is included in the range 22.2%–51.3% (with $\sigma_{dB} = 0.254$ dB). Therefore, one understands here that the RS test can lead to opposite conclusions (EMC compliance or not) according to the conditions of the test (for instance if the position of the EUT or of the antenna within the RC is changed). To get more certainty for these low p values (i.e. large $|E_S| / \langle |E_T| \rangle_d$ ratios), another strategy, developed in Section 4.3, has to be carried out.

From these results, it is clear that the value of $n = 12$ recommended by the IEC procedure [4] above $6f_s$ does not give strong insurance about the EMC compliance of the EUT in the case of absence of susceptibility during the RS test, including for $p = 20\%$. It is, therefore, recommended to increase this value of n to increase the reliability of the test. It is important to remind here that the assessment of the EM field strength in [4] is different, the average of the maximum magnitude of the rectangular components of the electric field, i.e. $\langle |E_R|_{max} \rangle$, obtained during the RC calibration process being used.

4.2.5 Experimental results

The probabilistic model presented in this chapter is illustrated in this subsection from experimental results obtained in the RC. The aim is to reproduce "virtually" an RS test using a homemade generic EUT using S-parameter measurements. The experimental setup depicted in Figure 4.8 has been implemented.

The emitting antenna is a horn antenna connected to the port 1 of a vector network analyzer (VNA). The mode stirrer is made of eight rectangular metallic blades of 60 by 40 cm^2 dimensions as described in the first chapter of this book.

The homemade generic EUT shown in Figure 4.9 is placed on a dielectric support within the RC. It is made of a closed metallic enclosure of 15.5 cm length, 12.5 cm width and 8.5 cm height. In order to illustrate that the probabilistic model is independent of the EUT directivity, two versions of the EUT are available. In the so-called isotropic version, four monopoles of different lengths (2.5, 3, 3.5 and 4 cm) are installed on different faces of the enclosure with at least one monopole oriented in each axis of a Cartesian coordinate system. Within the enclosure, each monopole is connected to a four inputs toward one output RF combiner. The output of the combiner, linked to the port 2 of the VNA, is, therefore, the sum of the collected power by each monopole. In the so-called directive version of the EUT, only one monopole is used, the output of the monopole is directly linked to the port 2 of the VNA.

The S_{21} parameter, representing the coupling between the emitting antenna and the EUT, is collected for $n = 50$ equidistant positions of the rotating mode stirrer at 3,001 frequencies linearly spaced between 1 and 4 GHz ($\delta_f = 1$ MHz). Measurements are repeated for eight different positions of both the antenna and the EUT within the RC in order to increase the number of independent samples measured. This process leads to obtain 24,008 sets of $n = 50$ samples when discarding the measurement frequency. Indeed, instead of comparing the probabilistic model from results collected at one single frequency (which would require to collect an unreasonable number of samples

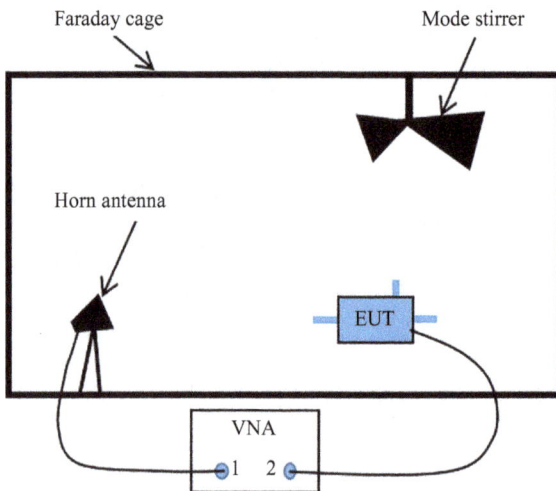

Figure 4.8   Schematic description of the experimental setup

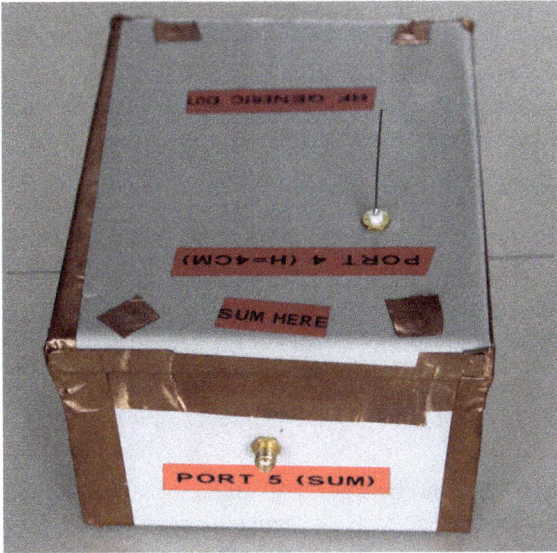

Figure 4.9   *Picture of the generic EUT. The 4 cm monopole is visible on the top of the EUT as well as the EUT output in the foreground*

at this particular frequency), the collected results are gathered together by considering each set of $n = 50$ samples (at any frequency) as independent. This is possible as the frequency shift $\delta_f = 1$ MHz between two sets of measurements is greater than the average mode bandwidth Δ_f of the RC, Δ_f being the ratio between the frequency f and the average composite quality factor $Q(f)$. Moreover, to confirm the absence of effect of the Q-factor provided that the RC is well stirred, the RC is left empty (i.e. when no absorber is inserted in the RC) during the measurements and also loaded with a block of 36 PAs, each PA having a squared ground surface of 100 cm^2 with a height of 30 cm. Δ_f is, respectively, 0.2 and 0.5 MHz for the "empty" and the "one block" loading configurations as shown in the first chapter.

The frequency range considered here is well above the frequency when the "well-stirred" condition is achieved (the so-called f_{wsc} as defined in the first chapter of this book) in order to be sure that the 50 samples are uncorrelated in each set. Indeed, f_{wsc} is achieved, respectively, around 560 and 400 MHz for the "empty" and the "one block" loading configurations in our RC. For each position of both the antenna and the EUT, the measurement time was around 11 minutes under these conditions.

Figure 4.10 shows $\langle |S_{21}| \rangle$ as a function of the frequency for the four tested configurations. Before analyzing the results, the data have to be corrected in order to cancel the frequency dependence of S_{21}. Indeed, $\langle |S_{21}| \rangle$ is not constant with the frequency and depends in particular of the frequency response of the different monopoles. $\langle |S_{21}| \rangle$ is also lower for the "isotropic" version of the EUT which is counter-intuitive. This is due to the RF combiner effect including its insertion losses and the attenuation related to the 12 dB splitter.

To correct the data, the EUT received power P_r for a frequency f, a position p_n of the mode stirrer and a measurement configuration i (a particular position of both the antenna and the EUT) is defined as follows

$$P_r (f, p_n, i) = |S_{21} (f, p_n, i)|^2. \tag{4.3}$$

This yields to the average received power at any frequency f

$$\langle P_r (f) \rangle = \langle |S_{21} (f, p_n, i)|^2 \rangle, \tag{4.4}$$

where the operator $\langle \cdot \rangle$ denotes averaging over the n stirring configurations.

Finally, the received power at any frequency is corrected as follows

$$P_r^{cor} (f, p_n, i) = \frac{P_r (f, p_n, i)}{\langle P_r (f) \rangle} \cdot \langle P_r \rangle, \tag{4.5}$$

where $\langle P_r \rangle$ is the average received power of all the collected samples. This leads to 24,008 sets of $n = 50$ samples, each set having the same average value.

For each set of data, it is verified if at least one sample is greater than a power susceptibility level P_S calculated with respect to $\langle P_r^{cor} \rangle$ and a probability of susceptibility p. The ratio between both values indicated in Table 4.3 for different p values is calculated from the exponential distribution, the induced power on an antenna being exponentially distributed as discussed in Section 4.2.3. The end of the post-treatment process is the same as described in Section 4.2.4. It is worth noting that no uncertainty computation is shown here as $\langle P_r^{cor} \rangle$ is known.

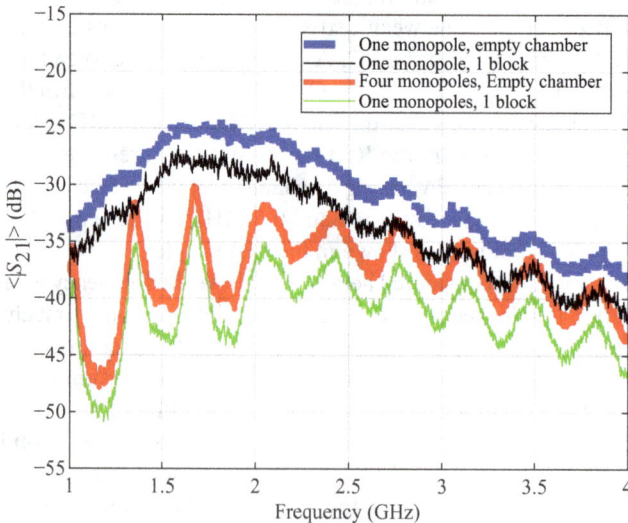

Figure 4.10   Plot of $\langle |S_{21}| \rangle$ measured in the RC for both versions of the generic EUT and two loading conditions of the RC

As expected, the experimental results presented in Figures 4.11 and 4.12 for two different experimental conditions fit the theoretical results presented in Section 4.2.4. This makes sense as this probabilistic model is independent of the EUT directivity and of the RC Q-factor provided that the samples are uncorrelated and the RC is "well stirred."

At this stage, two other important conclusions have to be reminded in the case of a real RS test

- the susceptibility level (and therefore p) is different for each EUT (this is the case here for the isotropic and the directive version of the EUT). Moreover, this value is unknown and frequency dependent;
- the power to inject to the antenna in the RC in order to obtain $\langle |E_T| \rangle_d$ is greater when the Q-factor is decreased in order to compensate the losses due to the presence of the absorber.

Table 4.3   *Relationship between P_S and $\langle P_r^{cor} \rangle$ for different probabilities of susceptibility p*

p	1%	5%	10%	20%
$P_S / \langle P_r^{cor} \rangle$	4.55	2.99	2.30	1.61

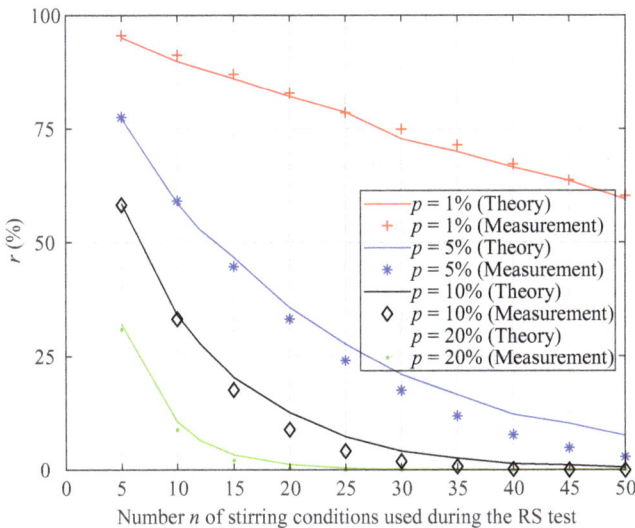

Figure 4.11   *Risk r to observe no susceptibility of faulty EUTs for different probability of susceptibility p according to n. Comparison with experimental results obtained with the EUT equipped with four monopoles when the RC is unloaded*

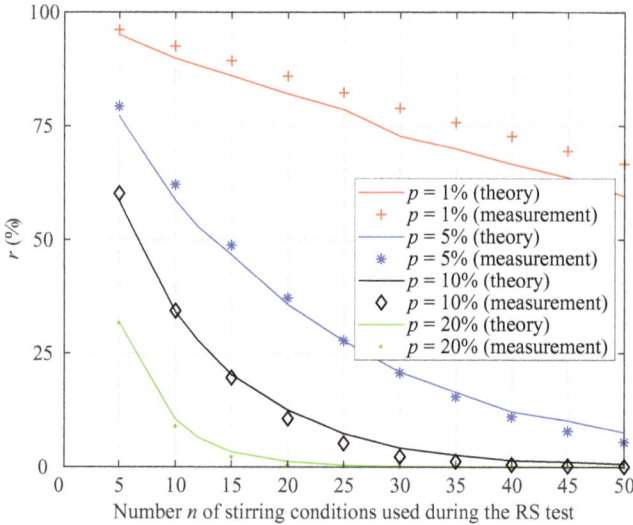

Figure 4.12   Risk r to observe no susceptibility of faulty EUTs for different probability of susceptibility p according to n. Comparison with experimental results obtained with the EUT equipped with one monopole when the RC is loaded with one block of absorber

4.3 Overtesting

4.3.1 Principle

As shown previously in Section 4.2.4, it is impossible to draw strong conclusions in the case of a weakly susceptible EUT (i.e. in the case of small values of p), even for large n. The open question is how to improve the reliability of an RS test in such cases.

The aim of this section is to illustrate with the same Monte Carlo approach that overtesting may be a solution increasing the confidence about the EMC compliance of the EUT. Overtesting consists in testing an EUT at a higher level (here $\langle|E_{OT}|\rangle_d$, the subscript "*OT*" meaning "overtesting") than the nominal disturbance level ($\langle|E_T|\rangle_d$). This concept has been introduced recently [14] for bulk current injection (BCI) purposes in EMC testing area. Therefore, if the EUT does not exhibit any susceptibility when exposed to $\langle|E_{OT}|\rangle_d$ even for a lower n, stronger conclusions can be drawn with respect to an RS test performed at $\langle|E_T|\rangle_d$. This constitutes the interest of the overtesting approach in such context.

The overtesting coefficient c_{OT} is introduced as follows

$$c_{OT} (\%) = 100 \left(\frac{\langle|E_{OT}|\rangle_d}{\langle|E_T|\rangle_d} - 1 \right). \tag{4.6}$$

Through this straightforward definition, if $\langle|E_T|\rangle_d$ equals 50 V/m, an overtesting coefficient c_{OT} of 25% leads to perform the overtest at $\langle|E_{OT}|\rangle_d = 62.5$ V/m.

With respect to the model presented in this chapter, the method is applied similarly with this unique exception: the EUT is tested during the RS test at the condition $\langle|E_{OT}|\rangle_d$, while the probability of susceptibility p is computed according to $\langle|E_T|\rangle_d$. Thus, $p = 1\%$ for $c_{OT} = 25\%$ and $\langle|E_T|\rangle_d = 50$ V/m still refers to 87.2 V/m. Therefore, the risk r to have a maximal constraint applied on the EUT lower than 87.2 V/m for a given n decreases when the EUT is tested at $\langle|E_{OT}|\rangle_d$.

4.3.2 Example of numerical results

The Monte Carlo approach is repeated (with $m = 10^4$ runs) in order to compute the risk r to observe no susceptibility of a faulty EUT for any (p, n) combination and for any c_{OT} coefficient. Figures 4.13 and 4.14 present respectively the results obtained for $c_{OT} = 25\%$ and 50%.

In the case of an overtesting coefficient of 50%, the risk r for $p = 5\%$ is cancelled when n equals 25. In other terms, if an EUT tested under the condition $\langle|E_{OT}|\rangle_d = 62.5$ V/m for $n = 25$ does not exhibit any susceptibility, $|E_S|$ is for sure greater than 75.5 V/m. As a reminder, no certainty could be obtained for this susceptibility level when testing the EUT at $\langle|E_T|\rangle_d = 50$ V/m, even for $n = 100$.

In the case of an overtesting coefficient of 50%, the risk r is null for any value of $p \geq 1\%$ if $n \geq 20$. Therefore, if the same EUT tested under the condition $\langle|E_{OT}|\rangle_d = 75$ V/m for $n = 20$ does not exhibit any susceptibility, $|E_S|$ is for sure greater than 87.2 V/m.

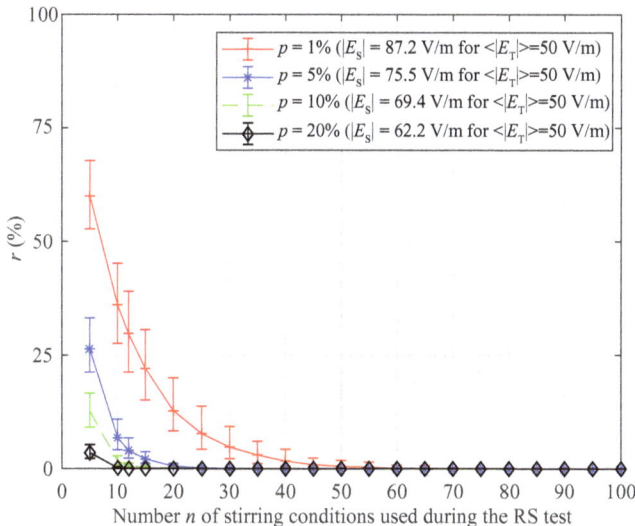

Figure 4.13 *Risk r obtained for different probability of susceptibility p according to n in the case of an overtesting coefficient c_{OT} of 25% (for $m = 10^4$ repeated runs). Uncertainty bars correspond to the uncertainty $\sigma_{dB} = 0.254$ dB on $\langle|E_T|\rangle_m^{s=100}$ obtained numerically in Section 4.2.2*

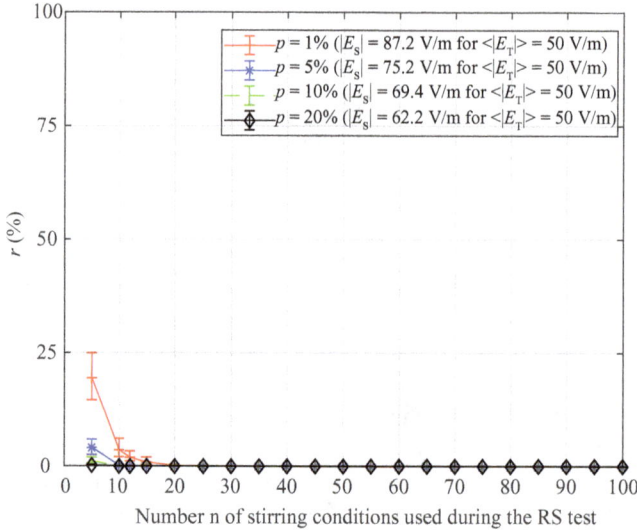

Figure 4.14   *Risk r obtained for different probability of susceptibility p according to n in the case of an overtesting coefficient c_{OT} of 50% (for $m = 10^4$ repeated runs). Uncertainty bars correspond to the uncertainty $\sigma_{dB} = 0.254$ dB on $\langle |E_T| \rangle_m^{s=100}$ obtained numerically in Section 4.2.2*

As a conclusion, these results illustrate that the overtesting approach leads (in absence of any EUT susceptibility during the test) to decrease the RS test duration (in reason of the reduction of n) while improving in parallel the reliability of the RS test itself. This is also illustrated in Figure 4.15 which presents the decrease of r as a function of the overtesting coefficient c_{OT} for $n = 20$. In such conditions, r becomes null for $p = 1\%$ with $c_{OT} = 45\%$.

4.3.3   A two-step approach using overtesting

It is worth noting that a parallel can be made between the overtesting approach introduced here and the method which consists in increasing gradually the level of the disturbance, an approach investigated in [3] for instance.

If an EUT susceptibility occurs during a simple overtest, it is necessary to repeat the RS test at the nominal level $\langle |E_T| \rangle_d$ in order to check if the susceptibility also occurs or if the EUT susceptibility is related to the overtest.

To overcome this problem, a two-step approach described as follows is considered to be more relevant than a simple overtest. In a first step, the RS test is performed at $\langle |E_T| \rangle_d$ in order to handle accurately the highest range of values of p. In a second step, an overtest is performed at $\langle |E_{OT}| \rangle_d$ in order to obtain reliable information about the EUT susceptibility for lower p values. Therefore, an EUT susceptibility occurring during the overtest can be attributed to this latter range of p values.

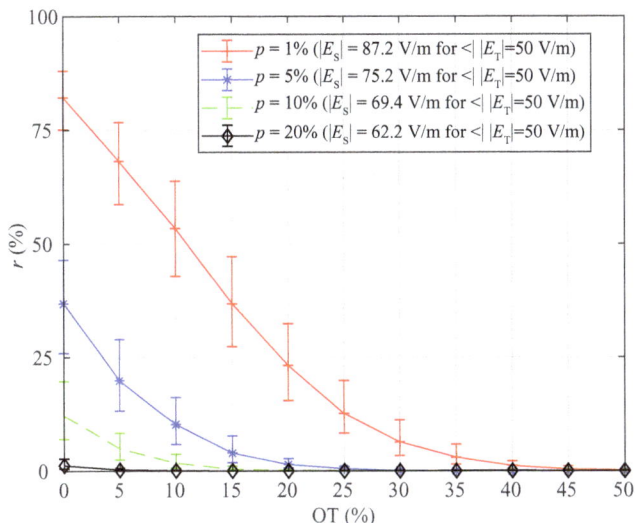

Figure 4.15 *Risk r to observe no susceptibility of faulty EUTs of different*
probability of susceptibility p according to the overtesting coefficient
c_{OT} *for n = 20. Uncertainty bars correspond to the uncertainty*
$\sigma_{dB} = 0.254$ *dB on* $\langle |E_T| \rangle_m^{s=100}$ *obtained numerically in Section 4.2.2*

4.4 Determination of the minimum susceptibility level of the EUT (after a successful test)

The probabilistic model in this chapter describes the *a priori* influence of n on the risk r to declare EMC compliant a faulty EUT.

In this section, it is shown that it is also possible to obtain *a posteriori* information on the minimum susceptibility level $|E_S|_{\min}$ of the EUT when no susceptibility is observed during the test (the same approach dealing with the received power would lead to the same results for the reasons given in Section 4.2.3). Indeed, the success of the RS test (i.e. no susceptibility observed during the test) implies that the EUT is not tested for the condition $|E_T| > |E_S|$. It is, therefore, possible to compute from a similar probabilistic approach the probability $p(|E_T|)$ that a given value of $|E_T|$ has been applied on the EUT during the test. The maximum value of $|E_T|$ satisfying the condition $p(|E_T|) = 100\%$ corresponds, therefore, to $|E_S|_{\min}$, the real value of $|E_T|$ being for sure higher or at least equal.

This approach allows the "power" of the RS test to be compared with respect to the selected test conditions, i.e. n and c_{OT}. If $|E_S|_{\min}$ increases (this metric being computable before the test for each set of testing conditions), the power of the test is better. Therefore, the users obtain more information about the EUT susceptibility in case of absence of EUT susceptibility during the test.

This reasoning is illustrated in the results shown in Figure 4.16. This figure presents the probability $p(|E_T|)$ (calculated from $m = 10^4$ repeated runs) that a given

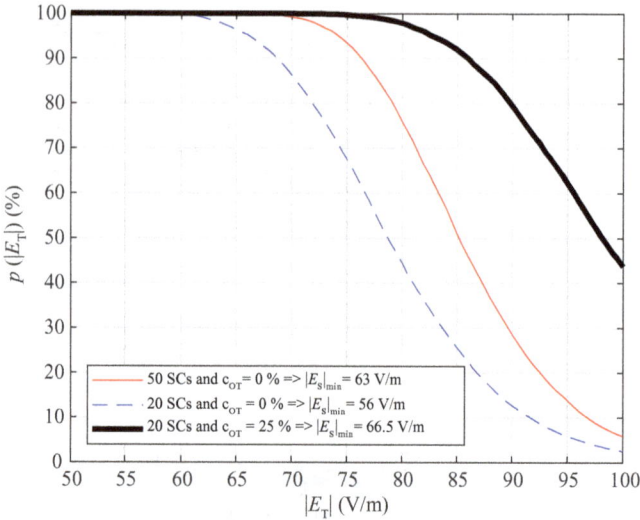

Figure 4.16   Probability $p(|E_T|)$ of the EUT to be exposed at a level $|E_T|$ at least once during a test performed over n stirring configurations for $\langle|E_T|\rangle = 50$ V/m. The value of $|E_S|_{min}$ is given in the legend for each testing configuration. $p(|E_T|) = 50\%$ means that a test performed with such value of n has 50% of chances to expose the EUT to the corresponding value of $\langle|E_T|\rangle$

exposure level $|E_T|$ has been imposed on the EUT for a given n and $\langle|E_T|\rangle = 50$ V/m. The interesting value to compare for the three different configurations is the value of $|E_S|_{min}$ extracted from the test conditions. As expected, the increase of n allows more accuracy about $|E_S|_{min}$ to be obtained, $|E_S|_{min}$ being 63 V/m for $n = 50$ but only 56 V/m for $n = 20$ in the absence of any overtest.

The interest to proceed to an overtest is emphasized clearly. With only $n = 20$ and an overtesting coefficient c_{OT} of 25% (i.e. $\langle|E_{OT}|\rangle = 62.5$ V/m), the absence of EUT susceptibility during the RS test leads to $|E_S|_{min} = 66.5$ V/m. Thus, in these conditions, the overtest leads, in addition to a reduction of the test duration, to obtain more reliable information about the minimum susceptibility level $|E_S|_{min}$ of the EUT.

The probabilistic approach presented in this section (which can also take into account the uncertainty related to the applied field) allows, therefore, the power of the test to be compared according to the selected test conditions by studying the value of $|E_S|_{min}$ obtained for each configuration.

4.5   Conclusion

A probabilistic model based on a Monte Carlo approach (requiring a few seconds of computation) which aims at helping RC users to understand the consequences of the number n of stirring configurations taken into account during an RS test performed in a "well-stirred" RC has been presented in this chapter.

This approach computes the risk r to observe no susceptibility of a faulty EUT (or more exactly of EUTs having different susceptibility levels) as a function of n, the choice of n being generally a trade-off between the expected RS test duration and the reliability of the test (the risk r increasing for a lower n). The model may help RC users to select the value of n fitting best their requirements and constraints. The model also deals with the uncertainty of the amplitude of the EM field disturbance applied on the EUT, which is directly related to the number of samples s used during the RC calibration process.

The field of application of the model is general, in the sense of the method is applicable for any EUT independently of its size, its receiving polarization pattern and its susceptibility level (for each frequency). Results of Figure 4.7 are, therefore, directly usable for any EUT. To take a concrete example, if the test is performed at $\langle|E_T|\rangle = 50$ V/m with $n = 50$, the user is sure to observe the EUT susceptibility during the test if $|E_S| = 69.4$ V/m but the risk r to miss the EUT susceptibility reaches 60% if $|E_S| = 87.2$ V/m (without considering the uncertainty on the EM field disturbance).

This probabilistic approach is also shown to give information about the minimum susceptibility level $|E_S|_{min}$ of the EUT as a function of the testing conditions (i.e. n and c_{OT}). This authorizes to compare the "power" of the RS test for different set of testing conditions. Thus, if $|E_S|_{min}$ increases (this metric being calculated before the RS test), the users know that they will obtain more information on the EUT in the case of absence of susceptibility observed during the test.

In such context, overtesting is able to improve the test reliability and to reduce the test duration. Overtesting is especially interesting if the required instrumentation can remain unchanged, if for instance the same couple of generator and amplifier can be used to generate either $\langle|E_T|\rangle_d$ or $\langle|E_{OT}|\rangle_d$ in the RC.

In real life, RS tests are often performed at frequencies where the "well-stirred" condition of the RC is not matched (i.e. below f_{wsc}). An interesting way of improvement of the model consists in investigating the influence of this imperfect reverberant environment on the risk r. This could probably be achieved by reproducing the experiments shown in this chapter at lower frequencies, in particular below the frequency when the "well-stirred" condition is reached.

References

[1] Andrieu G, Ticaud N, and Lescoat F., On the Risk to Declare EMC Compliant a Faulty EUT During Radiated Susceptibility Tests in Reverberation Chambers. IEEE Transactions on Electromagnetic Compatibility. 2019:1–9.
[2] Höijer M., Maximum Power Available to Stress Onto the Critical Component in the Equipment Under Test When Performing a Radiated Susceptibility Test in the Reverberation Chamber. IEEE Transactions on Electromagnetic Compatibility. 2006;48(2):372–384.

[3]   Amador E, Krauthäuser HG, and Besnier P. A Binomial Model for Radiated Immunity Measurements. IEEE Transactions on Electromagnetic Compatibility. 2013;55(4):683–691.

[4]   Reverberation Chamber Test Methods. Int Electrotech Commiss Standard IEC 61000-4-21:2011; 2011.

[5]   Hill DA. Electromagnetic theory of reverberation chambers. National Institute of Standards and Technology (US) Technical Note 1506. 1998;36.

[6]   Lehman TH. A statistical theory of electromagnetic fields in complex cavities. EMP Interaction Note 494. 1993.

[7]   Kostas JG and Boverie B. Statistical Model for a Mode-Stirred Chamber. IEEE Transactions on Electromagnetic Compatibility. 1991;33(4):366–370.

[8]   Corona P, Ferrara G, and Migliaccio M. Reverberating Chambers as Sources of Stochastic Electromagnetic Fields. IEEE Transactions on Electromagnetic Compatibility. 1996;38(3):348–356.

[9]   Hill DA. Plane Wave Integral Representation for Fields in Reverberation Chambers. IEEE Transactions on Electromagnetic Compatibility. 1998;40(3): 209–217.

[10] Arnaut LR, Krauthauser HG, and Höijer M. Comparison of Different Definitions of Field Strength Used in Reverberation Chamber Standards. In: 2007 IEEE International Symposium on Electromagnetic Compatibility. IEEE, Honolulu, USA; 2007. p. 1–5.

[11] Orjubin G. Maximum Field Inside a Reverberation Chamber Modeled by the Generalized Extreme Value Distribution. IEEE Transactions on Electromagnetic Compatibility. 2007;49(1):104–113.

[12] Gradoni G and Arnaut LR. Generalized Extreme-Value Distributions of Power Near a Boundary Inside Electromagnetic Reverberation Chambers. IEEE Transactions on Electromagnetic Compatibility. 2010;52(3):506–515.

[13] Hu D, Wei G, Pan X, *et al.* Investigation of the Radiation Immunity Testing Method in Reverberation Chambers. IEEE Transactions on Electromagnetic Compatibility. 2017;59(6):1791–1797.

[14] Badini L, Spadacini G, Grassi F, *et al.* A Rationale for Statistical Correlation of Conducted and Radiated Susceptibility Testing in Aerospace EMC. IEEE Transactions on Electromagnetic Compatibility. 2017;59(5):1576–1585.

Chapter 5

Over-the-air testing of wireless devices in heavily loaded reverberation chambers*

Kate A. Remley[1], Chih-Ming Wang[2] and Robert D. Horansky[1]

In many ways, what makes a "good" reverberation-chamber configuration for wireless over-the-air (OTA) tests that require signal demodulation is the opposite of what makes a good chamber configuration for electromagnetic compatibility (EMC) and electromagnetic interference (EMI) applications. For EMC/EMI tests, we strive for a highly reverberant, high-Q environment with as little unstirred energy as possible to efficiently expose the device under test (DUT), on average, to the same isotropic, Gaussian-distributed field using a minimum number of mode-stirring states.[†] Even though the DUT is not instantaneously exposed to the same field on a frequency-by-frequency basis, with a well-stirred chamber setup the fields should, theoretically, be identically distributed when averaged over a mode-stirring sequence. The isotropy of the fields can be verified with a goodness-of-fit test to confirm that the magnitude of the fields is Rayleigh distributed. As well, providing the minimum loading to obtain the well-stirred condition [1] usually ensures excellent spatial uniformity when measured samples are averaged over a mode-stirring sequence, reducing the criticality of device placement in the chamber (subject to the constraints of the working volume for the specific reverberation-chamber configuration at a given frequency of interest).

On the other hand, for OTA testing, we often intentionally load the chamber with significantly more RF absorber than that required to meet the well-stirred condition in order to replicate the flat-fading channel conditions that wireless-device equalizers are designed to accommodate. Figure 5.1 illustrates this, where we see that a significant amount of absorbing material is needed to measure the correct value of receiver sensitivity, as indicated by the plateau in the value of total isotropic sensitivity (TIS). For smaller amounts of loading, the chamber setup itself introduces distortion into the measurement [2–4]. This distortion obscures the goal of the OTA test, which is

*Work of the U.S. government, not subject to copyright in the United States.
[†]This chapter focuses exclusively on stepped mode stirring (mode tuning).
[1]Communications Technology Laboratory, U.S. National Institute of Standards and Technology, Boulder, CO, USA
[2]Information Technology Laboratory, U.S. National Institute of Standards and Technology, Boulder, CO, USA

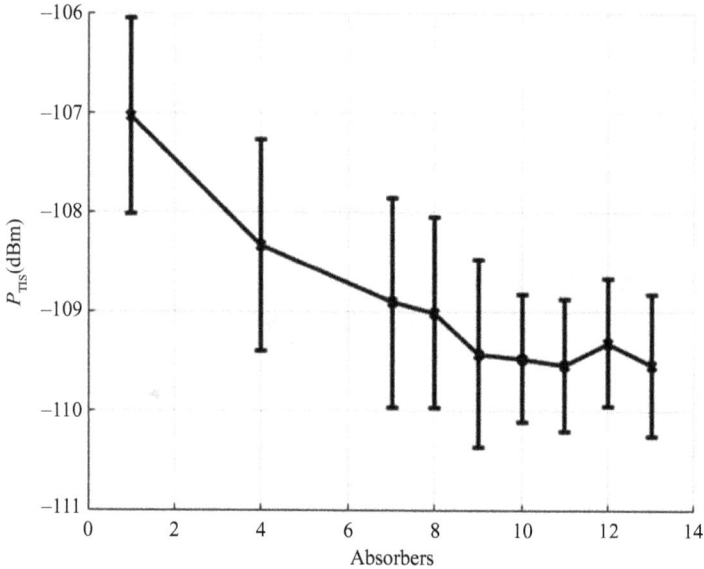

Figure 5.1   *Received power corresponding to the receiver's estimated total*
isotropic sensitivity (P_{TIS}) as a function of chamber loading [17]. The
metric P_{TIS} will be described in Section 5.2.1. Loading with eight
absorbers corresponds to approximately 1% of the chamber volume,
whereas the well-stirred condition may be achieved with a much lower
amount (e.g., 0.11% in [1]). Even though absorbers will perform
differently, these values provide an order-of-magnitude difference

to assess the receiver's performance. Once the channel is sufficiently flat, adding additional amounts of absorber does not significantly affect the mean value of the measured TIS, although the uncertainty due to lack of spatial uniformity may increase.

Seminal work in this area can be found in the literature, including [2–17]. Such loading is necessary to study device performance under conditions for which the device was designed to operate and where we wish to demodulate a communication signal having a finite instantaneous bandwidth (as opposed to the CW signals that are typically used in EMC/EMI tests). Loading reduces the Q and, because of an increase in unstirred energy, necessitates the use of position or antenna stirring. While it may be more complicated than for unloaded chambers, the goal of a stirring sequence design in loaded reverberation chambers is to provide, on average, Gaussian-distributed fields in three dimensions, just as for the EMC/EMI test [16].

As we will discuss next, the significant amount of loading needed for wireless-device tests increases the correlation between measured frequency samples to provide the flat-fading channel. However, this also increases correlation between the positions of mechanical mode stirrers, the locations of antennas, and even antenna orientations. Correlation reduces the number of independent samples in a reverberation-chamber measurement, which can significantly complicate the development of a stirring sequence that provides low uncertainty in the measurement of a quantity of interest. As such, quantifying correlation in the development of a stirring sequence and its effect on measurement uncertainty will be a large focus in this chapter.

In addition to chamber characterization and measurement uncertainty, we will also discuss types of wireless-device tests that are currently performed in loaded reverberation chambers. Here, we need to distinguish between two types of wireless-device OTA tests. The first is used by wireless industry organizations such as the CTIA and 3GPP [15] to assess specific metrics associated with device performance such as total radiated power (TRP) or TIS (also called total radiated sensitivity). These are standardized tests that allow labs having a variety of setups to obtain results comparable to an isotropic environment provided by an anechoic chamber. It is indeed counter intuitive to imagine that a highly reflective environment such as the reverberation chamber can provide channel conditions similar to an anechoic environment, but that is the goal of these tests. While the focus in this chapter is on cellular applications, the chamber characterization and tests discussed here are generally applicable to a wide variety of wireless technologies and to coexistence testing, where multiple devices are present within the chamber [18–21].

The second type of OTA wireless-device test conducted in heavily loaded reverberation chambers uses the reflective properties of the reverberation chamber to replicate specific multipath conditions. Loading is used to tune multipath decay times to those seen in real-world channels or to provide certain spatial channel characteristics specified by standards groups. Examples of these include replication of specific channels [8,10,22,23,63]; proposed tests for cellular handsets that use multiple-input, multiple-output (MIMO) antennas in which a channel emulator is used to replicate a specific multipath channel model and the MIMO antenna gain is measured [24–27]; and the implementation of 3D spatial channel models such as in [28].

The two types of tests have many similarities because they allow users to replicate specific channel conditions (e.g., a specific power-delay profile, power-angle spectrum, or an isotropic channel). The distinction is that the former, isotropic channel, is designed to test wireless-device performance under conditions for which it was designed. The latter, more realistic, channel conditions may stress the wireless device in unexpected ways. After a discussion of the chamber characterization steps that must be considered for both isotropic and more realistic multipath OTA test conditions, we will discuss each of these applications with a focus on the former because these tests are currently standardized.

5.1   Chamber characterization for OTA tests

For many EMC/EMI tests, the spatial uniformity of the chamber setup is first evaluated by performing measurements of orthogonal field components at multiple locations throughout the working volume of the chamber. Samples of the fields measured at a specific location are averaged over a mode-stirring sequence, which is often carried out by the movement of mechanical mode stirrers. If the variation of the averaged fields measured at the locations within the working volume is sufficiently low, then the user may perform the measurement at a single location within the working volume. For a chamber setup satisfying the well-stirred condition, having Gaussian-distributed real and imaginary field components, the uncertainty related to the mode-stirring process ideally decreases by $1/\sqrt{N}$, where N is the number of uncorrelated mode-stirring samples [29] in a stepped (static-channel) mode-stirring sequence.

Let us distinguish here between the small amount of loading that may be used under the "well-stirred condition" described in [1] and Chapter 2 of this book, and the (typically) heavier loading required for demodulating communication signals with defined instantaneous bandwidths. For clarity, we shall refer to the latter as "heavily loaded" reverberation chambers throughout this chapter. The amount of loading needed to demodulate the communications signal without significant chamber-induced distortion depends on the bandwidth and transmission format of the signal to be demodulated. "Spread spectrum" signals cover the entire communications channel during transmission. In these cases, loading is used to create a coherence bandwidth (CBW) (a metric defined in Section 5.1.3) that meets or exceeds the signal bandwidth [15,17,30]. Examples include the signals utilized in the wideband code-division multiple access (W-CDMA) transmission format, which have a bandwidth of 3.84 MHz, and the 22 MHz 802.11b wireless local-area-network waveform, which contains a modulated signal that is spread over the entire communications channel. Such waveforms require that the chamber is loaded with RF absorbers in order to achieve a CBW of, approximately, the channel bandwidth (e.g., 3.84 and 22 MHz for the previous examples). On the other hand, the orthogonal frequency-division multiplexing (OFDM) transmission format is designed to mitigate frequency selective fading by the use of many separate subcarriers, where each subcarrier is sufficiently narrowband to experience frequency-flat fading in high multipath environments. For the OFDM case, little loading may be needed [15,30,31].

In such heavily loaded reverberation chambers, we typically must use a combination of mode-stirring mechanisms that both alter the boundary conditions (e.g., mechanical-paddle stirring) and move the DUT to various locations within the chamber (e.g., antenna-position or "platform" stirring) to account for the decreased spatial uniformity caused by the RF absorbing elements. Antenna-position stirring may consist of physical movement of the antenna on, for example, a rotating platform or linear translation stage or the use of multiple antennas at various locations and/or polarizations within the reverberation chamber. The need to perform averaging over many locations within the working volume is one differentiator of wireless test from traditional EMC/EMI test applications. Intimately related to this is the need to account for the increased correlation between mode-stirring samples in both the estimate of a quantity of interest and the corresponding measurement uncertainty.

In this section, we focus on methods for identifying correlations and optimizing stirring sequences to minimize correlation between samples. While correlated samples may be used to estimate a quantity of interest, designing a stirring sequence to minimize correlation between mode-stirring samples increases the efficiency of a measurement by requiring the acquisition of fewer overall samples.

We start with an introduction to the basic setup used for testing wireless devices, including a derivation of the important chamber metric called the "reference power transfer function," G_{Ref}. The metric G_{Ref} characterizes the complex loss in the chamber setup. It is written as a gain so that losses take on negative values on a decibel scale. An accurate estimate of the chamber loss is important for obtaining calibrated power-based metrics such as TRP. While G_{Ref} itself may be derived from S-parameter measurements made with a vector network analyzer (VNA), the power-based DUT metrics themselves require calibration with a power-measurement instrument such as a power meter, spectrum analyzer, or base-station emulator (BSE).

G_{Ref} is also a convenient metric for assessing the correlations between mode-stirring samples in a chamber setup, as well as for obtaining various components of uncertainty. Our derivation of G_{Ref} is followed by a description of the effects of loading on the electromagnetic properties within the chamber, such as spatial uniformity and the distribution of samples under loaded conditions. This is followed by a discussion of methods to identify the correlation between samples, both in terms of frequency and location within the chamber. Finally, we discuss the metrics of K-factor, which describes the level of unstirred energy relative to stirred energy in a chamber, and isotropy, including methods for characterizing isotropy in heavily loaded reverberation chamber setups. These two metrics, K-factor and isotropy, may be used to assess different characteristics of a chamber setup.

The techniques presented in this section are intended to provide straightforward methods for optimally configuring heavily loaded reverberation chambers for wireless-device tests, in terms of both accuracy and efficiency in a measurement.

5.1.1 Configuring a reverberation chamber for wireless-device testing

While some parameters such as antenna efficiency may be determined from relative measurements of gain or loss, power-based metrics such as TRP and TIS (described later in Sections 5.2.1 and 5.2.2, respectively) report an absolute power value. For these measurements, it is essential to have a good estimate of the chamber's reference power transfer function G_{Ref}. G_{Ref} is usually measured with a VNA in a separate step from the DUT measurement. It is then corrected for (calibrated out) during the calculation of power-based performance metrics. For example, to measure the TRP from a cellular-enabled device, the device is placed in the reverberation chamber and commanded OTA by the BSE to transmit at full power. For each step in the stirring sequence, the BSE will measure the DUT's power minus the chamber's loss. Thus, to obtain the correct value of TRP in post processing, it is necessary to account for the power that was lost in transmission through the chamber, G_{Ref}.

The reference measurement, G_{Ref}, is intended to estimate the chamber's transfer function as experienced by the DUT's transmitted and received signals. Because most wireless devices have integrated antennas that are not removable from the DUT, a direct measurement of G_{DUT} is typically not possible. Thus, the reference measurement is generally performed with a reference antenna that is different from the DUT's antenna. However, the radiation pattern of the reference antenna is typically selected to be similar to that of the DUT (e.g., both azimuthally omnidirectional antennas) so that it is exposed to a similar set of mode-stirred fields as the DUT during the stirring sequence.

5.1.1.1   Chamber setup

A commonly used [4,14,15] two-step procedure for testing wireless devices in reverberation chambers is illustrated in Figure 5.2(a) and (b). The procedure consists of the measurement of the chamber's reference power transfer function [the "reference measurement," shown in Figure 5.2(a)] and the DUT measurement [shown in Figure 5.2(b)]. Samples are acquired by the measurement antenna at each stepped mode-stirring state and are averaged to yield an estimate of the quantity of interest, to within a desired uncertainty.

The two setups look very similar to each other, with the primary difference being the measurement instruments used: a VNA is used for the reference measurement and a BSE is used for the DUT measurement. In Figure 5.2(a) and (b), two metallic mode-stirring paddles change the boundary conditions in the chamber and a rotating turntable, on which both the DUT and the reference antenna sit, steps through various angles. At each step, a mode-stirring sample is acquired via the measurement antenna.

Because the average spatial uniformity of the mode-stirred fields is less constant in a heavily loaded chamber than in a chamber designed to meet the well-stirred condition, it is important that the chamber configuration for the reference and DUT measurements be as similar to each other as possible to minimize uncertainty in the measurement. For example, the placement of the turntable, RF absorber, fixtures, and measurement antenna should be the same for both measurements. The reference and

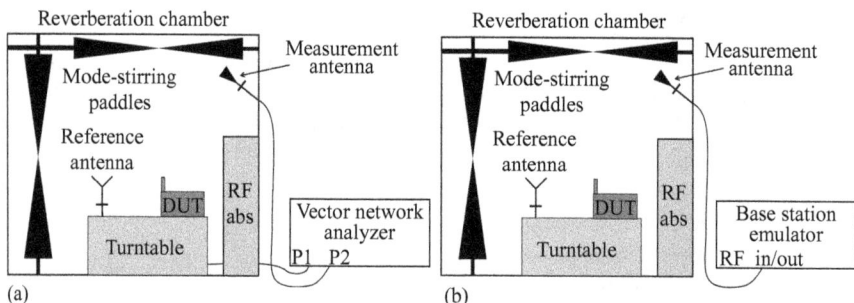

Figure 5.2   Common configuration for wireless-device measurements in a reverberation chamber, including two rotating paddles, RF absorber to broaden the coherence bandwidth, and a rotating platform on which the DUT and the reference antenna are placed. In (a), a VNA measures the reference power transfer function between the reference and the measurement antennas and in (b), a base-station emulator measures the performance of the DUT. The reference planes of the VNA in (a) are denoted by the line at the base of each antenna. In (b), the reference antenna is terminated in a 50-Ω load in order to provide the same loading condition in both the cases

DUT antennas should be exposed to the same nominal stirring sequence, although they may be placed anywhere within the working volume of the chamber (i.e., the region that has been bounded by a separate set of characterization measurements), which is why Figure 5.2 shows them both placed concurrently for both measurement setups on the turntable, but not in exactly the same physical locations.

The reference measurement shown in Figure 5.2(a) is carried out with a VNA that acquires a full set of S-parameters between the reference antenna and the measurement antenna at the measurement reference planes indicated by a short perpendicular line on each antenna.

In addition to the rotating paddles and platforms shown in Figure 5.2, other types of mode-stirring might be employed, including horizontal and/or vertical translation of mechanical paddles, translation of the DUT itself, multiple sampling antennas located at various locations and/or at different polarizations within the chamber, and movable walls.

Because an OTA test requires frequency averaging over the band of the modulated signal, frequency stirring, per se, is not typically considered as a separate stirring mechanism. While frequency stirring and frequency averaging consist of the same procedure (averaging mode-stirred samples over a specified frequency band), the wireless community often views frequency averaging as a necessary part of computing a power-based metric—by averaging over the channel bandwidth—rather than as part of the stirring sequence.

All mode-stirring mechanisms are not created equal in terms of efficient stirring of the fields, as we will discuss in Section 5.1.4. However, the use of rotating paddles and platform stirring, as shown in Figure 5.2, are common stirring mechanisms and, without loss of generality, we will use them to illustrate concepts related to OTA wireless-device test.

5.1.1.2 The reference power transfer function

The chamber's reference power transfer function, G_{Ref} is intended to provide an estimate of the chamber loss associated with the measurement of the DUT, G_{DUT}, which typically cannot be measured directly for devices with integrated antennas. G_{Ref} is determined by averaging over a stepped mode-stirring sequence during the reference measurement. This power loss is then corrected for during the calculation of the power-based metric. The value of G_{Ref} may be estimated from S-parameter measurements as [14,24]

$$G_{Ref} = \frac{\left\langle |S_{21,Ref}|\right\rangle^2_{N_W}}{\eta_M \eta_R \left(1 - |\Gamma_M|^2\right)\left(1 - |\Gamma_R|^2\right)}, \tag{5.1}$$

where G_{Ref} is the estimate of the reference power transfer function, η_x is the efficiency of the measurement ($x = M$) or reference ($x = R$) antenna [see Figure 5.2(a)], and the term Γ_x corresponds to the free-space reflection coefficient of each of the two antennas, which may be measured in an anechoic or reverberation chamber. If measured in a reverberation chamber, the chamber must be unloaded (or lightly loaded) condition in order to maximize uniformity and minimize loss (see [32] for additional detail). Note that the implicit frequency dependence is suppressed to simplify the expression. To compute G_{Ref}, the ensemble average is taken over N_W-stepped mode-stirring samples (N_W mode-stirring samples *within* a stirring sequence). We refer the reader to [14] for a detailed derivation of this expression, where expressions for choices of reference planes other than those shown in Figure 5.2(b) are derived.

Multiple *independent realizations* of G_{Ref} (each denoted next as $G_{Ref,p}$) are often acquired to estimate the uncertainty due to the lack of spatial uniformity. The independent realizations consist of mode-stirring sequences whose mode-stirring samples are uncorrelated from sequence to sequence, to be discussed in Section 5.1.4.

The independent realizations may be averaged to improve the estimate of the chamber's reference power transfer function. The average may be written as

$$\widehat{G}_{Ref} = \frac{1}{N_B} \sum_{p=1}^{N_B} G_{Ref,p}, \tag{5.2}$$

where p denotes an independent realization, and measurements are made over N_B-independent realizations of the mode-stirring sequence (with N_W mode-stirring samples and N_B samples *between* stirring sequences). The multiple independent realizations are primarily used to obtain the component of uncertainty due to lack of spatial uniformity, derived from the standard deviation taken over the N_B positions as

$$\sigma_{G_{Ref}} = \sqrt{\frac{1}{N_B - 1} \sum_{p=1}^{N_B} \left(G_{Ref,p} - \widehat{G}_{Ref}\right)^2}. \tag{5.3}$$

The value $\sigma_{G_{Ref}}$ is often used as a metric for evaluating the lack of spatial uniformity in a heavily loaded reverberation chamber, with the component of uncertainty corresponding to $\sigma_{G_{Ref}}/\sqrt{N_B}$ This will be discussed further in Section 5.4 on uncertainty.

Typically, the DUT antenna will not be identical to the reference antenna, in terms of radiation pattern. This means that the ratio of unstirred energy to stirred energy (the K-factor) may be different for the reference and DUT measurements, over and above the expected spatially dependent uncertainty that is captured by $\sigma_{G_{Ref}}$. This, in turn, may result in different correlations and additional uncertainty. To minimize this effect, current practice is to select a reference antenna that produces a K-factor that is similar to that of the DUT. For example, omnidirectional reference antennas are often used to estimate G_{Ref} for testing a cellular device having, essentially, an omnidirectional pattern. A discussion of uncertainty related to this effect is discussed in [33], which is summarized in Section 5.4.3.

The reference power transfer function G_{Ref}, derived from the complex S_{21}, may be used to determine the power delay profile (PDP) (power-based response in time) of the reverberation chamber setup. For the nth stepped mode-stirring position, which can be considered to have been measured at delay, τ_n, the impulse response of the chamber setup may be estimated from the inverse Fourier transform of the measured

S-parameters $h_n(t) \cong \text{IFT}\{S_{21}(f, n)\}$. The magnitude squared of the impulse response corresponds to the *PDP* [10]:

$$PDP(t) = \langle |h(t, \tau_n)|^2 \rangle, \tag{5.4}$$

where $h(t, \tau_n)$ is the nth mode-stirring sample of the linear, time-varying impulse response of the channel, and the brackets denote the ensemble average. The root-mean-square (RMS) delay spread is found from the square root of the second central moment of the *PDP* as

$$\tau_{\text{RMS}} = \sqrt{\frac{\int_0^\infty (t - t_0)^2 PDP(t) dt}{\int_0^\infty PDP(t) dt}}, \tag{5.5}$$

where t_0 is the mean delay of the propagation channel for step n given by

$$t_0 = \frac{\int_0^\infty t PDP(t) dt}{\int_0^\infty PDP(t) dt}. \tag{5.6}$$

This time-domain representation of the chamber's reference power transfer function can be useful in several applications, as will be discussed in Section 5.1.3 on CBW and Section 5.3 on replicating specific multipath channels in reverberation chambers.

5.1.2   Distribution of mode-stirring samples in loaded chambers

Both the reference power transfer function and the S-parameters themselves will, ideally, follow specific distributions. In chambers loaded to meet the well-stirred condition, the samples often follow these distributions without the need for position stirring. In heavily loaded chambers, assessing the degree to which the distribution of samples follows the theoretical value is a good diagnostic to determine whether a stirring sequence contains a sufficient number of independent mode-stirring samples.

As discussed in previous sections, there are three complex, orthogonal components of the electric field vector, $E_i = E_{i1} + jE_{i2}$, where $i = x; y; z$. When sampled over many mode-stirring samples, each component of the complex electric field vector may be represented by a complex Gaussian distribution. For "non-invasive" antennas (typically those that are physically small relative to the chamber volume), the chamber may be characterized by measuring the S-parameters over the stepped mode-stirring sequence with a VNA, instead of measuring the electric field. If we measure the chamber's complex transfer function (proportional to S_{21}) with a VNA, the real and imaginary components will also, ideally, both be Gaussian distributed. Further, the magnitude of S_{21} will be Rayleigh distributed and the magnitude squared ($|S_{21}|^2 \propto G_{\text{Ref}}$) will be exponentially distributed [1,29,34,35].

As we have seen in previous chapters, in a well-stirred chamber, the distribution of measured mode-stirring samples can be nearly ideal if it is obtained under conditions that include a sufficient number of mode-stirring samples; little direct coupling between antennas; a high-Q chamber setup; and the use of effective, uncorrelated mechanical mode-stirring mechanisms [1,34].

However, for a heavily loaded chamber, obtaining the desired distribution can be more complicated. When a large amount of RF-absorbing material is present in the chamber, some of the energy introduced into the chamber is not randomized through mechanical mode-stirring. This increases correlation between measured mode-stirring samples, including those acquired at different frequencies, antenna orientations, locations, and paddle positions. This, in turn, reduces the effective number of samples in the chamber. Use of an inadequate number of uncorrelated samples in a mode-stirring sequence can lead to an incorrect estimate of the quantity of interest and an underestimate of its uncertainty.

Physically, in a heavily loaded chamber, the lack of spatial uniformity of the measured, averaged fields due to the absorbers increases the correlation between the mode-stirring samples. If measurements are made in one location within a heavily loaded chamber, the distributions may be less ideal, for example, resulting in a nonzero mean in the Gaussian distribution of the real and imaginary components of the measured S_{21}. Such a nonzero mean can be visualized as an offset from the origin when the real and imaginary components of the complex transmission coefficient are plotted in polar form (see, e.g., [8,36]). However, such an offset is negligible under well-stirred conditions, which may be different for lightly and heavily loaded chamber setups because of the need for position stirring in the heavily loaded chamber.

An illustration of the importance of position stirring for heavily loaded chambers is shown in Figure 5.3, which plots $\mathrm{Im}(S_{21})$ vs. $\mathrm{Re}(S_{21})$ samples from a stepped mode-stirring sequence for a heavily loaded chamber. The loading was chosen to provide a CBW of approximately 4 MHz. This corresponds to a volume of 0.46 m^3 of the RF absorber in a 45.2-m^3 chamber, which represents roughly 1% of the chamber's volume. In Figure 5.3(a), antenna-position stirring of the reference (transmit) antenna was used, while in Figure 5.3(b) antenna-position stirring was not used. Measurements were made for $N_B = 8$ [Figure 5.3(a)] or $N_B = 9$ [Figure 5.3(b)] independent realizations of the stirring sequence. Independent realizations correspond to those for which the same nominal stirring sequence is used (e.g., a stirring sequence of $N_W = 100$ paddle positions performed at $N_B = 9$ antenna locations), and for which correlation between the mode-stirring samples has been minimized, as described in sections that follow.

As can be seen in Figure 5.3(a), the measured values of $\mathrm{Re}(S_{21})$ and $\mathrm{Im}(S_{21})$ are closely distributed around the origin, with the mean values of the eight independent realizations of the stirring sequence clustered at the origin. For this plot, the stirring sequence consisted of 120 mode-stirring states involving a combination of platform, paddle, and height stirring.

The case in Figure 5.3(b) utilizes similar loading conditions as the case in Figure 5.3(a), but no position stirring was used. For this plot, the stirring sequence consisted of 72 mode-stirring states with paddle stirring only. We see that mean values of $\mathrm{Re}(S_{21})$ and $\mathrm{Im}(S_{21})$ are less-well clustered around the origin in Figure 5.3(b), illustrating the reduction in spatial uniformity of the averaged fields within the chamber's working volume. This highlights the importance of position stirring in loaded reverberation chambers.

The magnitude of the real and imaginary components of S_{21} in Figure 5.3 is smaller than it would be in an unloaded chamber because fields within the heavily

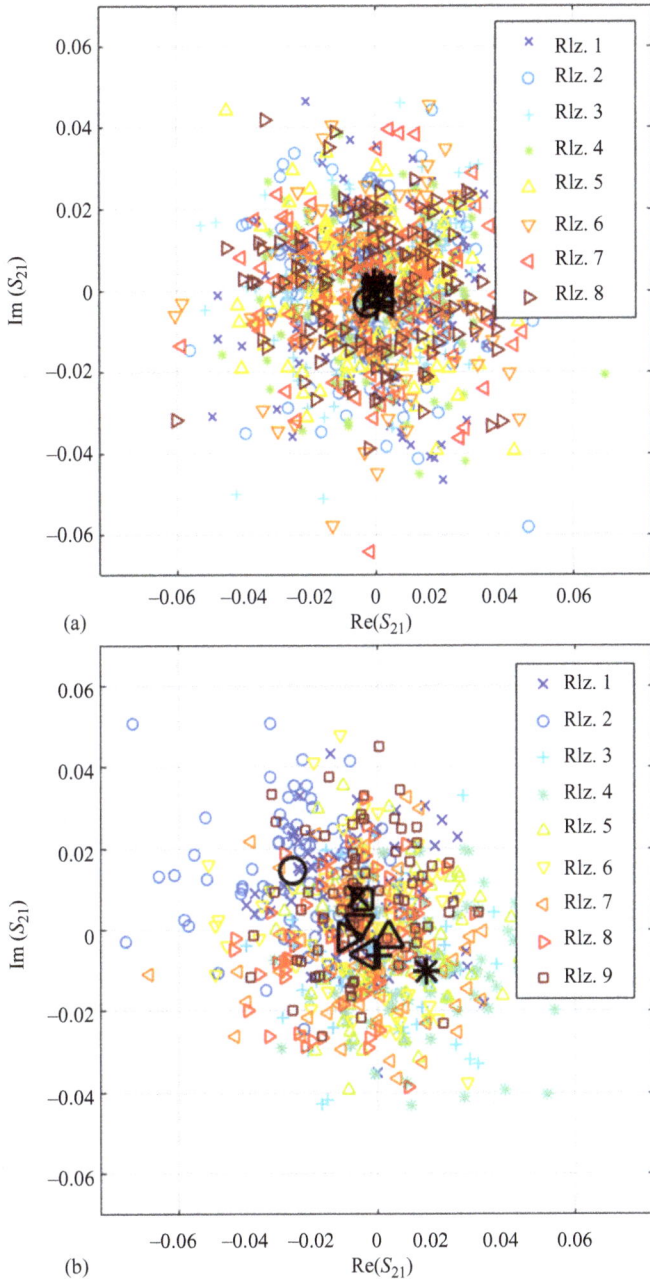

Figure 5.3   *Scatter plot of complex S_{21} data at 2 GHz measured in a chamber that was heavily loaded for a CBW of approximately 4 MHz (a) with position stirring and (b) without position stirring. Note that the mean values, given by the thick, black symbols, are clustered at the origin for (a) but not (b)*

loaded chamber interact with the RF absorber and decay before they arrive at the receive antenna. This decrease in received signal can be calibrated out with a reference measurement made under the same loading conditions and with the same (or similar) antennas as are used in the DUT measurement [33].

Physically, the decrease in spatial uniformity can be attributed in part to a faster decrease in the "stirred energy" relative to the "unstirred energy" in the heavily loaded chamber. That is, the "stirred" fields that reflect off of multiple surfaces within the chamber as they undergo mode-stirring have a greater chance of interacting with an absorber as they travel between transmit and receive antennas, resulting in a reduction in measured power of the stirred components. In contrast, the "unstirred" fields, such as those with direct coupling between antennas or that bounce off of a single wall, do not experience a reduction in measured power with loading. Thus, the unstirred components, which remain fixed for specific frequencies and positions within the chamber, become more significant with loading relative to the stirred components. This introduces correlation between measured mode-stirring samples. This effect of decreased stirred energy relative to the unstirred energy as a function of loading is quantified by the Rician K-factor, sometimes referred to as simply K-factor. This metric is discussed in Section 5.1.6.

The decrease in the magnitude of the reference power transfer function with increased chamber loss is further illustrated in Figure 5.4 [37]. The top set of curves shows the "relative mean power," defined as $|S_{21}|^2$, for chamber loading with an increasing number of large, stacked metallic boxes having a maximum surface area of approximately 5.8 m^2. The bottom set of curves shows $|S_{21}|^2$ for stacked RF absorber of similar physical dimensions (maximum surface area of approximately 6.2 m^2). Measurements were made over the frequency range from 1.5 to 2.5 GHz, utilizing 72 rotating-paddle mode-stirring positions. In the experiment, eleven monopoles used as receive antennas were located throughout the chamber. The results in Figure 5.4 are for one of these locations. The transmit antenna was a broadband dual-ridge horn antenna (DRHA) aimed at the rotating paddle.

Figure 5.4 shows that for loading with reflective material, the power insertion loss does not change significantly even for the maximum surface area. However, when RF absorbers having a surface area similar to the metal boxes are placed within the chamber, the values of $\langle |S_{21}|^2 \rangle$ significantly decrease as a function of loading.

To quantify the decrease in the spatial uniformity of the chamber, we consider the standard deviation of the relative power samples $|S_{21}|^2$ measured at the 11 monopole antennas for the different loading configurations. The absolute standard deviation is not independent from the mean. For this reason, we use a normalized measure, called the coefficient of variation

$$C_v = \frac{\sigma}{\mu} \times 100\%, \tag{5.7}$$

where σ is the standard deviation and μ is the mean received power over the number of antenna locations for a given loading configuration (11 in this example). Figure 5.5 shows an increase of the coefficient of variation with RF absorber loading, unlike loading with metallic objects.

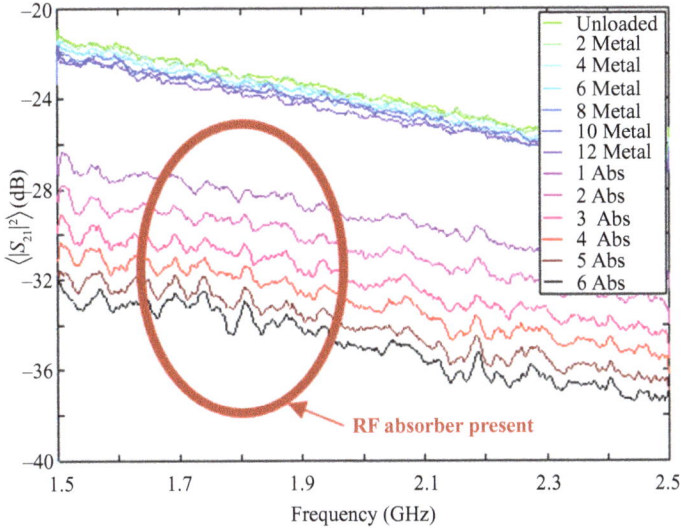

Figure 5.4 *Relative mean power* $\langle |S_{21}|^2 \rangle$ *as a function of loading. The top curves show a chamber with metallic boxes (labeled "2 Metal" through "12 Metal"). The lower curves (circled) show the chamber in the same configuration but with RF absorber labeled "1 Abs" through "6 Abs." RF absorber loading increases the average insertion loss in the chamber, which can be calibrated out*

The increased variance as a function of frequency with loading—fundamentally caused by increased correlation between mode-stirring samples—must be accounted for when optimizing the stirring sequence in order to minimize uncertainty and maximize efficiency in determining power-based metrics such as TRP or TIS. In the following sections, we discuss methods for identifying and quantifying correlation as a function of frequency and in terms of spatial locations within a loaded reverberation chamber.

5.1.3 Coherence bandwidth

Understanding the correlation between various mode-stirred samples in a heavily loaded reverberation chamber is the key to performing OTA tests with low uncertainty. Much effort has been expended by the community to understand the effects of frequency correlation and to account for it in the determination of uncertainty [2,4,13,38], especially for standardized certification tests such as those whose goal is to emulate an isotropic environment [15].

As mentioned earlier, loading is necessary to create correlation between frequencies within the chamber in order to provide a channel for which the device was designed to operate. If the channel is not sufficiently flat over the bandwidth of the

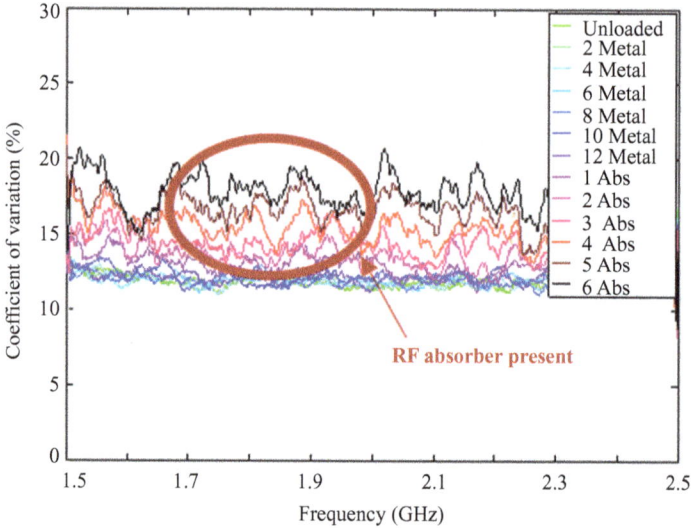

Figure 5.5   The coefficient of variation calculated for 11 monopoles placed throughout a loaded reverberation chamber for metallic and absorbing material

modulated communication signal, the chamber setup itself will artificially introduce distortion, masking the potential distortion from the DUT that the OTA test has been developed to uncover.

The effect of increased loading as a function of frequency is illustrated in Figure 5.6. In particular, we show the chamber's loss, G_{Ref}, vs. frequency across a 4-MHz communication channel bandwidth for an increasing number of RF absorbers and for a single mode-stirring sequence step (i.e., a static channel with specific paddle orientations, turntable locations, and antenna polarizations). The figure shows a decrease in fluctuation as a function of frequency as more loading is added to the chamber. The chamber response for each loading case has been offset by arbitrary values to make the graph easier to read. Note that this effect was not visible in Figure 5.4 due to the expanded frequency axis scale.

A common metric that allows assessment of the amount of the frequency flattening provided by a given loading case is the CBW. The CBW was originally used in RF propagation channel modeling, describing the frequency separation necessary for two signals to be considered statistically independent [39]. This metric assesses the frequency selectivity of a channel. A wider CBW corresponds to a smoother frequency response. The CBW has an inverse relationship to the quality factor of a reverberation chamber and is similar to the average mode bandwidth metric $f/Q(f)$ described in previous chapters. That is, the more that a reverberation chamber configuration stores energy, the more frequency selective the chamber setup is.

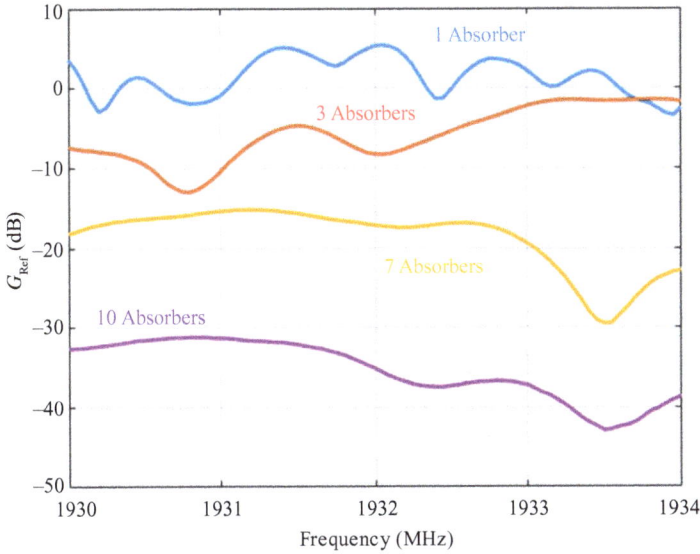

Figure 5.6   The chamber loss, G_{Ref}, across a 4-MHz communication channel bandwidth for an increasing number of absorbers [17]. Curves are offset by arbitrary decibel values to clearly illustrate the variation as a function of frequency

The CBW may be defined by the autocorrelation function, R, as given by

$$R(i, n) = \frac{\sum_{j=1}^{M-i} S_{21}(f_j, n) S_{21}^*(f_{j+i}, n)}{\sum_{j=1}^{M-i} S_{21}(f_j, n) S_{21}^*(f_j, n)}, \tag{5.8}$$

where $S_{21}(f_j, n)$ corresponds to the measured complex S_{21} at frequency step f_j with M frequency points measured within the bandwidth of interest, BW, so that $f_1 = f_c - (BW/2)$ and $f_M = f_c + (BW/2)$. The index n is the mode-stirring sample (out of N_W). The index i corresponds to one of several frequency-step offsets (lags) over the bandwidth of interest (here $BW = 100$ MHz [15]) where $-(M-1) \leq i \leq (M-1)$. The frequency lag for a set of measured data is given by $\Delta f = i((f_M - f_1)/(M-1))$. The asterisk denotes complex conjugation. The bandwidth, BW, is taken over 100 MHz to provide stable results [40]. The calculation of CBW is similar to that of the "first-order autocorrelation" coefficient of [1]. For lightly loaded chambers, assessing correlation between two adjacent frequency steps provides a metric to judge the effectiveness of a mode-stirring process in a given chamber setup. For the heavy loading required in wireless-device testing, correlation is studied on further adjacent frequencies.

A representative set of CBW plots for various amounts of RF absorber is shown in Figure 5.7. In current practice, e.g., [15], the CBW would be chosen to meet or

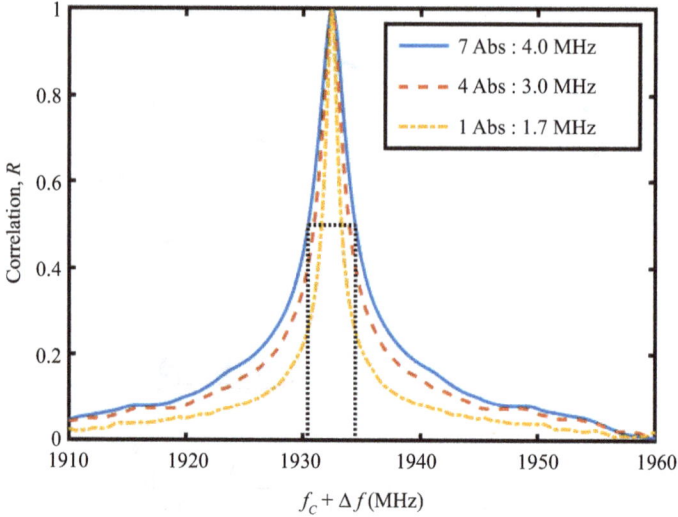

Figure 5.7   The frequency correlation function for a chamber loaded with three different amounts of RF absorber. The corresponding CBW is given in the legend, illustrating that CBW increases with the amount of loading inserted in the reverberation chamber

exceed the modulation bandwidth of the communication signal being tested. The CBW is computed for a defined threshold of the autocorrelation function. A value of 0.5 for the threshold is used within standards for wireless-device testing in reverberation chambers [15], although other values, such as $1/\sqrt{2}$, can also be used. The width for a threshold of 0.5 is illustrated by the dotted line in Figure 5.7 for the seven-absorber case.

This choice of threshold can be derived by understanding that the CBW is the Fourier-space equivalent of the RMS delay spread τ_{RMS}, which is a measure of the decay of power in the chamber as a function of time. It was defined in (5.5) in Section 5.1.1 and is inversely equivalent to the quality factor. If the temporal power in the chamber, P_τ, follows an ideal exponential decay, it may be written as [41,42]

$$P_\tau = S_0 e^{-\tau/\tau_{RMS}},\tag{5.9}$$

for $\tau \geq 0$, where τ is the time where the power is measured after injection at $\tau = 0$ and S_0 is an arbitrary magnitude. In this case, the frequency autocorrelation is the symmetric two-sided Fourier transform, which is

$$R = \frac{S_0 \tau_{RMS}}{1 + j2\pi \Delta f \tau_{RMS}}.\tag{5.10}$$

In this case, the full-width at 0.5 threshold of the magnitude of R (the CBW) relates to the RMS delay spread as [2,41,42]

$$CBW = \frac{\sqrt{3}}{\pi \, \tau_{RMS}}. \tag{5.11}$$

Other threshold values may be used to define correlation depending on the application. For example, in the next subsection, we choose a threshold of 0.3 to define spatial correlation in order to provide "independent realizations" of the stirring sequence.

5.1.4   Spatial correlation and the mode-stirring sequence

As discussed earlier, loading is necessary for unimpaired demodulation of communication signals during OTA testing. Loading not only increases correlation between frequency components (as quantified by the CBW discussed in Section 5.1.3) but also increases spatial correlation between mechanical paddle positions, antenna orientations, and platform positions [4,13,38,43]. Correlation between mode-stirring samples can reduce measurement efficiency by requiring additional samples to obtain a desired level of uncertainty [6,12,34,35,44]. In the worst case, if the number of independent samples provided by the stirring sequence is too small, it may not even be possible to measure the quantity of interest with sufficient accuracy. Therefore, this section focuses on techniques to develop and verify optimized mode-stirring sequences.

Prior work in the literature has discussed identification of spatially uncorrelated samples. For assessing spatial correlation between stepped mode-stirring samples *within* a stirring sequence, sample linear autocorrelation is often used [4,12,35,44]. For assessing the correlation *between* two mode-stirring sequences samples, the cross correlation (such as Pearson's cross correlation), is often used [2,15,44]. This latter statistic estimates the linear dependence between two realizations of a given mode-stirring sequence and is sometimes used to determine if these realizations may be considered "independent." Multiple independent realizations are used to quantify the uncertainty due to lack of spatial uniformity for a given chamber setup. These two techniques are next described, with examples of their use in OTA test applications.

5.1.4.1   Correlation within a measurement

To identify the physical step size for a specific mode-stirring mechanism beyond which we deem samples to be uncorrelated, the linear autocorrelation is often used. This metric assesses the relative change in mode-stirring samples within a proposed stirring sequence (e.g., are samples less correlated if we use one large paddle rotating in 3° steps or two smaller paddles rotating with 2° steps?). From the sample linear autocovariance, we may identify, for example, the "coherence angle" for platform stirring, or the "coherence length" for linear translation. Correlation may also be determined for the case of multiple mode-stirring mechanisms used simultaneously. The better the stirring mechanisms are at randomizing (or, more accurately, statistically altering) the fields in the chamber, the smaller the coherence distance between samples.

The sample linear autocorrelation across mode-stirring samples may be found from S-parameter measurements as

$$R_{CA}(f,i) = \frac{\sum_{j=1}^{N_w-i} \left(S_{21}(f,n_j) - \langle S_{21}(f,n)\rangle_{N_w}\right)\left(S_{21}(f,n_{j+i}) - \langle S_{21}(f,n)\rangle_{N_w}\right)^*}{\sum_{j=1}^{N_w} \left(S_{21}(f,n_j) - \langle S_{21}(f,n)\rangle_{N_w}\right)\left(S_{21}(f,n_j) - \langle S_{21}(f,n)\rangle_{N_w}\right)^*},$$

(5.12)

where $\langle S_{21}(f,n)\rangle_{N_w}$ is the mean of the S_{21} measurement at frequency, f, over the mode-stirring samples, N_w. As with CBW, the normalization term in the denominator provides a maximum value of 1. By subtracting the mean, we may evaluate the relative effectiveness of a stirring mechanism on the stirred energy. The effect of unstirred energy on the measurement is evaluated from the uncertainty due to the lack of spatial uniformity, described briefly in Section 5.1.1, and in more detail in Section 5.4.2.

Linear autocorrelation (5.12) operates on the entire mode-stirring sequence by calculating the correlation coefficient from a copy of itself shifted one step (lag) away, which is repeated until the entire sequence has been shifted from the first to the last point in the sequence. The minimum step size between "uncorrelated samples" corresponds to the set of samples physically spaced farther than those for which the correlation coefficient falls below the specified threshold. This technique is very similar to the calculation of CBW discussed earlier. And, as with CBW, correlation between samples in a reverberation chamber is a function of the loading and the frequency of operation.

Linear autocorrelation (5.12) characterizes the spatial correlation between mode-stirring samples within a stirring sequence and, thus, may be used to judge the relative effectiveness of various mode-stirring mechanisms. This makes it useful in developing optimized stirring sequences in which spatial correlation is minimized. The concept is illustrated in Figure 5.8(a) and (b), which shows correlation between mode-stirring samples for various stirring mechanisms as a function of frequency in an unloaded chamber [Figure 5.8(a)] and a loaded chamber [Figure 5.8(b)]. The various stirring mechanisms are evaluated separately and in groups, allowing the user to identify the spatial step requirements of each.

For this large (4.27 m × 3.65 m × 2.90 m) chamber, the available mechanical mode-stirring mechanisms included two rotating paddles and a rotating platform. We conducted measurements of each mode-stirring mechanism separately, collecting samples over uniform angular steps for stirring sequences in which (1) only the horizontal paddle was stepped; (2) only the vertical paddle was stepped; (3) only the rotating turntable was stepped; (4) all the three were stepped (the "All stepped" case in the legend); and (5) both paddles were stepped together (only for the loaded case). The spatial acquisition step size was chosen to be well below the expected coherence angle in order to develop correlation curve (similar to Figure 5.7): 1° for all mechanisms, except for the all stepped case which was 3°.

At each center frequency of interest, we applied (5.12) to the complex S_{21} data over a 100-MHz bandwidth and applied a threshold of 0.3 to this curve to estimate the minimum step size for each mode-stirring mechanism. At least four independent

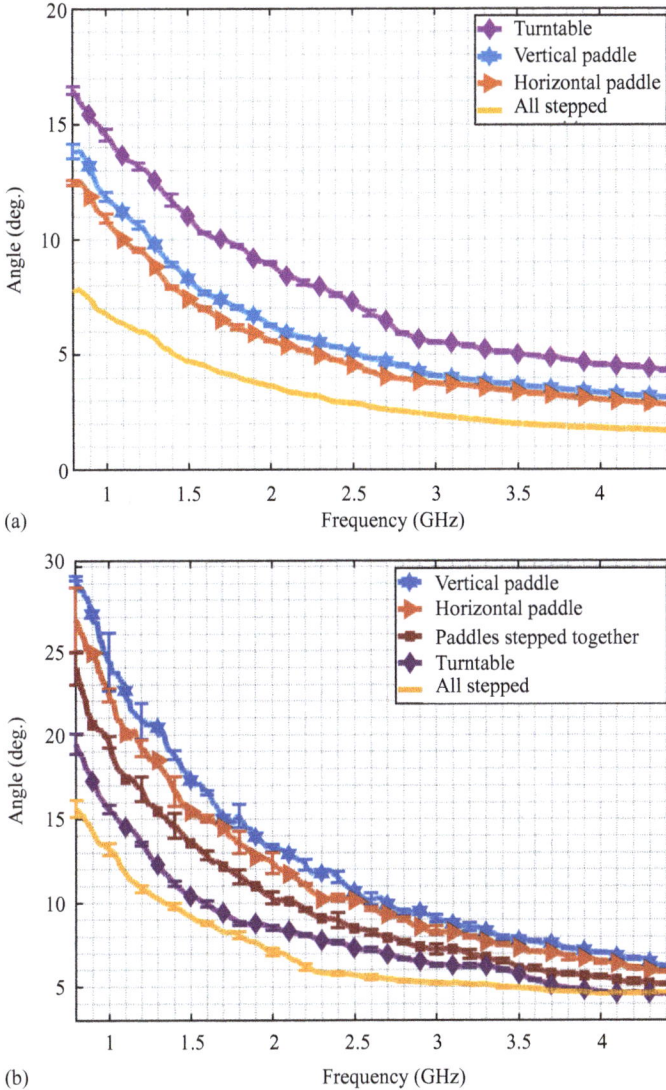

Figure 5.8 *Coherence angle determined as a function of frequency for a large reverberation chamber which was (a) unloaded (CBW approximately 0.6 MHz) and (b) loaded (CBW approximately 3.2 MHz). Individual stirring mechanisms and combinations are utilized for the mode-stirring sequences, including the vertical paddle only; horizontal paddle only; two paddles together; turntable only; and all three of these mechanisms stepped simultaneously ("All stepped")*

realizations were collected for each mode-stirring mechanism (except the unloaded All-stepped case) to create the error bars.

Note that the threshold of 0.3 provides a more stringent requirement for samples to be deemed uncorrelated than would a threshold of 0.5 (commonly used for CBW). As the threshold value decreases, the coherence-angle values become larger. This not only reduces potential correlation between samples but also can limit the number of uncorrelated samples that is possible to obtain within a chamber's working volume. That is, if the turntable coherence angle for a heavily loaded chamber is 10°, only 36 turntable samples may be used in the stepped mode-stirring sequence. This can affect the user's ability to develop multiple independent realizations for the uncertainty analysis and is a topic of current research.

Several effects may be noted for the unloaded chamber configuration in Figure 5.8(a). First, for all frequencies, the smaller coherence angle for the horizontal paddle indicates that it provides more efficient mode-stirring than the vertical paddle. This makes sense due to its larger physical size (a cylinder tracing out a volume of 2.97 m^3 as compared to 2.27 m^3 for the vertical paddle). A second effect to note is that the wider coherence angle for the turntable means that the turntable provides the least-effective individual stirring mechanism of the three. This is due to both the absence of significant boundary condition changes as compared to paddle stirring and also the high value of spatial uniformity in the large, unloaded chamber. That is, the mode-stirring effectiveness of an antenna moving on the turntable through the relatively uniform spatial environment of the unloaded chamber is less than the mode-stirring effectiveness provided by the paddles.

Finally, the smallest coherence angles occur between samples for the All-stepped case, indicating, as expected, that this is the most efficient mode-stirring sequence. For example, an optimized stirring sequence at 2 GHz for the unloaded reverberation chamber could consist of mode-stirring samples in which the two paddles and the platform each rotate in approximately 4° steps, providing approximately 90 uncorrelated stepped mode-stirring samples.

For the loaded chamber case shown in Figure 5.8(b), the correlation between mode-stirring samples is significantly higher. Again, the large (horizontal) paddle is shown to be more effective at mode-stirring by its smaller coherence angle (relative to the other mode-stirring mechanisms). The figure clearly illustrates the importance of position stirring in loaded-chamber configurations because the coherence angle for the turntable is now the single most effective mechanism. Physically, position stirring allows the antenna to sample the peaks and nulls within the reduced-uniformity environment.

Finally, note that the All-stepped case at 2 GHz indicates that angular steps of approximately 7° would be required to obtain uncorrelated samples, as compared to 4° for the unloaded chamber.

Measurements such as these provide a straightforward method for assessing the correlation between mode-stirring mechanisms within a chamber configuration in order to derive an optimal stirring sequence. More advanced methods such as principle component analysis have been investigated in [45] and are the subject of current research.

5.1.4.2 Correlation between measurements

While autocovariance approaches may be used to evaluate samples within a mode-stirring sequence, cross-correlation methods are typically used for evaluating spatial correlation between independent realizations of a mode-stirring sequence. Sample cross correlation, sometimes referred to as Pearson's cross-correlation function [2], may be determined from S-parameter measurements made for two independent realizations a and b of the same mode-stirring sequence as

$$R_{AB}(f) = \left| \frac{\sum_{n=1}^{N_W} \left[\left(S_{21,a} - \langle S_{21,a} \rangle_{N_W} \right) \left(S_{21,b} - \langle S_{21,b} \rangle_{N_W} \right)^* \right]}{\sqrt{\sum_{n=1}^{N_W} \left[\left| S_{21,a} - \langle S_{21,a} \rangle_{N_W} \right|^2 \right]} \sqrt{\sum_{n=1}^{N_W} \left[\left| \left(S_{21,b} - \langle S_{21,b} \rangle_{N_W} \right) \right|^2 \right]}} \right|.$$

(5.13)

Pearson's cross-correlation function illustrates spatial correlation between pairs of mode-stirring sequences as a function of frequency, averaged over the N_W values in the mode-stirring sequence. Generally, (5.13) is applied to all pairs of realizations of a mode-stirring sequence to determine the correlation between measurements or independent realizations, verifying their independence.

As an example of the application of (5.13) in OTA wireless test, in the standardized tests of [15], Pearson's cross correlation is applied to the measurements of G_{Ref} to determine whether the reference antenna is too close to a lossy DUT. Close placement of an antenna near an RF-absorbing material can block the radiation pattern of the antenna, causing a systematic underestimate in the determination of G_{Ref} [46,47]. The problem is that, when pre-characterizing a chamber, there is no way to know in advance how lossy a DUT may be. For example, a metal cellular-enabled parking meter in close proximity may not alter the radiation pattern of the reference antenna, whereas a lossy section of a foam-enclosed car dashboard may cause a significant change in the pattern. The test plan in [15] was designed to be sufficiently general to allow chamber pre-configuration and pre-characterization to measure both of these devices accurately.

The procedure described in [15] was developed to verify that the reference antenna used for chamber characterization is placed far enough from the DUT to ensure a sufficiently accurate estimate of G_{Ref}. The procedure requires that the user insert a block of RF absorber having a surface area as large (or larger) as the largest DUT to be measured. Because very few wireless devices will be as lossy as RF absorber, this scenario is expected to provide a worst-case-simulated lossy DUT. The user then performs a measurement of G_{Ref} with the reference antenna placed at the location in the chamber where it is anticipated that the reference antenna will be located for chamber characterization prior to DUT measurement. This measurement is followed by a second, *spatially uncorrelated* measurement of G_{Ref} made closer to the block of RF absorber. As long as the change in G_{Ref} due to its closer proximity to the simulated DUT does not exceed a specified threshold, its nominal location is deemed to be sufficiently far from any DUT having a surface area of that size or smaller.

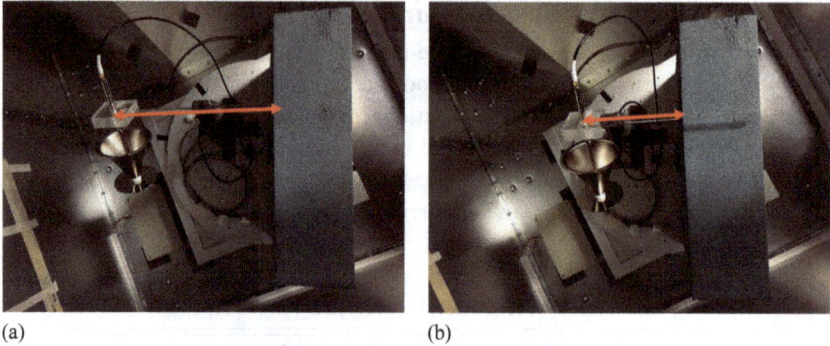

(a)   (b)

*Figure 5.9   Top view looking down on a discone reference antenna placed (a)
34.5 cm and (b) 19.5 cm from a block of RF absorber to study the
"proximity effect." The block of absorber is intended to simulate a
worst-case lossy DUT*

This procedure is illustrated in Figure 5.9(a) and (b), where an azimuthally omnidirectional reference antenna was placed at a distance of 34.5 cm (2.3 λ at 2 GHz) [Figure 5.9(a)] and 19.5 cm (1.3 λ at 2 GHz) [Figure 5.9(b)] from the block, respectively.

Figure 5.10 shows the results of application of (5.13) to the measured complex S_{21} data for the frequency band covering 1.85–2.0 GHz (Personal Communications Service (PCS) Band in the United States). A threshold of 0.3 correlation has been plotted as a horizontal red bar on the graph. The cross correlation between the two locations for nine different turntable positions is shown, where the stirring sequence consists of mechanical paddle stirring only (a nonideal stirring sequence used to illustrate the spatial correlation effect).

The figure shows that the turntable positions exhibit lower correlation at higher frequencies, where the data lie below the 0.3 threshold line (with one exception). At lower frequencies, there are more frequencies where correlation between antenna locations exists. The black vertical lines indicate a 3.84-MHz W-CDMA cellular communication channel. For this example, the user would be recommended to place the reference antenna somewhat farther from the simulated lossy DUT for the reference measurement to ensure that the reference antenna does not significantly couple directly into the DUT. The need to assess and minimize spatial correlation is a clear illustration of the statistical nature of the reverberation chamber, and, again, the importance of position stirring in heavily loaded chambers.

5.1.5   Lack of spatial uniformity due to loading

The metric $\sigma_{G_{Ref}}$, derived in (5.3) (Section 5.1), may be used to quantify the lack of spatial uniformity in a heavily loaded reverberation chamber [4,40]. The localized placement of RF absorber dampens the modes at specific locations, reducing the spatial uniformity of the stirred energy throughout the chamber. Because the

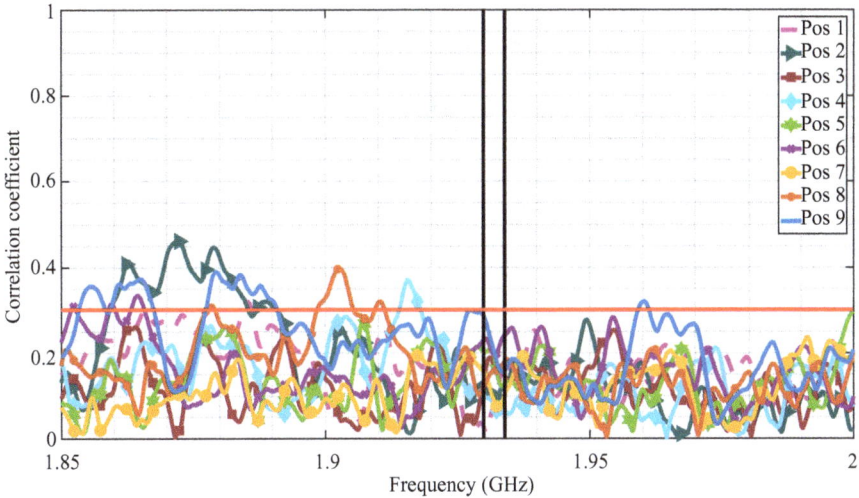

Figure 5.10 *Pearson's cross-correlation function applied to nine pairs of G_{Ref} measurements each made at two distances from an RF absorbing block to study the "proximity effect" of [15]*

TRP and TIS are derived from distributions of samples that should ideally have specific distributions (see Section 5.1.2), understanding the lack of spatial uniformity is essential for designing an adequate stepped mode-stirring sequence.

For OTA measurements of wireless devices, the uncertainty in the estimate of TRP or TIS is often dominated by the lack of spatial uniformity [14], as will be discussed in Section 5.4.1. In fact, uncertainty derived from (5.3) has been adopted by the wireless community in standardized test methods that utilize heavily loaded reverberation chambers [15].

To assess the uncertainty due to the lack of spatial uniformity for a given chamber configuration and stirring sequence, one may perform measurements of multiple independent realizations of the proposed stirring sequence, where independence can be evaluated by use of the cross-correlation procedure in (5.13) of Section 5.1.4. The procedure is similar to, but somewhat more complicated than, the method for assessing spatial uniformity in [48]. For a wireless-device test, the chamber is first loaded for a desired CBW and, typically, a stirring sequence is developed to minimize correlation between mode-stirring samples. The chamber's reference power transfer function is then measured with several different independent realizations of the same stirring sequence, where each sequence uses the same number of paddle angles, antenna positions, etc., but the samples are nominally spatially uncorrelated from one implementation to the next, as confirmed using (5.13). The standard deviation between these independent realizations is then computed from (5.3) with $N_B - 1$ degrees of freedom [49]. The value of $\sigma_{G_{Ref}}$ will depend on the chamber configuration and the types of mode-stirring used in the stirring sequence.

Often nine or more independent realizations are required for standardized OTA tests [15]. To increase the number of independent realizations of stirring sequences, users often elect to use mechanisms such as polarization stirring (where the antenna is moved to one of three orthogonal polarizations during the multiple realizations of the stirring sequence), linear position stirring (where a linear translation stage moves the antenna up-and-down or side-to-side within the chamber), and/or multiple-antenna stirring (where multiple measurement antennas are placed at uncorrelated locations within the chamber, with correlation defined by the cross-correlation function in (5.13)). A method for optimizing the number of independent samples is presented in Section 5.4. The stirring sequence must enable a sufficiently low value of $\sigma_{G_{Ref}}$ so that the combined uncertainty (see Section 5.4.3) is below a specified value for the chamber configuration (including the stirring sequence) to be deemed acceptable for use in OTA tests.

5.1.6   Isotropy for loaded chambers

An ideal reverberation chamber setup will theoretically provide an "isotropic" electromagnetic environment [34], where the fields impinge on the receive antenna from all angles of incidence equally when averaged over the collection of mode-stirring samples in a mode-stirring sequence. Real reverberation chamber environments will naturally be anisotropic, and characterizing the level of anisotropy is often a key figure of merit in standardized test procedures. Typically, to characterize anisotropy through measurement, the magnitude squared of either the electric field or the transmission parameters S_{21} is considered [48,50,51]. To simplify notation, we use the field representation here.

The statistics used for current standardized isotropy validation tests [48] are based on measurements of the three orthogonal components of the electric field, E_i with $i = x, y, z$ for each mode-stirring state. The real and imaginary components of the electric field may be represented by complex Gaussian distributions. We denote

$$E_i = E_{i_1} + jE_{i_2}. \tag{5.14}$$

Then, $E_{i_1} \sim N(\mu_{i_1}, \sigma_{i_1}^2)$ and $E_{i_2} \sim N(\mu_{i_2}, \sigma_{i_2}^2)$ each follow a normal distribution with mean μ_{i_x} and variance $\sigma_{i_x}^2$ for $x = 1, 2$, respectively, and E_{i_1} and E_{i_2} are independent. If we further assume that the in-phase and quadrature components have equal variances, i.e., $\sigma_{i_1}^2 = \sigma_{i_2}^2 = \sigma_i^2$, the magnitude squared of E_i is written as

$$|E_i|^2 = (\mu_{i_1} + Z_1\sigma_i)^2 + (\mu_{i_2} + Z_2\sigma_i)^2 \tag{5.15}$$

$$= \sigma_i^2 \left[\left(Z_1 + \frac{\mu_{i_1}}{\sigma_i} \right)^2 + \left(Z_2 + \frac{\mu_{i_2}}{\sigma_i} \right)^2 \right]$$

$$= \sigma_i^2 W_i, \tag{5.16}$$

where Z_1 and Z_2 are independent standard Gaussian random variables, and, hence, W_i is distributed as a noncentral χ^2 with 2 degrees of freedom and noncentrality parameter (e.g., see [52])

$$\lambda_i = \frac{\mu_{i_1}^2 + \mu_{i_2}^2}{\sigma_i^2}. \tag{5.17}$$

For stirred fields in a reverberation chamber, a definition for the statistical isotropy of the field may be given by the following equality

$$\langle |E_x|^2 \rangle = \langle |E_y|^2 \rangle = \langle |E_z|^2 \rangle, \tag{5.18}$$

where x, y, and z represent any three orthogonal orientations, and the average is taken over all mode-stirring samples. Given the result in (5.16), (5.18) reduces to

$$2\sigma_x^2 + \mu_{x_1}^2 + \mu_{x_2}^2 = 2\sigma_y^2 + \mu_{y_1}^2 + \mu_{y_2}^2 = 2\sigma_z^2 + \mu_{z_1}^2 + \mu_{z_2}^2.$$

If we further assume that the unstirred energy is sufficiently small, i.e., both μ_{i_1} and $\mu_{i_2} \ll \sigma_i$ for $i = x, y, z$, then (5.18) is equivalent to

$$\sigma_x^2 = \sigma_y^2 = \sigma_z^2 = \sigma^2. \tag{5.19}$$

Note that most stirring sequences used in wireless-device tests utilize position stirring. Thus, when averaged over the stirring sequence, (5.19) will often be satisfied even for chamber configurations with significant amounts of unstirred energy at a single location. With (5.19), the distribution of $|E_i|^2$ in (5.16) is

$$|E_i|^2 \sim \sigma^2 \chi_2^2, \tag{5.20}$$

where χ_2^2 is a central χ^2 distribution with 2 degrees of freedom. This distributional property of $|E_i|^2$ in (5.20) can be used to develop statistical hypothesis tests for isotropy by comparing the observed and theoretical distributions of $|E_i|^2$ using, for example, a chi-square goodness-of-fit test (e.g. see [53]). Note that, while the Anderson–Darling goodness-of-fit test may converge more quickly for continuous samples [35], the chi-square formulation is easy to use in standardized tests where specific distribution binning must be utilized between labs.

Since the distribution of $|E_i|^2$ involves an unknown parameter σ, a common approach is to use ratios of $|E_i|^2$ to develop test statistics. To simplify the notation, denote $X = |E_x|^2$, $Y = |E_y|^2$, and $Z = |E_z|^2$. Then the ratios of X, Y, and Z are ratios of independent χ^2 random variables and are free of any unknown parameters. IEC 61000-4-21 [48] denotes anisotropy coefficients based on the ratios of X, Y, and Z. Specifically, planar and total field anisotropy coefficients A_{ij} with $i \neq j$; $i, j = x, y, z$ and A_t are used

$$A_{xy} = \frac{X - Y}{X + Y}, \quad A_{yz} = \frac{Y - Z}{Y + Z}, \quad A_{zx} = \frac{Z - X}{Z + X}, \tag{5.21}$$

$$A_t = \frac{1}{3}\sqrt{\left(A_{xy}^2 + A_{yz}^2 + A_{zx}^2\right)}. \tag{5.22}$$

Under (5.20), A_{xy}, A_{xz}, and A_{yz} are uniformly distributed ranging from -1 to 1 and are correlated, that is, no two A_{ij} pairs are independent.

The distribution of A_t can be derived numerically from knowledge of the planar field-anisotropy-coefficient distributions, but it is also conveniently obtained by simulation. Figure 5.11 displays the CDF of the total anisotropy coefficient A_t based on 1,000,000 Monte Carlo samples. The difference between the measured CDF of A_t and ideal CDF may be computed within the framework of the chi-square test to determine if the chamber is isotropic.

It can be shown by simulation that an isotropy validation test based on A_t tends to pass the test even when the chamber exhibits a substantial level of anisotropy. To correct this deficiency, one may use any two coefficients A_{ij} to carry out the isotropy validation test. The problem is that these coefficients are not independent, and hence the exact significance level of such a test cannot be determined. Consequently, validation tests based on two independent test statistics that are functions of the ratios of X, Y, and Z are desirable so that the correct significance level of the tests can be determined and components that cause anisotropy can be identified. For example, test statistics $X/(X + Y)$ and $Z/(X + Y + Z)$ are independent under the hypothesis that the chamber is isotropic and hence can be used for this purpose. The development of isotropy validation tests based on a pair of independent statistics that produce adequate powers in all anisotropy conditions and correct significance levels is a topic for future research.

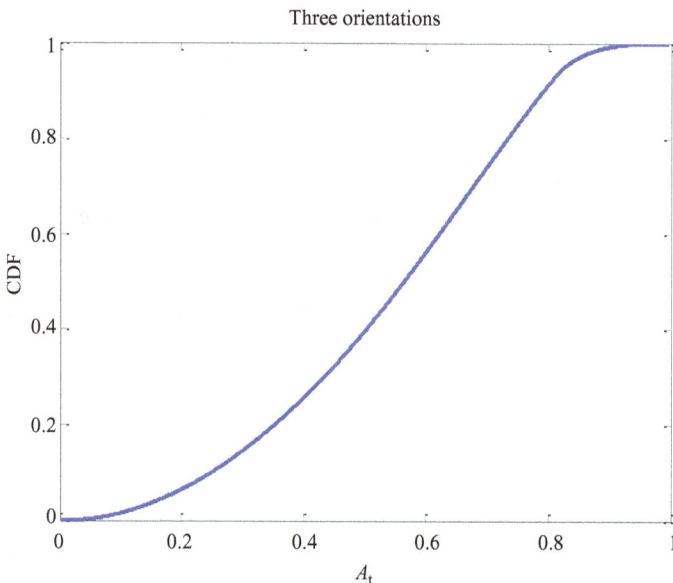

Figure 5.11　Cumulative distribution function of A_t calculated with (5.22)

5.1.7 K-factor

The Rician K-factor, or simply K-factor, is a metric that is commonly used in the area of wireless communications to describe a reflective multipath channel. Specifically, it is the ratio of "discrete" multipath, in which a certain number of paths between the transmitter and receiver consist of specular, deterministic reflections, to the number of randomly scattered, "diffuse" multipath reflections [39]. For communication channels, specularly reflected signal paths are typically more desirable than diffuse multipath both because the former usually provide stronger received signals and because they can be modeled analytically, allowing adaptive protocols in receivers to adjust appropriately. Consequently, high K-factor is often a desirable figure of merit for wireless communications.

Conversely, for traditional reverberation-chamber measurements, stirred energy is often equated with a well-performing setup and, effectively, diffuse multipath, and is deemed desirable. In [8], the K-factor is defined as the ratio of the unstirred to stirred energy. That is

$$K \approx \frac{|\langle S_{21}\rangle|^2}{|\langle S_{21} - \langle S_{21}\rangle\,|^2\rangle}. \tag{5.23}$$

In this definition, the unstirred energy consists of the power after averaging over the mode-stirring sequence, denoted by the angled brackets. The denominator is the variance of the received power and yields an approximation of the stirred energy [8].

There will always be a finite K-factor in a reverberation chamber due to, as examples, nonideal direct coupling between antennas, unstirred components of the measured signal, losses in the metal and, if utilized, RF absorber. However, minimization of the K-factor will often minimize measurement errors and uncertainties. The value of the K-factor will depend on a number of measurement setup parameters, including the dimensions of the chamber, the reflectivity of the chamber walls, the types of antennas used and their orientation with respect to each other, the number of mode-stirring samples, and, of course, the amount of RF absorber, as well as its placement within the chamber.

One can take simple steps to minimize the K-factor by eliminating the line-of-sight path and simple unstirred reflected paths between the measurement antenna and the DUT. For example, one can point the measurement antenna toward the mechanical mode stirrers within chamber and cross-polarize the measurement antenna with the DUT and the reference antenna. Additionally, the use of a directional measurement antenna can help to reduce line-of-sight interactions between this antenna and the DUT [4,40].

Figure 5.12 shows that the K-factor increases as a function of loading. That is, for a given mode-stirring sequence, as the relative level of unstirred energy increases due to loading, so does the K-factor. In this figure, the K-factor was averaged over 72 paddle positions in a chamber with a single rotating paddle and no platform stirring. The legend refers to the "p" positions and the "or" orientations of an azimuthally omnidirectional reference antenna.

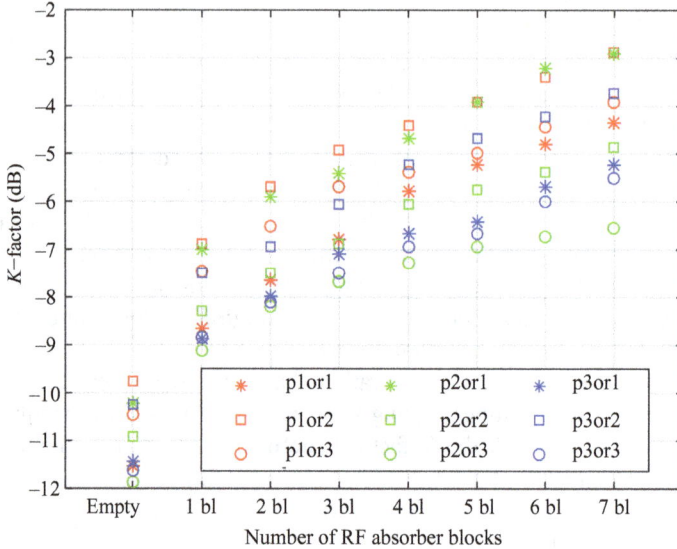

Figure 5.12   Rician K-factor calculated from measurements made for three omnidirectional transmit antenna positions ("p") and three orientations ("or") for each location. Results were averaged over the PCS band, 1.85–2.0 GHz (from [40])

In Figure 5.12, it is interesting to note that the K-factor values can differ by 2–4 dB for various locations within the chamber due to the lack of spatial uniformity (indicating different amounts of spatial correlation between locations). As well, for different frequency averaging bandwidths, different K-factors will be obtained for a particular chamber setup, again due to the correlations between mode-stirring samples. Understanding the K-factor and the parameters that increase or decrease its value can help in the optimization of a stirring sequence and troubleshooting an unanticipated measurement result.

5.1.8   Chamber characterization: a summary

For wireless-device tests in which the signal must be demodulated, use of a loaded reverberation chamber is generally required to provide a channel that is flat enough to match that in which the device was designed to operate. Because loading degrades the spatial uniformity of the averaged fields in the chamber, measured samples are often correlated. We have presented several methods for assessing this correlation in terms of the necessary increase in frequency correlation (CBW) and the less-desirable spatial correlation between mode-stirring samples. With these techniques, users may develop optimized mode-stirring sequences in which each sample contributes maximally to an improvement in the estimate of a quantity of interest. In the next section, we discuss the estimation through the measurement of power-based metrics that may be used to evaluate the OTA performance of many cellular-based wireless devices.

5.2 Over-the-air tests for radiated power and receiver sensitivity

Given a well-devised mode-stirring sequence based on the knowledge of the correlations between mode-stirring samples, users can perform measurements and derive uncertainties for metrics such as those used by cellular-device certification groups, obtaining results comparable to anechoic chambers, which have traditionally been used for these tests [54]. Commonly used metrics are TRP, TIS, data throughput, and channel capacity [40,55,56]. Here, we focus on the two most basic tests that have been standardized used for single-input, single-output wireless devices: TRP and TIS [15].

5.2.1 Total radiated power

TRP is one of the most fundamental metrics that can be obtained from OTA testing of a wireless device, as reported in the literature [48,57,58]. As mentioned previously, heavy chamber loading is not necessary for TRP measurements in reverberation chambers because there is no need to demodulate the signal. However, establishing a communication link by a BSE is typically necessary in cellular applications to control the DUT power output. Also, many labs do load so that they can use the same chamber setup and pre-characterization for both TIS and TRP measurements.

The description of TRP here is consistent with the OTA test methodology that is utilized by the cellular industry [14,15]. It accounts for impedance mismatch between the antennas and the chamber, as well as the antenna efficiencies. As for the chamber characterization steps of Section 5.1, these methods are based on "stepped mode-stirring" with samples collected under static conditions.

The objective in a TRP measurement is to estimate the total power radiated by the DUT in free space and the associated measurement uncertainty. Typically, G_{Ref} is first measured with a VNA, as illustrated in Figure 5.2(a). We then define P_{Meas} as the average power measured by the receiver over the occupied bandwidth of the modulated signal. The measurement is carried out at the reference plane indicated in Figure 5.2(b), where the power measurement instrument is the receiver section of a BSE.

In practice, for cellular device testing, the BSE establishes a communication link with the wireless DUT and instructs it to radiate at full power. Samples of the received power are measured at the reference plane of the receive antenna by the BSE or another receiver. This is P_{Meas} in (5.24) and (5.25). Each measured sample is corrected for the chamber setup's reference power transfer function, which is estimated from G_{Ref}, the measurement antenna characteristics (mismatch and efficiency), the impedance mismatch between the measurement antenna and the receiver, and the cable loss (if applicable). As discussed in Section 5.1.1, the reference measurement, G_{Ref} is intended to provide an estimate of the chamber loss associated with the measurement of the DUT, G_{DUT}, which typically cannot be measured directly for devices with integrated antennas. Thus, the reference measurement typically involves a different antenna than that of the DUT, but generally one with a similar radiation pattern [33]. Uncertainties related to this will be discussed in Section 5.4.3.

The TRP may then be estimated from the ensemble average of the P_{Meas} samples acquired over all N_W mode-stirring positions and N_B independent realizations, as given by

$$P_{\text{TRP}} = \frac{\langle P_{\text{Meas}} \rangle_{N_W N_B} |1 - \Gamma_M \Gamma_{\text{RX}}|^2}{G_{\text{Ref}} \, \eta_M \left(1 - |\Gamma_M|^2\right)}, \tag{5.24}$$

where η_M is the efficiency of the measurement antenna, Γ_M is the free-space reflection coefficient of the measurement antenna, and Γ_{RX} is the reflection coefficient of the BSE or other receiver assembly (including cable, if used). Again, note that if the antenna efficiencies are measured in a reverberation chamber, it must be unloaded [32].

In terms of measured quantities P_{Meas} and $S_{21,\text{Ref}}$, we have

$$P_{\text{TRP}} = \frac{\langle P_{\text{Meas}} \rangle_{N_W N_B} |1 - \Gamma_M \Gamma_{\text{RX}}|^2 \eta_R (1 - |\Gamma_R|^2)}{\langle |S_{21,\text{Ref}}|^2 \rangle_{N_W N_B}}, \tag{5.25}$$

where $\langle |S_{21,\text{Ref}}|^2 \rangle$ is the ensemble average of the measured transmission parameter obtained during the reference measurement, frequency averaged over the bandwidth of the channel being tested. Note that the measurement antenna's mismatch-and-efficiency terms from (5.1) in Section 5.1.1 and (5.24) cancel, leaving only the reference antenna-efficiency and mismatch terms. In practice, the DUT measurement $\langle P_{\text{Meas}} \rangle_{N_W N_B}$ is often made for a single independent realization of the stirring sequence. That is, typically $N_B = 1$ for this term.

It is typically necessary to average each measurement sample over many transmitted symbols (or bursts, blocks, or frames, depending on the type of communication signal) to obtain a valid estimate of the power corresponding to that sample. The duration required for each measurement sample will depend on the modulation and transmission format of the radio-access technology for the particular measurement. These measurement parameters are described in OTA test plans such as [15,54]

As mentioned, loading is not critical for TRP measurements, but if the chamber is loaded, then averaging over position is critical. We have seen the importance of position stirring in Section 5.1.2.

In [40], the TRP of a wireless router transmitting with W-CDMA was measured in the Cellular (800–900 MHz) and PCS (1.85–2.0 GHz) bands. Agreement between the reverberation chamber measurement and two different anechoic chambers was within 2 dB, which is within the CTIA limit for TRP.

5.2.2   Total isotropic sensitivity

TIS is a measure of receiver sensitivity of a wireless device derived from OTA measurements. This metric quantifies the amount of spatially averaged RF power that must be incident on a device's receive antenna to achieve a defined minimum standard of data fidelity in an isotropic environment. As with TRP, this measurement may be made in both anechoic or reverberation chamber setups. The BSE first establishes a communication link with a wireless device and sends a known sequence of bits as a preamble. The device then sends back the bits it reads from the preamble (typically at full DUT power) and the error rate is computed at each mode-stirring step. This "error rate" may be the bit error rate (BER), block error rate, frame error rate, or other, depending on the transmission protocol. The BER tends to increase gradually as the power level of the BSE is lowered.

To find the device's receiver sensitivity for each stepped mode stirring sample, the RF power emitted by the BSE is reduced in successive measurement steps and the error rate is again computed. This procedure continues until the minimum power required at the device is reached that is necessary to maintain a specified error rate below a given threshold. For example, in [54], this threshold is 1.2% BER for the W-CDMA transmission protocol.

A schematic representation of the TIS measurement concept is shown in Figure 5.13. Since the estimated value of TIS is based on the power incident on the DUT's receive antenna, the measurement requires a determination of the power emitted by a BSE, P_{BSE}, and corrections for the power lost between the BSE and the DUT. As depicted in Figure 5.13, these losses could be due to cable loss, $G_{CableLoss}$, power that is not transmitted by the antenna due to impedance mismatch at the antenna port, and reference power transfer function, G_{Ref} (5.1) from Section 5.1. Procedures for performing TIS tests have been developed for anechoic [54,59] and reverberation chambers [15,60]. We will present the reverberation-chamber method here.

The procedure for estimating the TIS of a wireless device from a reverberation-chamber measurement is similar to that for an anechoic-chamber measurement, except that instead of averaging over angular orientations of the DUT as in the

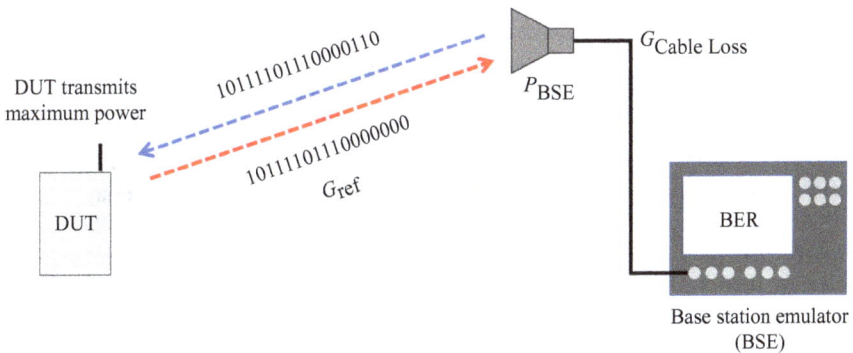

Figure 5.13 *Schematic representation of a basic over-the-air measurement of TIS. The base-station emulator transmits a known code sequence that is received by the DUT and transmitted back to the BSE at full power. The BSE transmit power (P_{BSE}) is reduced until incorrect bits are sent back at a specified rate (shown in the rightward-directed red arrow). The TIS is estimated from the power incident on the DUT, so P_{BSE} is combined with any losses between the BSE and the DUT, which are shown as cable loss ($G_{CableLoss}$) and average path loss (G_{Ref}) (from [17])*

anechoic-chamber case, the measurements are averaged over the changing boundary conditions provided by the mode-stirring sequence.

For a given stepped mode-stirring sample n, the power emitted from the BSE that yields the threshold error rate, $P_{\mathrm{BSE}}(n)$, is recorded. Normally, the measurements for G_{Ref} are averaged over the stirring sequence in advance. If, instead, the reference and P_{BSE} samples were taken in precisely the same configuration (e.g., by use of a switch between the reference and DUT antennas as was done in [17]), then the TIS power, P_{TIS}, may be given by

$$P_{\mathrm{TIS}} = \frac{\eta_{\mathrm{M}}(1 - |\Gamma_{\mathrm{M}}|^2)}{N_{\mathrm{W}} |1 - \Gamma_{\mathrm{M}}\Gamma_{\mathrm{RX}}|^2} \sum_{n=1}^{N_{\mathrm{W}}} G_{\mathrm{Ref}}(n)P_{\mathrm{BSE}}(N). \tag{5.26}$$

Here, G_{Ref} defined in (5.1) has not yet been averaged, N_{W} is the number of samples in the mode-stirring sequence, and corrections are made for the efficiency and free-space reflection coefficient of the measurement antenna and the impedance mismatch between the measurement antenna and the receiver. As for TRP, we assume that the measurement antenna is identical for the measurement of P_{BSE} and G_{Ref} and that G_{Ref} provides a reasonable estimate of G_{DUT}. Even though the reference and DUT measurements are made with different antennas, when the radiation patterns are similar, the measured results will be similar [4,33].

The TIS measurement is intended to estimate the true, intrinsic receiver sensitivity of a wireless device. Since the sensitivity is intrinsic to the DUT, the value of P_{BSE} is dependent on the value of the chamber's reference power transfer function [here termed $G_{\mathrm{Ref}}(n)$] for each mode-stirring sample. That is, at step n in the stirring sequence, if the sample of G_{Ref} goes up, P_{BSE} should ideally compensate by going down. This dependence is illustrated in Figure 5.14, which shows a plot of $P_{\mathrm{BSE}}(n)$ overlaid with a plot of $G_{\mathrm{Ref}}(n)$ averaged over a 4-MHz bandwidth, centered at 1,932.4 MHz. The G_{Ref} values have been offset and the inverse taken, allowing the two quantities to be compared. Qualitatively, the two parameters track each other very well. The uncertainty on the plotted quantities is approximately 0.5 dB [17]. Note that if the chamber loading was not sufficient, the two curves would not overlay as well.

Given the inverse dependence of P_{BSE} on G_{Ref}, as evidenced by the data in Figure 5.14, the separate averaging of the two parameters must be complimentary. To illustrate this, we recast (5.26) as

$$\frac{P_{\mathrm{TIS}}(n)}{P_{\mathrm{BSE}}(n)} = \frac{(1 - |\Gamma_{\mathrm{M}}|^2)\,\eta_{\mathrm{M}}}{|1 - \Gamma_{\mathrm{M}}\Gamma_{\mathrm{RX}}|^2} G_{\mathrm{Ref}}(n). \tag{5.27}$$

Often, the chamber response and base-station power are measured at different times and with different realizations of the mode-stirring sequence. Representing these by $N_{\mathrm{W}1}$ and $N_{\mathrm{W}2}$, we can calculate P_{TIS} from (5.26) as

$$P_{\mathrm{TIS}} = \frac{(1 - |\Gamma_{\mathrm{M}}|^2)\,\eta_{\mathrm{M}}}{|1 - \Gamma_{\mathrm{M}}\Gamma_{\mathrm{RX}}|^2} \frac{1}{N_{\mathrm{w}1}} \left(\sum_{n_1=1}^{N_{\mathrm{w}1}} \frac{1}{P_{\mathrm{BSE}}(n_1)} \right)^{-1} \frac{1}{N_{\mathrm{w}2}} \sum_{n_2=1}^{N_{\mathrm{w}2}} G_{\mathrm{Ref}}(n_2). \tag{5.28}$$

Equation (5.28) utilizes the arithmetic mean of the chamber's reference power transfer function, as was done in (5.1) in Section 5.1.2, and the harmonic mean (with inverse

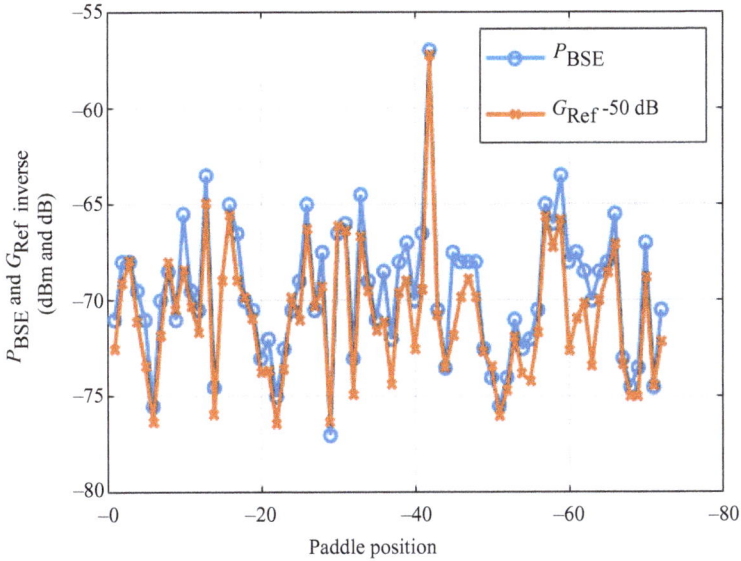

Figure 5.14   *Plot of individual samples of P_{BSE} (blue circles) and of G_{Ref} (red ×'s) vs. paddle angle for a single stepped mode-stirring sequence. The chamber was sufficiently loaded with RF absorber to produce the ideal inverse behavior between the two parameters. For comparison, the G_{Ref} has had 50 dB subtracted and has been inverted in the y-axis (from [17])*

averaging) for the receiver sensitivity power. This is not a unique choice of averaging. That is, theoretically, the assignment of arithmetic and harmonic mean could be swapped, or the median of each quantity used. The measurement and uncertainty dependence on these choices was discussed in [60], where it was shown that the combination in (5.28) yields the most stable results.

As was mentioned, in practice [15], the measurement of G_{Ref} is carried out in a pre-characterization step of the reverberation-chamber configuration to determine proper loading and to calculate the uncertainty due to lack of spatial uniformity. This step is seldom performed again. The P_{BSE} measurement is then performed quickly on customer-provided DUTs and combined with the pre-characterization parameters to yield the TIS of each device.

As stated in the introduction, the measurement of metrics such as TIS in which the communication signal must be demodulated motivates the heavy loading of the reverberation chamber for testing of wireless devices, because the CBW must be sufficiently wide to provide a frequency flat channel. Figure 5.15 shows measured examples of TIS as a function of chamber loading for two different reverberation chambers. The chamber with the larger error bars utilized a single mode-stirring paddle and no position stirring. The other chamber utilized two rotating paddles,

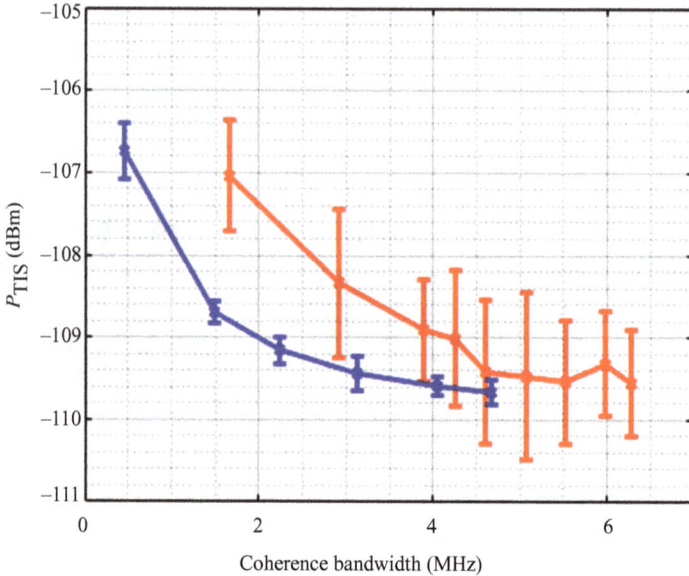

Figure 5.15   TIS power of a cellular-enabled wireless router operating in the 1.85–2.0 GHz US PCS Band measured in two different reverberation chambers. The error bars represent nine independent realizations of each stirring sequence. Position stirring was not used in the chamber with the larger error bars

and a turntable for position stirring, resulting in smaller error bars. Both chambers demonstrate a "plateau" in TIS for CBW values greater than the 3.84 MHz bandwidth of the W-CDMA communications signal. This again illustrates the need for heavy loading of the chamber for OTA measurements of wireless devices.

One key advantage of the reverberation chamber over the anechoic chamber method is that a significant fraction of the volume of the chamber is the working volume. Thus, the same chamber setup can be used to measure devices as small as printed circuit boards or as large as trash compactors [2,40]. The key limitation is that this setup is only useful for spatially averaged metrics of interest since angle-of-arrival information is lost. Obtaining a sufficiently wide CBW may be another limitation, depending on the characteristics of the chamber setup and the bandwidth of the communications signal to be measured [61].

5.3   Replicating specific multipath channels for OTA test

Many OTA tests for wireless devices require the creation of specific multipath conditions, in terms of the timing between multipath components incident upon the DUT [given by the delay spread that was defined in (5.5)], the angles from which multipath components arrive, or both. The high number of reflections in the reverberation chamber can simulate a dense multipath environment, and loading can tune the decay time, for example. For channels with longer delay spreads and/or discrete multipath components, a channel emulator may be coupled to the reverberation chamber. Experimental work on using heavily loaded reverberation chambers to simulate spatial channels is ongoing. Each of these applications is briefly described in this section.

5.3.1 Highly reflective power delay profiles

Some channels present a short decay time to the device. Examples include outdoor-to-indoor building penetration scenarios, within a building, or some urban environments in which the decay time may be on the order of a few microseconds or tens to hundreds of nanoseconds. In these cases, the chamber itself can sometimes be tuned to provide a controlled OTA test environment having this same decay time. This configuration is especially appropriate when testing in a highly reflective, dense multipath condition. An example involving tests of an LTE cellular base station in which an urban environment is replicated by a loaded reverberation chamber is presented in [31].

Another example is presented in Figure 5.16 [10] that compares measurements of the PDP in an oil refinery to those in a loaded reverberation chamber. In all of these

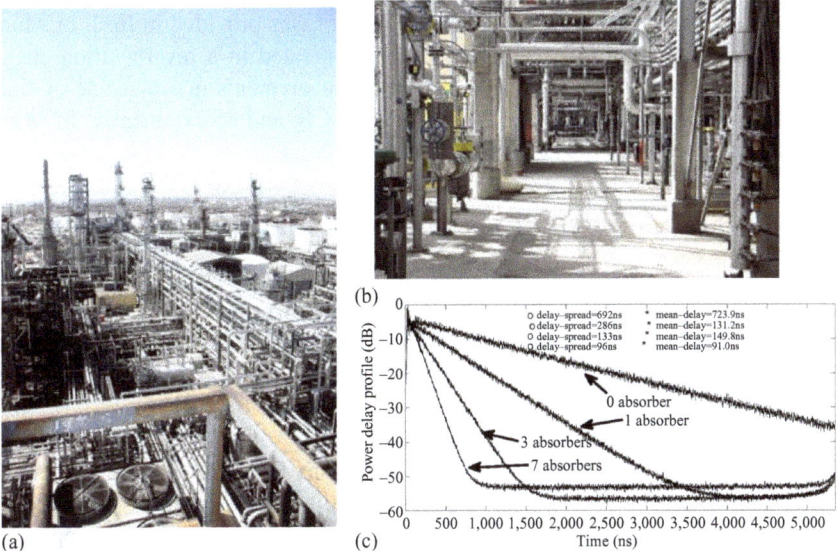

Figure 5.16 *The power delay profile was measured in an oil refinery and replicated in a reverberation chamber. Photographs of the oil refinery are shown in (a) and (b), while (c) shows the chamber measurements averaged over a stirring sequence (from [10])*

examples, the number of reflections is sufficiently high that the densely reflective PDP of the reverberation chamber captures the characteristics of the environment.

A third example of the use of a reverberation chamber to directly simulate a specific multipath environment is an outdoor-to-indoor channel model investigated by the CTIA and 3GPP. The model is based on measurements made in several representative urban environments. RMS delay spread values of less than 90 ns in the 700 MHz band were extracted from measured power-delay profiles [62]. An interlaboratory comparison of reverberation chambers configured to emulate this channel model was presented in [27] with 0.7 dB standard deviation between laboratories.

5.3.2   Longer power delay profiles

When a longer delay spread, or a PDP with discrete multipath components is to be emulated to better match measured propagation conditions, a common technique is to connect a channel emulator to the reverberation chamber (this configuration is sometimes termed "RC + CE"). The channel emulator creates multiple reflected copies of the transmitted signal, each delayed in time, amplitude, and phase according to a specified channel model. Two examples of these channel models are the "Isotropic Urban Micro (IS UMi)" and the "Isotropic Urban Macro (IS UMa)," which consist of the same PDP as the Spatial Channel Model Extended UMi and UMa, although in an isotropic environment. These RC + CE models were emulated in the interlaboratory comparison of [27], with 1.25 dB standard deviation between three laboratories for the UMi model and 1.88 dB for the UMa model.

Another example of an RC + CE measurement was provided in [63], in which the PDP corresponding to an urban area was replicated in a reverberation chamber. Figure 5.17 shows a comparison between measurements in a non-line-of-sight environment in downtown Denver, CO, the RC + CE, and an exponential fit to the measured data.

5.3.3   Emulating spatial channels

The previous examples illustrate the use of the reverberation chamber for emulating certain isotropic-channel timing characteristics. Over the years, researchers have also investigated the use of reverberation chambers to create specific non-isotropic, directional channels [23,28,64,65].

As an example, in [28], a reverberation chamber was heavily loaded to create a channel having specific spatial characteristics. RF absorber placed along one wall of the reverberation chamber damped out reflections from that surface, creating a channel measurement system with a shorter, spatially controlled PDP.

Figure 5.18 shows results from a synthetic aperture technique that was applied to visualize the channel. These figures show the power received at the synthetic aperture array as a function of time (on the *x*-axis) and azimuthal angle of arrival (on the *y*-axis). In these figures, the received power is indicated on the scale to the right.

The unloaded chamber shown in Figure 5.18(a) shows several reflection incidents on the receive antenna from a wide range of azimuthal angles. In Figure 5.18(b),

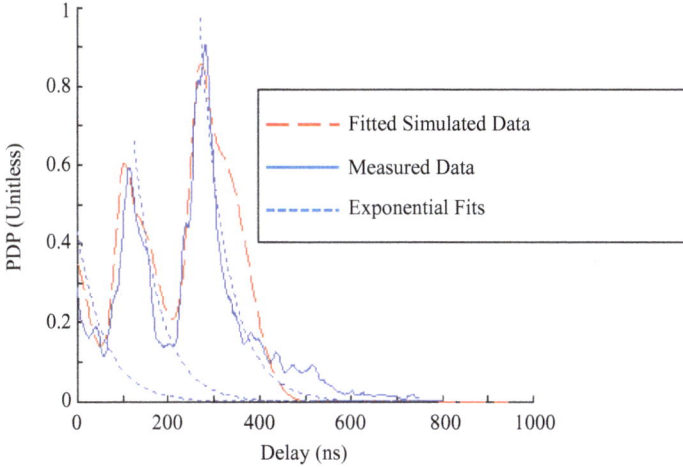

Figure 5.17 *Power delay profile based on measurements performed in downtown Denver, CO (solid blue line), a reverberation-chamber-plus channel emulator (dashed red line), and an exponential fit to the measured data (dotted blue line). Figure extracted from [63]*

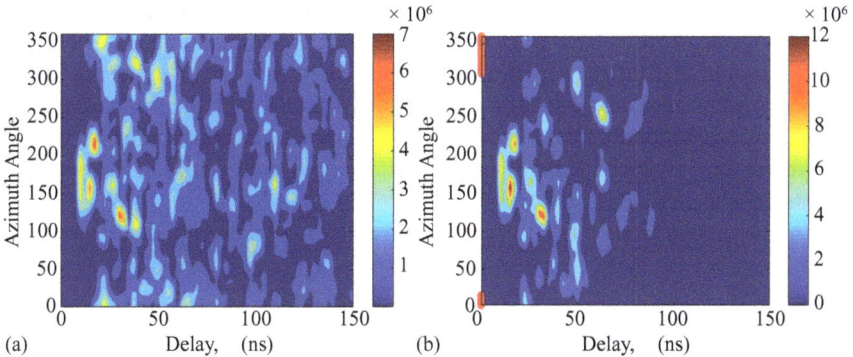

Figure 5.18 *Synthetic aperture results showing reflected energy incident on a receive antenna in (a) an unloaded reverberation chamber and (b) a reverberation chamber with RF absorber placed along the left-hand wall in the locations indicated by the thick vertical lines at the left of the plot (from [28])*

the thick vertical lines on the left side of the plot show the placement of the RF absorber along that wall. Reflections are damped out in that azimuthal direction and die out sooner than for the unloaded case. A controlled spatial-and-temporal channel test environment such as this could be useful for repeatable testing of multiple-antenna wireless devices.

5.4　Uncertainty in OTA measurements with heavily loaded reverberation chambers

5.4.1　*Number of samples vs. spatial uniformity: which uncertainty mechanism dominates?*

Two important sources of uncertainty in the estimate of power-based OTA metrics such as TRP and TIS derived from reverberation-chamber measurements concern the number and type of stepped mode-stirring samples used in the measurement. Understanding the origins and impact of these measurement parameters on uncertainty for OTA tests conducted in heavily loaded reverberation chambers is the focus of this section. Spatial correlation due to loading reduces the effective number of mode-stirring samples. This can increase uncertainty if the mode-stirring sequence does not contain a sufficient number of uncorrelated samples. Likewise, the lack of spatial uniformity due to loading can contribute to increased uncertainty related to non-repeatable placement of the device within the working volume of the chamber. Both these effects—the limited number of mode-stirring samples and the lack of spatial uniformity—may impact the uncertainty in an estimate of OTA-derived metrics such as TRP and TIS.

For heavily loaded chambers, these two components interact, yet it is difficult to assess which dominates without further statistical analysis. For example, if a stirring sequence does not include a sufficient number of mode-stirring samples, the variance between independent realizations will be higher than if a large number of samples is used. However, it is difficult to attribute the increased variance to either the lack of mode-stirring samples or the lack of spatial uniformity. As well, understanding which source of uncertainty dominates may lead to a simplified expression for uncertainty in reverberation-chamber measurements, which is a desirable goal for standardized test methods that strive for efficiency.

In [14], a significance test was developed that allows a user to identify which of two expressions for these components of uncertainty is most appropriate to account for these two different, yet related, sources of uncertainty.

A significance test can be used to identify the dominant source of uncertainty for a given chamber setup and the corresponding form that the expression for uncertainty may take [49]. The significance test in [14] may be used to identify whether statistically significant differences exist between realizations of a stirring sequence. Based on commonly used procedures outlined in [49], it is based on an F distribution which is formed from the ratio of two variances. To carry out this test, we first compute the variance of the samples within the stirring sequence from the full measurement set consisting of N_W samples within each stirring sequence and N_B independent realizations of that stirring sequence.

$$s_W^2 = \frac{1}{N_B(N_W - 1)} \sum_{p=1}^{N_B} \sum_{n=1}^{N_W} (G_{\text{Ref}}(n,p) - G_{\text{Ref}}(p))^2, \qquad (5.29)$$

which has $N_B(N_W - 1)$ degrees of freedom. We then compute the variance of the samples *between* multiple realizations of the stirring sequence

$$s_B^2 = \frac{N_W}{N_B - 1} \sum_{p=1}^{N_B} \left(G_{\text{Ref}}(p) - \hat{G}_{\text{Ref}} \right)^2, \tag{5.30}$$

where $\hat{G}_{\text{Ref}}$ is defined in (5.2) and with $N_B - 1$ degrees of freedom. (Note that the term N_W in the numerator arises from the decomposition of the sum of squares [14].) The ratio of the two variances is then formed

$$F\left(s_B^2, s_W^2\right) = \frac{s_B^2}{s_W^2}, \tag{5.31}$$

where G_{Ref} is defined in (5.2) with $N_B - 1$ and $N_B(N_W - 1)$ degrees of freedom for the numerator and denominator, respectively. F_{α,n_1,n_2} is the α percentile (e.g., 95%) of the F distribution with n_1 and n_2 degrees of freedom.

If the test is *not significant*, that is

$$F\left(s_B^2, s_W^2\right) < F_{0.95,\, N_B-1,\, N_B(N_W-1)}, \tag{5.32}$$

then the uncertainty due to the number of mode-stirring samples and the uncertainty due to the lack of spatial uniformity both contribute to the uncertainty in the measurement of the reference. In this case, the uncertainty squared of the reference value may be given as the weighted average of s_B^2 and s_W^2 as

$$u_{\text{Ref}}^2 = \frac{1}{N_B N_W (N_B N_W - 1)} \sum_{n=1}^{N_W} \sum_{p=1}^{N_B} \left(G_{\text{Ref}}(n,p) - \hat{G}_{\text{Ref}} \right)^2, \tag{5.33}$$

with $N_B N_W - 1$ degrees of freedom. Note that in this case, the uncertainty essentially decreases as $\sqrt{1/N_B N_W}$. If the test *is significant*, then the uncertainty due to lack of spatial uniformity dominates, and the uncertainty squared in the reference value may be given as

$$u_{\text{Ref}}^2 = \frac{1}{N_B(N_W - 1)} \sum_{p=1}^{N_B} \left(G_{\text{Ref}}(p) - \hat{G}_{\text{Ref}} \right)^2$$

$$= \frac{1}{N_B} \sigma_{G_{\text{Ref}}}^2, \tag{5.34}$$

where the second equality is taken from (5.3) in Section 5.1.1 and with $N_B - 1$ degrees of freedom. Note that (5.33) may considerably underestimate u_{Ref}^2 when (5.34) applies. Essentially, what (5.34) states is that to correctly estimate the lack of spatial uniformity in the chamber, it is first necessary to ensure that an adequate number of mode-stirring samples are used in the stirring sequence. The reader is referred to [14] for more detail on this significance test to identify the component of uncertainty related to a specific reverberation chamber setup.

Equations (5.33) and (5.34) represent the component of uncertainty that accounts for the impact of the nonideal reverberation chamber setup on the estimate of G_{Ref} including the number of mode-stirring samples and lack of spatial uniformity, on the estimate of G_{Ref}. The more independent realizations that are made, the lower the uncertainty in the reference measurement. For example, in [15], 12 independent realizations of G_{Ref} are measured at the edges and center of the working volume ($N_B = 12$).

Both the reference measurement and the power-based measurement of the DUT (P_{meas} for TRP in Section 5.2.1, or P_{BSE} for TIS in Section 5.2.2) utilize the same OTA measurement setup, including chamber configuration, stirring sequence, and measurement antenna. Thus, this component should also be applied to the measurement of the DUT. In this case, the uncertainty squared for the DUT measurement should also account for the number of DUT measurements

$$u_{\text{DUT}}^2 \cong \frac{N_B}{N_{\text{B,DUT}}} u_{\text{Ref}}, \tag{5.35}$$

with the degrees of freedom corresponding to u_{Ref}^2 from (5.33) or (5.34), under the assumption of independence between measurements.

While uncertainty in the reference measurement is typically estimated from multiple measurements, in practice, typically only a single measurement of the DUT is performed to save time. The smaller value of $N_{\text{B,DUT}}$ for the DUT measurement will increase this component of uncertainty. This will be discussed further in Section 5.4.3 on combined uncertainty.

5.4.2   Relative uncertainty due to the type of stirring mechanisms

As discussed in Section 5.4.1, increased spatial correlation between samples and between mode-stirring sequences due to loading can increase the related uncertainty due to the use of a nonideal reverberation-chamber setup. Thus, this subsection is focused on methods for reducing the reverberation-chamber-specific components of uncertainty related to spatial correlation, rather than describing a specific component of uncertainty.

As illustrated in Section 5.1.4, certain mode-stirring mechanisms more effectively stir the fields than others for a given reverberation-chamber configuration. It is possible to quantify how the selection of these mechanisms affects uncertainty using a trial-and-error empirical approach to minimize the value of $\sigma_{G_{\text{Ref}}}$ from (5.3). With this method, a user may design an optimal stirring sequence for a particular chamber configuration. The uncertainty due to the lack of spatial uniformity will be the lowest for a combination of stirring mechanisms that most efficiently stir the fields within the chamber.

More predictive approaches have been explored, as well [4,13,38]. Using a more predictive approach, in [38], a measurement-based model of the correlation within a loaded chamber was developed and compared to the empirical approach. The correlation-based model identifies the most effective combination of mode-stirring mechanisms, providing levels of uncertainty on the order of those for an unloaded chamber with the minimal number of mode-stirring steps.

To derive the correlation-based model, the correlation coefficients between spatial mode-stirring samples are first identified. To do this, the S-parameters are measured on a fine spatial grid that covers the range of stirring mechanisms to be used for a particular chamber setup. The estimated values of the correlation terms are obtained from a single large $M_{max} \times N_{max}$ measurement and applied to the qth $M \times N$ subset stirring-sequence, where Q indicates the number of unique realizations that can be extracted from the large measurement without using samples more than once. That is, $Q = M_{max}N_{max}/MN$ realizations of an $M \times N$ measurement that are collected from the large $M_{max} \times N_{max}$ measurement.

Two 2D correlation matrices $\mathbf{R}_{M_{max}}$ and $\mathbf{R}_{N_{max}}$, with dimensions $M_{max} \times M_{max}$ and $N_{max} \times N_{max}$ and elements $r_{mm'}$ and $r_{nn'}$, are calculated for each stirring mechanism, respectively. The correlation coefficients r_M and r_N are then identified.

The correlation-based model for uncertainity takes the form [38]

$$\hat{u} = \frac{1}{\sqrt{MN}}\{[1 + (M - 1)r_M][1 + (N - 1)r_N]\}^{1/2}, \qquad (5.36)$$

where a stirring sequence consists of M samples of one stirring mechanism (such as platform stirring), N samples of a second stirring mechanism (such as paddle stirring), r_M and r_N describe the correlation between samples, and $\hat{u}$ is the relative uncertainty based on the model. This uncertainty is defined as

$$\hat{u}^2 = \frac{\sigma^2 \left(\langle|S_{21}|^2\rangle_{M,N}^{(q)}\right)}{\left((1/Q)\sum_{q=1}^{Q}\langle|S_{21}|^2\rangle_{M,N}^{(q)}\right)^2}. \qquad (5.37)$$

Here, without loss of generality, our given stirring sequence consists of M antenna positions and N paddle positions.

The correlation coefficients are applied to yield the effective number of samples that should be used in the determination of uncertainty for a given stirring sequence. In [38], it was shown that the uncertainty could be reduced nearly to that of an unloaded chamber $1/\sqrt{MN}$ when the optimal configuration (ratio of paddle stirring to platform stirring) was used.

To verify the model, it was compared to the empirically derived uncertainty due to lack of spatial uniformity in (5.34). For the comparison, a normalized version of the uncertainty due to lack of spatial uniformity given in (5.34) was utilized, given by

$$\sigma_{G_{\text{Ref, norm}}} = \frac{\sigma_{G_{\text{Ref}}}}{\hat{G}_{\text{Ref}}} \qquad (5.38)$$

The F test described in Section 5.4.1 was applied to ensure that the uncertainty due to lack of spatial uniformity dominated, as opposed to the number of samples within the mode-stirring sequence.

As an example from the study in [38], 60 mode-stirring samples comprised a stirring sequence. The number of paddle and platform samples was selected randomly (without replacement) from a large set of measured data that covered the range of stirring mechanisms (15 platform positions $\times$ 64 paddle positions $\times$ 4 heights). Ten ratios of platform-to-paddle samples ranging from 60:1 to 1:60 were chosen from this larger set of data. Four loading conditions were considered in three chambers, two of which are shown here.

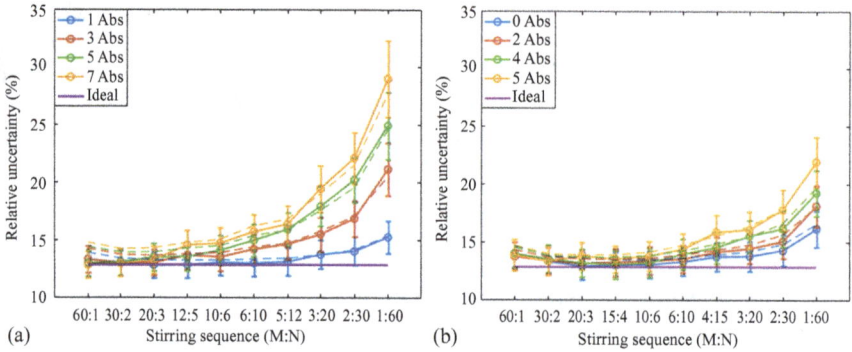

Figure 5.19   *Relative mode-stirring uncertainties corresponding to selected stirring sequences with a total of M × N = 60 measurements for (a) a large dual-paddle chamber, and (b) a small dual-paddle chamber. The modeled results are shown by the dashed lines (from [38])*

Figure 5.19(a) and (b) illustrate results for the empirical uncertainty method (solid lines) and the measurement-based model (dashed lines). The data error bars correspond to the standard error over a 150-MHz frequency band and the standard deviation over 20 repeat measurements of each realization of the empirical uncertainty analysis. The "ideal" uncertainty, if no correlation existed, is shown by the horizontal line that corresponds to $1/\sqrt{60}$ [29].

As shown in Figure 5.19(a) and (b), the predictive correlation-based model for uncertainty agrees well with the empirical approach. Further, the 20:3 ratio of platform-to-paddle stirring minimizes the relative uncertainty in the chamber configurations for all loading cases that were studied. The results clearly show that for both of these chambers, the use of only paddle stirring can lead to a significant increase in uncertainty. The appropriate combination of paddle and platform mode-stirring brings the uncertainty within a close range of that in a chamber with no correlation between mode-stirring samples, shown by the flat line in Figure 5.19(a) and (b). Design rules for optimizing a stirring sequence will be different for every chamber and set of mode-stirring mechanisms. However, it is clear that position stirring must be used in loaded chambers to obtain the lowest uncertainty.

5.4.3   Uncertainty due to difference in reference and DUT antennas

In OTA tests of wireless devices made using reverberation chambers, a reference measurement is used to estimate the reference power transfer function experienced by the DUT. That is, the measurement of G_{Ref} is intended to estimate G_{DUT} because G_{DUT} cannot be directly measured on a wireless device with an integrated antenna. However, differences in the radiation pattern of the reference antenna and DUT antenna can lead to differences between G_{Ref} and G_{DUT} (if it could be measured) in a loaded reverberation chamber. This is because the reduced spatial uniformity for loaded chambers combined with the potentially limited "field of view" of directional antennas may lead to different amounts of stirred vs. unstirred energy at the antenna element. As a result, the ratio of unstirred energy to stirred energy [the K-factor, defined in (5.23)] could be different for the reference and DUT measurements. This, in turn, could result in different values of the measurement-derived reference power transfer function, as well as different correlations and uncertainties. Note that the K-factor is spatially dependent in the heavily loaded reverberation chamber, and therefore, as with G_{Ref}, it is calculated as an average of measurements made in several locations.

This effect was studied in [33], where measurements of simulated DUTs with removable antennas allowed comparison of G_{Ref} and G_{DUT} for antennas having similar and different radiation patterns. The effect was studied for a "typical" OTA test configuration for cellular device testing in a large reverberation chamber. The test configuration included the following:

• A single-stepped rotating paddle (72 steps) and platform stirring (9 locations).
• A directional DRHA measurement antenna oriented to minimize direct coupling.
• Tests at Cellular-band (800–900 MHz), PCS-band (1.8–1.95 GHz), and mmWave frequencies (43–47 GHz). Results are shown here for the PCS-band frequencies.
• A reverberation chamber in both unloaded and loaded (with a 4-MHz CBW for the PCS band) configurations.

Several antenna types and configurations were studied, including two omnidirectional antennas ("Discone" and "Router") and a directional antenna that was randomly oriented with respect to the measurement antenna ("DRHA RPol"). These three antenna configurations had significant direct coupling to the measurement antenna and, hence, significant average K-factors (averaged over the stirring sequence). Two different directional antennas were cross polarized and pointed away from the measurement antenna at each platform location to minimize the unstirred coupling ("DRHA XPol" and a standard gain horn "Std Gain") resulting in lower average K-factors for these cases.

From this set of antennas, we denoted one antenna as a "DUT" antenna and measured the difference in the reference power transfer function for all other "reference" antennas. This difference was termed ΔG_{DUT} in [33]. For the results shown in Figure 5.20, we used the discone as a DUT antenna and measured the reference power transfer function using all other antenna types. In a second case, we used the standard gain horn as the DUT and all other antennas as the reference. In this way, we could study both "similar" reference/DUT antenna pairs (e.g., omni/omni) and "different" reference/DUT antenna pairs (e.g., omni/dir).

These studies showed several effects: (1) for unloaded chambers, differences in G_{Ref} and G_{DUT} are insignificant, as shown by the small value of ΔG_{DUT} for the "Absorber 0" cases in Figure 5.20 (the black xs). This is expected because the stirred energy dominates in unloaded chambers, which has the effect of eliminating antenna pattern [34]. (2) For loaded chambers in which the reference and DUT antennas had nominally similar radiation patterns, differences between G_{Ref} and G_{DUT} were small, as shown by the small value of ΔG_{DUT} for the "Router" antenna (as compared to the omnidirectional discone) in the "Absorber 7" case (first red × in Figure 5.20). (3) For

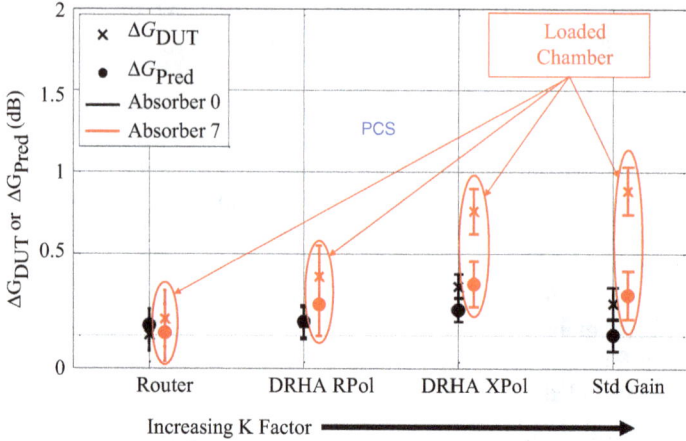

Figure 5.20   *The difference between a simulated device's power transfer function G_{DUT} and either (1) the reference power transfer function G_{Ref}, given by (ΔG_{DUT}) (xs), or (2) $G_{DUT, Pred}$ from (5.39) (ΔG_{Pred}) (dots). Two different loading cases are shown. 0 dB corresponds to the case where there is no error in the estimate of G_{DUT} (from [33])*

loaded chambers, if the antennas have different radiation patterns that lead to different K-factor values, the value of ΔG_{DUT} could be significant, ranging from 0.35 to 1 dB for the "Absorber 7" cases in Figure 5.20 (second through fourth red xs).

To minimize this effect, current practice is to select a reference antenna that produces a K-factor that is "similar" to that of the DUT. For example, azimuthally omnidirectional reference antennas are often used to estimate G_{Ref} for testing cellular devices, which nominally radiate in all directions azimuthally.

To account for differences in the (typically unknown) DUT antenna radiation pattern and the reference antenna, certain standard groups require inclusion of a component of measurement uncertainty [15]. As an example, for current below-6-GHz wireless devices in the PCS band, an omnidirectional-DUT/directional-reference pattern mismatch could lead to up to 0.35 dB error (corresponding to an difference in power of 0.084), as shown by the second-from-the-left red $\times$ in Figure 5.20. If we assume that a less-directional reference antenna would typically be used, we may consider 0.35 dB a worst-case bound. The systematic uncertainty component u_K would then be uniformly distributed, with a standard deviation of $u_K = \sigma_K \approx 0.084/\sqrt{12} \approx 0.1$ dB [49].

If the user has access to the DUT's antenna ports and the K-factor, κ, associated with a measurement of the wireless device in a specific chamber configuration, can be determined, a correction factor can be calculated to minimize the differences in transfer function values [33]. In this case, the DUT transfer function can be written as

$$G_{DUT, Pred} = G_{Ref}\frac{1 + \kappa_D}{1 + \kappa_R}, \tag{5.39}$$

where κ_D and κ_R are the K-factors for the DUT and reference measurements, respectively, averaged over a stirring sequence. One may obtain an improved estimate of the wireless metrics TRP and TIS by using $G_{\text{DUT, Pred}}$ in place of G_{Ref} if the K-factors are known. In either case, obtaining this value of uncertainty is complicated and still is a topic of research in the community. Its importance will likely become more significant as directional antennas will be used in future wireless systems.

5.4.4 Combined and expanded uncertainties

We have identified three components of uncertainty that are specifically due to the use of reverberation chambers in wireless OTA tests:

- The uncertainties u_{Ref} and u_{DUT} are due to the nonidealities associated with reverberation-chamber measurements, including both the limited number of samples in a stepped mode-stirring sequence and the lack of spatial uniformity in the chamber setup (Section 5.4.1). The difference between u_{Ref} and u_{DUT} is related to the number of reference and DUT measurements that are performed, respectively. In practice, several measurements are typically performed to assess u_{Ref} as part of the chamber pre-characterization, while often only one DUT measurement is performed, resulting in a higher value of u_{DUT}.
- The uncertainty u_{K} related to the difference in radiation patterns of the reference and DUT antennas (Section 5.4.3).

In the final uncertainty analysis, there will be many other components of uncertainty, such as those related to the measurement equipment and the estimated antenna efficiency. But, for an example, we illustrate combination of these three.

These components may be combined using a root-sum-of-square approach as

$$u_{\text{Combined}}^2 = \sqrt{u_{G_{\text{Ref}}}^2 + u_{\text{DUT}}^2 + u_{\text{K}}^2}$$

$$= \sqrt{\frac{1}{N_{\text{B}}}\sigma_{G_{\text{Ref}}}^2 + \frac{1}{N_{\text{DUT}}}\sigma_{\text{DUT}}^2 + \sigma_{\text{K}}^2}, \tag{5.40}$$

with the number of degrees of freedom v for the first two terms corresponding to (5.33) or (5.34) from Section 5.4.1. We assume that the degrees of freedom for the systematic error u_{K} are infinite [66]. Because the components have different degrees of freedom, we use the Welch–Satterthwaite equation to find the effective degrees of freedom due to the pooled variance as [66]

$$v_{\text{eff}} \cong \frac{\left(\sum_{i=1}^{n}\kappa_i u_i^2\right)^2}{\sum_{i=1}^{n}\left(\left(\kappa_i u_i^2\right)^2 / v_i\right)}, \tag{5.41}$$

where

$$\kappa_i = \frac{1}{v_i + 1}.$$

Because the systematic uncertainty term is assumed to have infinite degrees of freedom, this term drops out from the summations in (5.41). Substituting $v_1 = v_2 =$

$N_B - 1$ and $\kappa_1 = \kappa_2 = 1/N_B$ into (5.41) and simplifying, we end up with an effective number of degrees of freedom of

$$\nu_{\text{eff}} \cong \frac{((1/N_B) + (1/N_{\text{DUT}}))^2}{(1/N_B)^2 + (1/N_{\text{DUT}})^2} \, (N_B - 1) \,. \tag{5.42}$$

To find the expanded uncertainty U_{95} corresponding to a 95% confidence level, the coverage factor k_{95} is determined from the effective degrees of freedom from (5.42) using the procedure recommended in Appendix B3 of [49]. That is, $k_{95} = t_{95}(\nu)$, where $t_{95}(\nu)$ is the two-sided 95th percentile of the Student's t-distribution having ν degrees of freedom. Note that, depending on the number of measured samples of N_B and N_{DUT}, ν can be small. This would make the coverage factor larger than two. Finally, U_{95} is found as

$$U_{95,\text{Combined}} = k_{95} u_{\text{Combined}} \,. \tag{5.43}$$

5.5   Conclusion

In this chapter, we covered some of the key issues related to the use of heavily loaded reverberation chambers for OTA testing of wireless-device performance. Specifically, we addressed cases where the communication signal is spread over a wide modulation bandwidth and must be demodulated and cases in which a particular PDP is to be replicated. Metrics used to characterize the chamber configuration, such as the reference power transfer function, spatial uniformity, isotropy, and chamber decay time, are not significantly different from those used in prior work of the EMC/EMI community. However, their application to the case where significant correlation exists between frequencies and spatial samples requires additional consideration. As such, we paid a great deal of attention to the assessment correlation and its impact on uncertainty in the estimate of power-based metrics such as TRP or TIS.

Throughout, we attempted to provide the theoretical basis for current standardized test approaches of cellular wireless devices, including a summary of the OTA test methods currently used for TRP and TIS of single-antenna devices. We briefly touched on the use of reverberation chambers in the emulation of specific multipath channels. This is expected to be a growth area in the future as wireless devices more commonly use multiple antenna systems designed for optimal operation in multipath environments. Such environments are provided naturally by the reverberation chamber.

Finally, we provide a summary of components of uncertainty specifically related to the use of heavily loaded reverberation chambers for OTA test of wireless devices. While this development focused on cellular-enabled wireless devices, the techniques and uncertainty analyses can be extended to other types of wireless devices in a straightforward way. With the increased prevalence of wireless devices for Internet-of-Things and machine-to-machine communications, we anticipate that the use of reverberation-chamber methods for efficient OTA test will continue to grow in the years ahead.

5.6 Acknowledgments

The authors are grateful for the contributions of current and former NIST colleagues and affiliates, many of whom co-authored papers from which this material was developed, including Christopher Holloway, John Ladbury, Benjamin Jamroz, Dylan Williams, Michael Frey, Maria Becker, Ryan Pirkl, Damir Senic, Robert Jones, Stephan van de Beek, Joop Ann Den Toorn, Jos Dortmans, Anouk Hubrechson, Thomas Meurs, Matt North, Vincent Neylon, Christian Lotback, Haider Shah, Arvand Homer, and Catherine Weldon.

References

[1] Andrieu G, Ticaud N, Lescoat F, *et al.* Fast and Accurate Assessment of the "Well Stirred Condition" of a Reverberation Chamber From S_{11} Measurements. IEEE Trans Electromagn Compat. 2019;61(4):974–82.

[2] Chen X, Kildal P, Orlenius C, *et al.* Channel Sounding of Loaded Reverberation Chamber for Over-the-Air Testing of Wireless Devices: Coherence Bandwidth Versus Average Mode Bandwidth and Delay Spread. IEEE Antennas Wireless Propag Lett. 2009;8:678–81.

[3] Remley KA, Floris SJ, Shah HA, *et al.* Static and Dynamic Propagation-Channel Impairments in Reverberation Chambers. IEEE Trans Electromagn Compat. 2011;53(3):589–99.

[4] Kildal P, Chen X, Orlenius C, *et al.* Characterization of Reverberation Chambers for OTA Measurements of Wireless Devices: Physical Formulations of Channel Matrix and New Uncertainty Formula. IEEE Trans Antennas Propag. 2012;60(8):3875–91.

[5] Corona P, Ferrara G, and Migliaccio M. Reverberating Chamber Electromagnetic Field in Presence of an Unstirred Component. IEEE Trans Electromagn Compat. 2000;42(2):111–5.

[6] Hill DA and Ladbury JM. Spatial-Correlation Functions of Fields and Energy Density in a Reverberation Chamber. IEEE Trans Electromagn Compat. 2002;44(1):95–101.

[7] Orlenius C, Kildal P, and Poilasne G. Measurements of total isotropic sensitivity and average fading sensitivity of CDMA phones in reverberation chamber. 2005 IEEE Antennas and Propagation Society International Symposium. IEEE, Washington, DC, USA; 3–8 Jul. 2005. p. 409–12 Vol. 1A.

[8] Holloway CL, Hill DA, Ladbury JM, *et al.* On the Use of Reverberation Chambers to Simulate a Rician Radio Environment for the Testing of Wireless Devices. IEEE Trans Antennas Propag. 2006;54(11):3167–77.

[9] Delangre O, Doncker PD, Lienard M, *et al.* Delay Spread and Coherence Bandwidth in Reverberation Chamber. Electron Lett. 200;44(5):32–9.

[10] Genender E, Holloway CL, Remley KA, *et al.* Simulating the Multipath Channel With a Reverberation Chamber: Application to Bit Error Rate Measurements. IEEE Trans Electromagn Compat. 2010;52(4):766–77.

[11] Sanchez-Heredia JD, Gruden M, Valenzuela-Valdes JF, *et al.* Sample-Selection Method for Arbitrary Fading Emulation Using Mode-Stirred Chambers. IEEE Antennas Wireless Propag Lett. 2010;9:40–12

[12] Chen X. Experimental Investigation of the Number of Independent Samples and the Measurement Uncertainty in a Reverberation Chamber. IEEE Trans Electromagn Compat. 2013;55(5):816–24.

[13] Remley KA, Pirkl RJ, Shah HA, *et al.* Uncertainty From Choice of Mode-Stirring Technique in Reverberation-Chamber Measurements. IEEE Trans Electromagn Compat. 201;55(6):1022–30

[14] Remley KA, Wang CM, Williams DF, *et al.* A Significance Test for Reverberation-Chamber Measurement Uncertainty in Total Radiated Power of Wireless Devices. IEEE Trans Electromagn Compat. 2016;58(1):207–19.

[15] CTIA Certification. Test Plan for Wireless Large-Form-Factor Device Over-the-Air Performance. CTIA; 2019.

[16] Orlenius C, Franzen M, Kildal P, *et al.* Investigation of heavily loaded reverberation chamber for testing of wideband wireless units. 2006 IEEE Antennas and Propagation Society International Symposium. IEEE, Albuquerque, NM, USA; 9–14 Jul. 2006. p. 3569–72.

[17] Horansky RD and Remley KA. Flexibility in Over-the-Air Testing of Receiver Sensitivity With Reverberation Chambers. IET Microwaves Antennas Propag. 2019;13:2590–7.

[18] Wolfgang A, Orlenius C, and Kildal P. Measuring output power of Bluetooth devices in a reverberation chamber. IEEE Antennas and Propagation Society International Symposium Digest Held in Conjunction With: USNC/CNC/URSI North American Radio Sci Meeting (Cat No. 03CH37450). IEEE, Columbus, OH, USA; 22–27 Jun. 2003. p. 735–8 Vol. 4.

[19] Monebhurrun V and Letertre T. Total radiated power measurements of WiFi devices using a compact reverberation chamber. 2009 20th International Zurich Symposium on Electromagnetic Compatibility. IEEE, Zurich, Switzerland; 12–16 Jan. 2009. p. 65–8.

[20] Rudander JH, Ikram-e-Khuda, Kildal PS, *et al.* Measurements of RFID Tag Sensitivity in Reverberation Chamber. IEEE Antennas Wireless Propag Lett. 2011;10:1345–8.

[21] Svedjenäs P, Dong W, Arvidsson K, *et al.* OTA device sensitivity in the presence of interference measured in a reverberation chamber. 2014 International Symposium on Electromagnetic Compatibility. IEEE, Gothenburg, Sweden; 1–4 Sep. 2014. p. 328–31.

[22] Delangre O, Doncker PD, Horlin F, *et al.* Reverberation chamber environment for testing communication systems: Applications to OFDM and SC-FDE. 2008 IEEE 68th Vehicular Technology Conference. IEEE, Calgary, BC, Canada; 21–24 Sep. 2008. p. 1–5.

[23] Valenzuela-Valdes JF, Martinez-Gonzalez AM, and Sanchez-Hernandez DA. Emulation of MIMO Nonisotropic Fading Environments With Reverberation Chambers. IEEE Antennas Wireless Propag Lett. 2008;7:325–8.

[24] Kildal P and Rosengren K. Correlation and Capacity of MIMO Systems and Mutual Coupling, Radiation Efficiency, and Diversity Gain of Their Antennas: Simulations and Measurements in a Reverberation Chamber. IEEE Commun Mag. 2004;42(12):104–12.

[25] Valenzuela-Valdes JF, Garcia-Fernandez MA, Martinez-Gonzalez AM, *et al.* The Influence of Efficiency on Receive Diversity and MIMO Capacity for Rayleigh-Fading Channels. IEEE Trans Antennas Propag. 2008;56(5): 1444–50.

[26] Kildal P, Orlenius C, and Carlsson J. OTA Testing in Multipath of Antennas and Wireless Devices With MIMO and OFDM. Proc IEEE. 2012;100(7):2145–57.

[27] Mora-Andreu M and Sánchez-Hernández DA. Reverberation chamber results on 3GPP/CTIA LTE MIMO OTA 2013 round robin tests using different channel models—A comparison of performance. The 8th European Conference on Antennas and Propagation (EuCAP 2014). IEEE, The Hague, Netherlands; 6–11 Apr. 2014. p. 3684–7.

[28] Becker MG, Horansky RD, Senic D, *et al.* Spatial channels for wireless over-the-air measurements in reverberation chambers. 12th European Conference on Antennas and Propagation (EuCAP 2018). IET, London, UK; 9–13 Apr. 2018. p. 1–5.

[29] Kostas JG and Boverie B. Statistical Model for a Mode-Stirred Chamber. IEEE Trans Electromagn Compat. 1991;33(4):366–70.

[30] Hussain A and Kildal P. Study of OTA throughput of LTE terminals for different system bandwidths and coherence bandwidths. 2013 7th European Conference on Antennas and Propagation (EuCAP). IEEE, Gothenburg, Sweden; 8–12 Apr. 2013. p. 312–4.

[31] Micheli D, Barazzetta M, Carlini C, *et al.* Testing of the Carrier Aggregation Mode for a Live LTE Base Station in Reverberation Chamber. IEEE Trans Veh Technol. 2017;66(4):3024–33.

[32] Holloway CL, Shah HA, Pirkl RJ, *et al.* Reverberation Chamber Techniques for Determining the Radiation and Total Efficiency of Antennas. IEEE Trans Antennas Propag. 2012;60(4):1758–70.

[33] Remley KA, Pirkl RJ, Wang CM, *et al.* Estimating and Correcting the Device-Under-Test Transfer Function in Loaded Reverberation Chambers for Over-the-Air Tests. IEEE Trans Electromagn Compat. 201;59(6):1724–34.

[34] Hill DA. Electromagnetic Fields in Cavities: Deterministic and Statistical Theories. Piscataway, NJ, USA: Wiley-IEEE Press; 2009.

[35] Lemoine C, Besnier P, and Drissi M. Investigation of Reverberation Chamber Measurements Through High-Power Goodness-of-Fit Tests. IEEE Trans Electromagn Compat. 2007;49(4):745–55.

[36] Chen X, Kildal P, and Lai S. Estimation of Average Rician K-Factor and Average Mode Bandwidth in Loaded Reverberation Chamber. IEEE Antennas Wireless Propag Lett. 2011;10:1437–40.

[37] van de Beek S, Remley KA, Holloway CL, *et al.* Characterizing large-form-factor devices in a reverberation chamber. 2013 International Symposium

on Electromagnetic Compatibility. IEEE, Brugge, Belgium; 2–6 Sep. 2013. p. 375–80.

[38]   Becker MG, Frey M, Streett S, *et al.* Correlation-Based Uncertainty in Loaded Reverberation Chambers. IEEE Trans Antennas Propag. 2018;66(10): 5453–63.

[39]   Jakes WC. Multipath Interference. In: Jakes WC, editor. Microwave Mobile Communications. New York, NY, USA: IEEE Press; 1974.

[40]   Remley KA, Dortmans J, Weldon C, *et al.* Configuring and Verifying Reverberation Chambers for Testing Cellular Wireless Devices. IEEE Trans Electromagn Compat. 2016;58(3):66–72.

[41]   Durgin GD. Space-Time Wireless Channels. Saddle River, NJ, USA: Prentice Hall Press; 2003.

[42]   Holloway CL, Shah HA, Pirkl RJ, *et al.* Early Time Behavior in Reverberation Chambers and Its Effect on the Relationships Between Coherence Bandwidth, Chamber Decay Time, RMS Delay Spread, and the Chamber Buildup Time. IEEE Trans Electromagn Compat. 2012;54(4):714–25.

[43]   Chen X. Scaling Factor for Turn-Table Platform Stirring in Reverberation Chamber. IEEE Antennas Wireless Propag Lett. 2017;16:2799–802.

[44]   Pirkl RJ, Remley KA, and Patane CSL. Reverberation Chamber Measurement Correlation. IEEE Trans Electromagn Compat. 2012;54(3):533–45.

[45]   Arnaut LR. Optimizing Low-Frequency Mode Stirring Performance Using Principal Component Analysis. IEEE Trans Electromagn Compat. 2014;56(1):3–14.

[46]   Burger WTC, Remley KA, Holloway CL, *et al.* Proximity and antenna orientation effects for large-form-factor devices in a reverberation chamber. 2013 IEEE International Symposium on Electromagnetic Compatibility. IEEE, Brugge, Belgium; 5–9 Aug. 2013. p. 671–6.

[47]   Aan Den Toorn J, Remley KA, Holloway CL, *et al.* Proximity-Effect Test for Lossy Wireless-Device Measurements in Reverberation Chambers. IET Sci Meas Technol. 2015;9(5):540–6.

[48]   International Electrotechnical Commission. IEC 61000-4-21 Electromagnetic Compatibility (EMC) – Part 4-21: Testing and Measurement Techniques – Reverberation Chamber Test Methods. 2nd Edition; 2011.

[49]   Joint Committee for Guides in Metrology. Evaluation of Measurement Data— Guide to the Expression of Uncertainty in Measurement. Sevres, France: International Bureau of Weights and Measures (BIPM); 2008.

[50]   Arnaut LR. Operation of Electromagnetic Reverberation Chambers With Wave Diffractors at Relatively Low Frequencies. IEEE Trans Electromagn Compat. 2001;43(4):637–53.

[51]   Kildal P-S and Carlsson C. Detection of a Polarization Imbalance in Reverberation Chambers and How to Remove It by Polarization Stirring When Measuring Antenna Efficiencies. Microwave Opt Technol Lett. 2002;34(2): 145–9.

[52]   Morrison DF. Multivariate Statistical Methods. New York, NY, USA: McGraw-Hill Book Company; 1967.

[53] NIST/SEMATECH. e-Handbook of Statistical Methods, Section 1.3.5.15 "Chi-Square Goodness-of-Fit Test"; 2012. Available from: http://www.itl.nist.gov/div898/handbook/eda/section3/eda35f.htm.

[54] CTIA Certification. Test Plan for Wireless Device Over-the-Air Performance; 2018.

[55] Chen X, Kildal P, Carlsson J, *et al.* Comparison of Ergodic Capacities From Wideband MIMO Antenna Measurements in Reverberation Chamber and Anechoic Chamber. IEEE Antennas Wireless Propag Lett. 2011;10:446–9.

[56] Patané CL, Skårbratt A, Rehammar R, *et al.* On the use of reverberation chambers for assessment of MIMO OTA performance of wireless devices. 2013 7th European Conference on Antennas and Propagation (EuCAP). IEEE, Gothenburg, Sweden; 8–12 Apr. 2013.

[57] Krauthauser HG. On the Measurement of Total Radiated Power in Uncalibrated Reverberation Chambers. IEEE Trans Electromagn Compat. 2007;49(2): 270–9.

[58] Orlenius C, Lioliou P, Franzen M, *et al.* Measurements of total radiated power of UMTS phones in reverberation chamber. The Second European Conference on Antennas and Propagation, EuCAP 2007. IET, Edinburgh, UK; 11–16 Nov. 2007.

[59] Qi Y, Yang G, Liu L, *et al.* 5G Over-the-Air Measurement Challenges: Overview. IEEE Trans Electromagn Compat. 2017;59(6):1661–70.

[60] Horansky RD, Meurs TB, North MV, *et al.* Statistical considerations for total isotropic sensitivity of wireless devices measured in reverberation chambers. 2018 International Symposium on Electromagnetic Compatibility (EMC EUROPE). IEEE, Amsterdam, Netherlands; 27–30 Aug. 2018.

[61] Senic D, Remley KA, Wang CJ, *et al.* Estimating and Reducing Uncertainty in Reverberation-Chamber Characterization at Millimeter-Wave Frequencies. IEEE Trans Antennas Propag. 2016;64(7):3130–40.

[62] Matolak DW, Remley KA, Holloway C, *et al.* Outdoor-to-Indoor Channel Dispersion and Power-Delay Profile Models for the 700-MHz and 4.9-GHz Bands. IEEE Antennas Wireless Propag Lett. 2016;15:441–3.

[63] Fielitz H, Remley KA, Holloway CL, *et al.* Reverberation-Chamber Test Environment for Outdoor Urban Wireless Propagation Studies. IEEE Antennas Wireless Propag Lett. 2010;9:52–6.

[64] Rosengren K and Kildal P. Theoretical study of angular distribution of plane waves in a small reverberation chamber for simulating multipath environment and testing mobile phones. IEEE Antennas and Propagation Society International Symposium 2001 Digest Held in Conjunction With: USNC/URSI National Radio Science Meeting (Cat No. 01CH37229). IEEE, Boston, MA, USA; 8–13 Jul. 2001.

[65] Pirkl RJ and Remley KA. Experimental Evaluation of the Statistical Isotropy of a Reverberation Chamber's Plane-Wave Spectrum. IEEE Trans Electromagn Compat. 2014;56(3):498–509.

[66] NIST/SEMATECH. e-Handbook of Statistical Methods; 2019. Available from: https://www.itl.nist.gov/div898/handbook.

Chapter 6

From material absorption to dosimetry for exposure of animals in reverberation chambers

Philippe Besnier[1]

This chapter deals with energy absorption in reverberation chamber (RC) and discusses two specific but related applications. Losses clearly play a fundamental role with regard to RC performance. It controls the modal width, which is proportional to the modal overlap. This is defined as the ratio between average modal width and average distance (in Hz) between consecutive natural modes. A too small modal overlap (much smaller than 1) does not provide a well-operated RC [1,2]. In the case of limited modal density and weak modal overlap, adding absorbing materials may lower the frequency of operation of a well-operated RC. However, it lowers the energy available for the RC. This may be a strong limitation for tests requiring high power densities such as immunity or dosimetry tests. Moreover, it reduces the composite Q-factor of the RC and may induce a restriction of the working volume.

Nevertheless, throughout this chapter, we assume that the conditions of ideal random field are met despite some changes of absorbing conditions. This is a pre-requirement for the applications which are further described in this chapter. For these applications, the frequency range of operation corresponds to largely oversized cavities with high modal density, and we assume that the measurement setup has a negligible impact with regard to the working volume delimitation.

The first application deals with material characterization and more specifically discusses the efficiency evaluation of absorbing materials in a RC. It is interesting to note that any change of absorption level in a RC may be sensed through a receiving probe or a receiving antenna. Averaging energy over RC states (i.e. realizations) is virtually equivalent to sample energy over space according to the ergodicity hypothesis. It is, therefore, possible to detect the presence of a piece of absorbing material in a simple way. The sensitivity and uncertainty of measurements depends on the number of available states. These measurements depend on the quality factor or related physical features such as the decay time constant of the chamber or its coherence bandwidth. Given a correct estimation of the composite Q-factor we show how to retrieve the so-called absorbing area cross section (AACS) and prove that this AACS is consistent with the size, shape and intrinsic electromagnetic (EM) features of the

[1]INSA Rennes, CNRS, IETR – UMR 6164, Rennes, France

tested material. It is also consistent with the position of this piece of material within the chamber.

The second application deals with bioelectromagnetics. By virtue of the properties of loaded RCs highlighted for AACS characterization, we easily conclude that RCs have a significant potential for experimental dosimetry. It is specifically attractive to provide calibrated dosimetry dedicated to animal exposure since they are allowed to move freely in their cage, under EM stress. Laboratory experiments with animal models are indeed useful to assess cancer-related risks of exposure to various sources of EM fields. Moreover, at millimeter waves (mm-waves), a growing concern with regard to advanced and future communication standards (fifth generation (5G) and beyond), a RC offers a relatively compact setup to deliver high electric field strength amplitudes over its almost entire volume. Using a dedicated phantom, whose EM intrinsic properties are precisely adjusted to mimic the behavior of skins at the relevant frequency range, we show that the temperature rise at its surface is consistent with the solution of the bio-heat transfer equation. This consistence is checked from the knowledge of the Q-factor of the phantom-loaded chamber, or the knowledge of the Q-factor of the empty chamber and the AACS of the phantom. Imaging the temperature rise at the phantom's surface offers an efficient calibration procedure.

This chapter is organized as follows: the composite Q-factor playing a key role for both applications, we first recall the roles of losses and define the composite Q-factor within a RC. We then introduce a specific measurement procedure for the Q-factor using different combination of data from scattering parameters measured at two antenna ports. This enables to check the quality of estimation while limiting its statistical uncertainty. The second part of the chapter describes the effects of the insertion of an absorbing material in the RC. In particular, we relate the observed change of the Q-factor to the AACS coefficient. This coefficient is shown to be related to the average transmission coefficient of a thick absorber and is proportional to its surface for simple geometries. The third part of the chapter extends the second part in relating the AACS to the heat deposition and describe how a dedicated mm-wave RC provides a calibrated exposure system for in-vivo experiments.

6.1   Losses and Q-factor

6.1.1   Role of losses in a RC

A lossless cavity is a limit case never met in practice. This is not to be confused with an ideal case from the point of view of ideal random fields within a RC. Without accounting for losses, a source of EM field can only be coupled to a single mode giving rise to an increasing energy level in the RC, if the source signal is applied continuously. In these virtual conditions the EM field is real since any mode is described by real eigenvalues and eigenvectors. Introducing losses, a source of EM fields may be more easily coupled to a mode that may be significantly excited over some bandwidth. The EM field is then complex and associated with complex eigenvalues and eigenvectors. If the modal bandwidth become of the order of magnitude of the frequency spacing between modes, the probability of excitation of one or several modes increases. The ratio of the average modal bandwidth to the average frequency spacing defines the average modal overlap. Through mechanical or frequency stirring mechanisms, many more modes may be involved. Above a sufficient modal overlap, the number of these significantly excited modes is sufficient enough to reach, asymptotically, the conditions of an ideal random field. This ideal random field may be represented, asymptotically, by an infinite addition of incoming plane waves with random phase, polarization and incident angles, according to Hill theory [3].

It turns out that the modal overlap is inversely proportional to the quality factor of the chamber. Therefore, losses dictate the frequency range for which this asymptotic regime is reached. Much more details and rigorous analysis about the role of losses in RC may be found in [1,4,5]. The viewpoint of universal features of chaotic properties in the frame of random matrix theory also offers a deep understanding of statistic fluctuation of fields in a RC according to frequency and particular with regard to the role of losses and of the modal density that both determine the modal overlap [2,6]. Beyond standards for EMC tests, more advanced techniques to settle appropriate conditions that guarantee that the RC behavior is close enough to Hill statistical description were proposed in relevant publications [1,2,6–9]. These techniques include fast and accurate approaches that may replace current standard calibration techniques [10,11].

Losses are closely related to the estimation of the Q-factor. Therefore, its estimation is a key when assessing the RC performance. We recall some definitions and describe a technique to estimate it with reduced statistical fluctuations. This technique is then readily used for the applications described in the following sections.

6.1.2 Definition of the Q-factor

According to the above discussion, the composite or effective Q-factor [4,5] is a quantity that enables to predict the behavior of a RC according to modal density and the probability of excitation of multiple modes. Even more importantly, its estimation also enables to calibrate the field intensity in the chamber associated to a prescribed injected power at the input of a transmitting antenna in the RC or vice-versa.

The Q-factor is said to be "composite" (or sometimes "effective") for two main reasons that may be confused with one another. The primary reason is that losses may come from different mechanisms. The second reason is that several modes are combined in the chamber, and the Q-factor of single modes cannot be retrieved. As far as RCs are concerned, the usual definition of the composite Q-factor corresponds to the average EM energy stored in the cavity over all states of the chamber obtained during the stirring process, per unit of time, and divided by the total dissipated power in the chamber at steady state. The steady state is reached once the dissipated power balances the transmitted power in the chamber. The effective or composite Q-factor used throughout this chapter follows this definition. However, it should be pointed out that the Q-factor may also be defined, in a more physical sounding approach, as a statistic of individual Q-factors accounting for modes generated at each RC state. It then appears as a random variable whose properties were analyzed theoretically [12,13] and from a large set of experiments [14,15].

The composite Q-factor links the ensemble average energy over different states of the chamber and the overall dissipated power as follows

$$Q_g = 2\pi f_0 \frac{\mathbb{E}[W_E]}{\mathbb{E}[P_d]} \tag{6.1}$$

This definition of the quality factor Q_g applies to steady state harmonics fields at frequency f_0. The notation $\mathbb{E}$ stands for the expected value. The term W_E, stands for the EM energy in the chamber and P_d represents the active power consumed in the RC.

6.1.3   Origin of losses and their contribution to the Q-factor

The effective quality factor Q_g is said to be a composite factor since it is composed of several losses mechanisms. Should a chamber be absolutely empty, the first cause of losses are metallic losses spread all over the inner surfaces of the Faraday cage. These losses are associated with the so-called wall quality factor, Q_w, which describes losses due to the residual resistance of metallic materials to electromagnetically induced currents.

However, a RC cavity must be coupled to a transmitted antenna and, possibly, to a receiving antenna. Both antennas absorb some energy, a transmitting antenna in a RC being also a receiving antenna in the diffuse field generated by itself. Losses at these antennas come from their intrinsic metallic or dielectric losses. These losses are distinct from losses in free space conditions, especially due to the current induced from the diffuse field on the structural parts of the antenna (including any supporting device). However, if they are matched (according to their free-space impedance), highly efficient and not in line-of-sight of each other, losses come significantly from power consumption at their load. Since a transmitting antenna imposes local boundary conditions for the field, it couples to modes with the same local pattern. In case of linearly polarized antennas, the polarization is forced at the transmitting antenna, whereas it is random at the receiving antenna. Therefore, it turns out that the receiving antenna is sensitive to half of the incoming energy density. The transmitting antenna is sensitive to the total energy density and absorbs twice more energy than the receiving antenna. Their respective Q_{Tx} and Q_{Rx} quality factors are, therefore, related to one another by $Q_{Rx} = Q_{Tx}/2$.

The last contribution we may describe is that coming from any object, including a possible device under test, that absorbs energy. We limit ourselves to a single object and name as Q_{obj} its own quality factor. The composition of all these losses may be justified as follows: at steady state, the required and constant active power generated in the RC by the source of energy, is absorbed through the different losses mechanisms described before. Assuming these mechanisms are not dependent from each other, this dissipated power is the sum of the dissipated power in antennas, walls (i.e. Q_w), and the object (i.e. Q_{obj}). Therefore, we may write Q_g as

$$\frac{1}{Q_g} = \frac{1}{Q_{Tx}} + \frac{1}{Q_{Rx}} + \frac{1}{Q_w} + \frac{1}{Q_{obj}} \tag{6.2}$$

It obviously means that the component having the lowest Q-factor absorbs the highest part of the energy.

6.1.4 Q-Factor measurements

Two different types of measurements may be provided to determine the quality factor of an oversized cavity. The composite Q-factor is related to the decay time constant of the chamber. Its relationship with the Q-factor is linked to the charge or the discharge time of resonators to reach equilibrium. Suppose that the source of energy is switched off once the steady state is reached. The elementary rate of change of the dissipated power P_d over a small time interval dt is compensated by the elementary rate of change of the stored energy (dW_E) in the chamber

$$dW_E = -P_d dt \tag{6.3}$$

Therefore, according to the definition in (6.1), it implies that $dW_E(t)$ decays exponentially with a time constant τ such as

$$dW_E(t) = \exp \frac{-t}{\tau} = \exp \frac{-t}{Q_g/\omega} \tag{6.4}$$

A direct measurement of the decay time after switching off a constant wave source at pulsation ω is an option. A second option consists in retrieving the impulse response from a spectral measurement over some bandwidth, using an inverse Fourier transform. A nice application of such techniques was provided in [16] for antenna efficiency measurements. Hereafter, we focus on techniques dedicated to narrowband measurements.

6.1.4.1 Test setup for Q-factor measurements

The theory of measurements of the Q-factor of a RC using a vector network analyzer (VNA) is presented. It is based on a standard configuration of measurements in a RC where two antennas are put in the chamber. Their location in the RC is arbitrary but not in line-of sight of each other (negligible contribution of unstirred energy). Both antennas are basically matched and highly efficient. They are connected to two ports of a VNA through coaxial cables and adequate through-wall interfaces. We illustrate in Figure 6.1 a typical configuration of measurement in the IETR large RC. Calibration is performed at the output connector of cables plugged on the input connectors of the antennas.

We note P_{inj} the arbitrary level of injected power from the internal generator of the VNA to any one of its active ports. This power is supposed to be constant through all operations described hereafter.

We also assume that the transmitted power by the emitting antenna that would have been radiated in free space is identical in the RC. In other words, we suppose that the backscattered energy from the RC couples to the antenna at a time of arrival which exceeds the settling time of the radiation. This settling time corresponds to the steady state of the antenna current distribution, giving birth to radiation. Therefore, any signal coupled from the RC to the transmitting antenna is considered an additional contribution.

Figure 6.1   A typical test setup for the measurement of the Q-factor with two log-periodic antennas in the IETR large RC

6.1.4.2   *Q*-Factor measurements using the full scattering matrix

Let the port 1 be the active port where the power is supplied and let us name "antenna 1" the antenna connected to this port. The other antenna is connected to port 2 and both roles may be interchanged, the propagation medium being linear and reciprocal. According to the "free space" hypothesis mentioned above, the transmitted power in the chamber P_t is given by

$$P_t = (1 - |\langle S_{11} \rangle|^2)\eta_1 P_{inj} \qquad (6.5)$$

In this expression $\langle\ \rangle$ corresponds to an ensemble average over the states of the chamber during the stirring process. In the hypothesis of the ideal random field, the $\langle S_{11} \rangle$ parameter (that definitely includes backscattering effects) tends to the free space reflection coefficient of the transmitting antenna ($\langle S_{11} \rangle \approx S_{11}^{fs}$). The term η_1 is the radiation efficiency of the considered antenna and accounts for intrinsic metallic and dielectric losses of its constitutive materials.

At the port 2 connected to antenna 2, the received power at the terminal of that antenna ($P_{rec,T}$) is written as

$$P_{rec,T} = (1 - |\langle S_{22} \rangle|^2)\eta_2 P_{rec} \qquad (6.6)$$

where η_2 is the radiation efficiency of the receiving antenna 2 and P_{rec} its average received power, if it would be ideally efficient and perfectly matched.

According to the hypothesis of an ideal random field, the average received power by the antenna, $\langle P_{rec} \rangle$, writes:

$$\langle P_{rec} \rangle = \frac{\lambda^2}{8\pi} \frac{E^2}{Z_0} = \frac{\langle P_{rec,T} \rangle}{(1 - |\langle S_{22} \rangle|^2)\eta_2} \qquad (6.7)$$

In this expression Z_0 may be considered the free space impedance associated with each individual plane wave composing the whole spectrum, i.e. $Z_0 = 120\pi$. The effective area $\lambda^2/8\pi$ is that of a linearly polarized antenna in a diffuse field (i.e. of uniformly distributed polarization) and E is the average total electric field strength magnitude in the chamber, i.e. $\langle|E_T|\rangle$.

According to (6.7), the average power received at the terminals of the receiving antenna, $P_{rec,T}$, is now related to the average magnitude of the electric-field strength in the RC. The energy in the volume of the chamber in the hypothesis of the ideal random field is uniformly spread in the volume V of the RC and, therefore,

$$\mathbb{E}[W_E] = \varepsilon E^2 V \tag{6.8}$$

with ε the medium permittivity. At steady state the transmitted power in the chamber equals the dissipated power

$$\mathbb{E}[P_d] = P_t \tag{6.9}$$

Using a VNA, we relate the scattering parameter S_{21} to the received power at the terminals of the receiving antenna at port 2 which is proportional to the injected power at port 1

$$\langle|S_{21}|^2\rangle = \frac{\langle P_{rec,T}\rangle}{P_{inj}} \tag{6.10}$$

For compactness of the notation, we then introduce the mismatch factors m_x, $x = 1, 2$ of each antenna, where $m_x = (1 - |\langle S_{xx}\rangle|^2)$. From (6.1), (6.8) and (6.9), we obtain

$$Q_g = \frac{\omega \varepsilon E^2 V}{P_t} \tag{6.11}$$

Using expression (6.11) and the sequence of equations (6.5), (6.6), (6.7) and (6.10), we get the following expression for the estimation of the composite Q-factor as a function of the full scattering matrix parameters

$$Q_g = \frac{16\pi^2 V}{\lambda^3} \frac{\langle|S_{21}|^2\rangle}{m_1 m_2 \eta_1 \eta_2} \tag{6.12}$$

According to the ideal random field hypothesis, the real and imaginary components of the scattering parameter S_{21} follow a centered Gaussian distribution and the variance of this parameter is proportional to the energy in the RC. The sample size being limited, the estimation of the mean of S_{21} (i.e. $\langle S_{21}\rangle$) deviates from zero. The variance estimation of S_{21} in (6.12) may be corrected to provide a slightly revised version of the Q-factor estimation

$$Q_g^c = \frac{16\pi^2 V}{\lambda^3} \frac{\langle|S_{21}|^2 - |\langle S_{21}\rangle|^2\rangle}{m_1 m_2 \eta_1 \eta_2} \tag{6.13}$$

6.1.4.3   Q-Factor measurements using the reflection coefficient from each antenna

Using the same setup, it is also possible to retrieve the composite Q-factor of the chamber from the reflection coefficients of each antenna, since the variance of these parameters is also proportional to the average energy in the cavity. Both antennas provide additional estimations of the Q-factor. Hereafter, we refer to any of these two antennas with the letter "x," i.e. $x = 1$ or $x = 2$.

The transmitted power by one or the other antenna is provided in (6.5) where we replace S_{11} by S_{xx}. However, the S_{xx} complex parameter can be written in the form of an addition of two contributions. The first one is the free space parameter and the second one is the backscattering response of the chamber. The first parameter is estimated from the ensemble average of all complex-valued S_{xx} measurements. The second one is proportional to the complex transfer function of the chamber, denoted by H. Thus, a given S_{xx} parameter at any chamber state may be written as

$$S_{xx} = \langle S_{xx} \rangle + (1 - |\langle S_{xx} \rangle|^2)H\eta_x \tag{6.14}$$

Rearranging and taking the square modulus of this equation, we have

$$|S_{xx} - \langle S_{xx} \rangle|^2 = (1 - |\langle S_{xx} \rangle|^2)^2 |H|^2 \eta_x^2 \tag{6.15}$$

Then, evaluating the ensemble average of this equality, we obtain

$$\langle |S_{xx} - \langle S_{xx} \rangle|^2 \rangle = (1 - |\langle S_{xx} \rangle|^2)^2 \langle |H|^2 \rangle \eta_x^2 \tag{6.16}$$

The term $|H|^2$ is homogeneous to the ratio between the received power at an ideally efficient and perfectly matched antenna, denoted by $P_{rec,I}$, and the transmitted power in the chamber

$$|H|^2 = \frac{P_{rec,I}}{P_t} \tag{6.17}$$

The average of this quantity over an ensemble of chamber states writes

$$\langle |H|^2 \rangle = \frac{\langle P_{re,cI} \rangle}{P_t} \tag{6.18}$$

The received power at an ideally efficient and perfectly matched transmitting (simultaneously receiving) antenna $P_{rec,I}$ appears to be

$$\langle P_{rec,I} \rangle = \frac{\lambda^2}{4\pi} \frac{E^2}{Z_0} \tag{6.19}$$

The equivalent surface of the transmitting antenna in the chamber is indeed $\lambda^2/4\pi$, thus different from that of a receiving antenna [17]. The transmitting antenna imposes locally the field polarization where it is located. The incident plane wave spectrum at the transmitting antenna may be considered uniform in terms of angles of arrival but the polarization is that of the transmitting antenna. Its effective area is, therefore, twice that of a receiving antenna.

We may now establish the composite Q-factor estimation from the following expression

$$Q^c_{gsingle} = \frac{8\pi^2 V}{\lambda^3} \frac{\langle |S_{xx}|^2 - |\langle S_{xx}\rangle|^2 \rangle}{m_x^2 \eta_x^2} \quad (6.20)$$

This particular estimation of the composite Q-factor is noted $Q^c_{gsingle}$ where "single" stands for estimation from data of a single antenna.

This estimation is consistent with that of (6.12). In case of ideally efficient and perfectly matched antennas, identifying these two questions yields to the result also established in [17]:

$$\frac{\langle |S_{xx}|^2 \rangle}{\langle |S_{21}|^2 \rangle} = 2 = e_b \quad (6.21)$$

which corresponds to the enhanced backscattered coefficient e_b in the ideal case. A corrected version of e_b may also be provided as follows

$$e^c_b = \frac{\langle |S_{xx}|^2 - |\langle S_{xx}\rangle|^2 \rangle}{\langle |S_{21}|^2 - |\langle S_{21}\rangle|^2 \rangle} \quad (6.22)$$

Furthermore, we note that this enhanced backscattered coefficient can be estimated from the measurement of these different Q-factors, i.e. $Q^c_{gsingle}$ from a single antenna in (6.20) and Q^c_g from a pair of antennas in (6.13). We may indirectly estimate e^c_b as

$$e^c_b \approx 2\frac{Q^c_{gsingle}}{Q^c_g}, \quad (6.23)$$

using each particular antenna or an average estimation from both antennas.

6.1.4.4 Q_{Tx} and Q_{Rx} measurements

Ideal antennas

If the transmitting and receiving antennas are ideal ($m_x = 1$ and $\eta_x = 1$), the quality factor of these antennas are linked to the power received at their loads. The quality factor of the transmitting antenna is then defined as

$$Q_{Tx} = \frac{\omega\varepsilon E^2 V}{P_{rec,I}}. \quad (6.24)$$

The injected power being entirely transmitted in the RC ($P_t = P_{inj}$), the dissipated power at the load of the transmitting antenna writes:

$$\langle P_{rec,I}\rangle = \langle |H|^2\rangle P_{inj} = \langle |S_{xx}|^2\rangle P_{inj}. \quad (6.25)$$

Therefore, we find:

$$Q_{Tx} = \frac{8\pi^2}{\lambda^3} V \qquad (6.26)$$

We recall that $Q_{Rx} = 2Q_{Tx}$.

Non ideal antennas
If the transmitting and receiving antennas are not ideal ($m_x \neq 1$ and $\eta_x \neq 1$), the dissipated power at the antenna is different from $P_{rec,I}$. The correct description of losses at such nonideal antennas was recently provided in [18]. Beyond the received power at the load, we must account for material losses induced by the incoming wave as well as induced from the partially reflected wave due to mismatch. The expression of Q_{Tx} writes:

$$Q_{Tx} = \frac{\omega \varepsilon E^2 V}{P_{dant}} \qquad (6.27)$$

The term P_{dant} accounts for all loss mechanisms at the receiving (and nonideal) antenna

$$\frac{P_{dant}}{P_{rec,I}} = (1 - n_x) + n_x m_x + n_x(1 - m_x)(1 - n_x) \qquad (6.28)$$

The first term on the right-hand side of (6.28) represents the material losses due to the incoming waves at the antenna, the second term accounts for the power partially dissipated at the load and the last term is the material losses associated with the reflected wave from the antenna due to mismatch. After elementary calculations we easily come to the final expression for Q_{Tx}

$$Q_{Tx} = \frac{8\pi^2}{\lambda^3} V \frac{1}{1 - n_x^2(1 - m)} \qquad (6.29)$$

This result has no impact on the determination of the global Q-factor in (6.13) or (6.20). However, it is of utmost importance for the analysis of the behavior of (far from) nonideal antennas in a RC in order to retrieve, for instance, their intrinsic properties such as their efficiency.

6.1.4.5   Experimental results of Q-factor estimation
Experiments have been carried out in the IETR large RC whose dimensions are 2.9 m × 3.7 m × 8.7 m (see Figure 6.1). The theoretical frequency of its first natural mode is 43 MHz. Measurements are carried out with two similar log-periodic antennas in the 200–1,000 MHz frequency range. These antennas are installed in the chamber in such a way that their direct coupling is relatively low compared to the diffuse energy. Using a VNA with its two ports connected to each antenna input port, the frequency response is recorded into a sequence of 20, 001 points with a 40-kHz frequency step. The ensemble average is performed over a 10-MHz frequency band consisting of 250 equally spaced frequency points. The ensemble average in (6.12) and in (6.20)

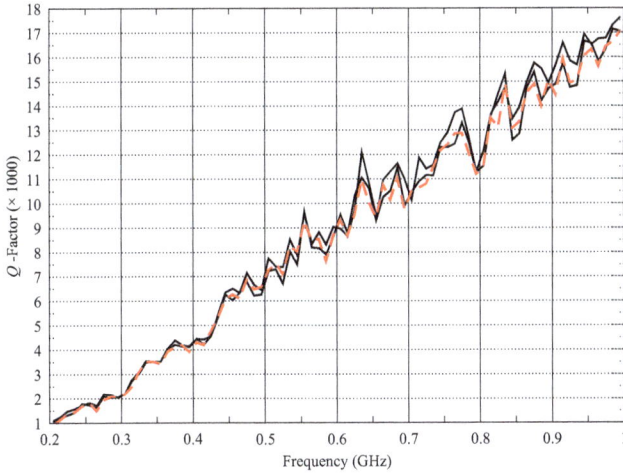

Figure 6.2   Three different RC Q-factor estimations with two antennas in the cavity. This estimation is performed using either antenna (both black curves) or both (red curve). Q-factors are first calculated at each frequency over 30 stirrer positions and then averaged over a 10-MHz bandwidth

is performed over 30 stirrer positions. Estimations are performed at each frequency step. Then, the reported Q-factor is determined from the average of 250 Q-factor estimations, each of these being taken over 30 stirrer positions, over a frequency window of 10 MHz.

First of all, the evaluation of the Q-factor is performed using the complete set of S parameters according to (6.13). Then, this estimation is compared to both estimations available from each antenna separately, estimated with (6.20). Results are reported in Figure 6.2. First of all, both estimations of the composite Q-factor with one or the other antenna only are very close to each other. Second, the estimation of the same factor using the full S parameter information yields to a very comparable result.

Two estimations of e_b^c according to (6.23) are provided in Figure 6.3. These estimations fluctuate around the value of two as expected and seems to slightly decrease close to 200 MHz.

This section of the chapter highlighted a method for evaluating the quality factor of a RC in the harmonic domain and through a set of two antennas. A satisfying level of accuracy may be obtained if many different states of the chamber are used through mechanical stirring and frequency averaging. Performing measurements at several positions is a complementary option. The accurate estimation of this coefficient opens the door for various applications of RC with regard to the insertion of objects. We focus in the next section on the characterization of absorbing performance of materials inserted in a RC.

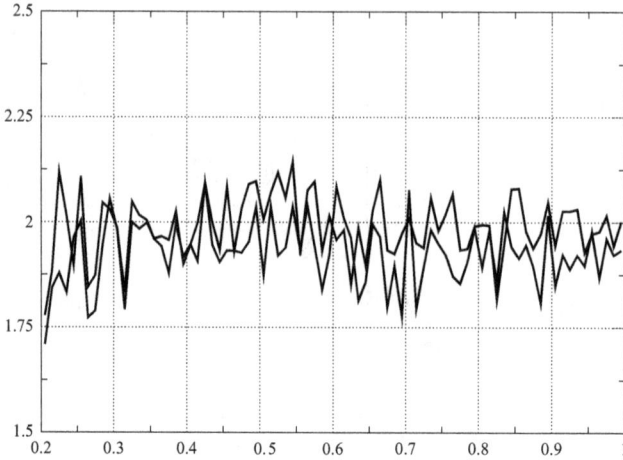

Figure 6.3   Enhanced backscattered coefficient from Q-factor estimations

6.2   The average absorption effective area of an object in a RC

When inserting any object in the RC and assuming that its absorbing properties does not modify those of other mechanisms, its own quality factor may be defined as

$$Q_{obj} = \frac{\omega \varepsilon E^2 V}{P_{d-obj}} \tag{6.30}$$

In this expression, P_{d-obj} is the power absorbed by the object due to the transduction of EM energy into thermal energy. This level of absorption of an object, and its associated quality factor depends on the geometry and on the intrinsic properties of its constitutive materials. In any case, this absorbed power is proportional to the available average power density in the chamber such as

$$P_{d-obj} = \sigma_{abs} \frac{E^2}{Z_0} \tag{6.31}$$

Since the power density has a unit of Watt per square meter, σ_{abs} is homogeneous to a surface. It is called the average AACS. It is a metric of absorption of an object when it is located in a diffuse field medium.

6.2.1   The Q-factor of an object in a RC and its average absorption cross section

According to (6.30) and (6.31), the Q-factor of an object inserted in the RC writes:

$$Q_{obj} = \frac{2\pi V}{\lambda} \frac{1}{\sigma_{abs}} \tag{6.32}$$

According to the definition of the composite Q-factor Q_g and given the independence of the different loss mechanisms, the addition of an object in the RC only changes the distribution of the dissipated power among the different loss mechanisms. Therefore, the quality factor, Q_g^0, evaluated if the object is removed from the RC (see (6.2)) is changed to

$$\frac{1}{Q_g^0} = \frac{1}{Q_{Tx}} + \frac{1}{Q_{Rx}} + \frac{1}{Q_w} \qquad (6.33)$$

Once the RC is loaded with the object under analysis, its AACS may be retrieved from the measurements of the loaded quality factor Q_g^L and the initial quality factor Q_g^0 such as

$$\sigma_{abs} = \frac{2\pi V}{\lambda} \left(\frac{1}{Q_g^L} - \frac{1}{Q_g^0} \right) \qquad (6.34)$$

The sensitivity of the measurement is related to the accuracy of the estimation of the difference between these two unloaded and loaded quality factors. The assessment of AACS has many practical applications such as (non-exhaustive list) analyzing absorbers performance [19] or material characterization [20], adjusting the characteristics of propagation channels [21] or predicting power losses for dosimetry applications as shown in Section 6.3. The estimation of AACS in the literature is available through different methods (with their own variants). A direct estimation from Q-factor estimations as performed in the Section 6.1.4.5 is proposed in this chapter. However, indirect methods using insertions losses [19,20], decay time constant [22] of the chamber or the relationship between the coherence bandwidth and the Q-factor are other interesting alternatives [21,23].

6.2.2 *Theoretical absorbing cross section of a rectangular piece of absorbing material*

We suppose that the inserted object is a homogeneous parallelepiped absorber of size $L_1 \times L_2 \times L_3$ much larger than the skin depth (δ) at the working frequency. Its total external surface is denoted S_{tot}. For such a canonical shape, the diffuse field in the chamber is impinging on the piece of absorber in the same way on each of its faces. It implicitly means that it is located at some distance from any walls or other object. Therefore, the AACS of this piece of absorbing material is approximately the sum of the partial AACS of each elementary surface of the parallelepiped absorber. Assuming a negligible skin depth, the AACS is proportional to the power density that is transmitted through the surface of the material. However, considering a planar surface, the total flux of energy in the direction of this surface is half the scalar power density in the RC [19]. This scalar power density corresponds (asymptotically) to the plane wave spectrum model as described by [3]. Any particular face of surface A_f of the parallelepiped absorber is supposed to be illuminated by a set of plane waves with uniformly spread angles of incidence. With respect to the considered surface, we define θ as the elevation angle and ϕ as the azimuthal angle. The flux of energy only depends on the elevation angle, so that we may relate it to a projected surface equals to $A_f \cos(\theta)$. A part of this flux is transmitted and absorbed through the material as a function of θ. Eventually, the total absorbed power associated with the elementary surface A_f is

$$P_{d-obj}^{A_f} = \frac{E^2}{Z_0} \int\limits_{\theta=0}^{\pi/2} \int\limits_{\phi=0}^{2\pi} A_f \cos(\theta) T(\theta) \sin(\theta) d\theta d\phi \tag{6.35}$$

This equation may be arranged as follows

$$P_{d-obj}^{A_f} = \frac{E^2}{Z_0} \frac{A_f}{2} \langle T \rangle \tag{6.36}$$

The symbol $\langle T \rangle$ stands for the average transmission coefficient over all incidence and polarization of the EM field with $0 \leq \langle T \rangle \leq 1$. From the above expression we determine the AACS for this elementary surface A_f as

$$\sigma_{abs}^{A_f} = \frac{A_f}{2} \langle T \rangle \tag{6.37}$$

It means that the maximum AACS of a planar surface A_f in the diffuse field is the half of this surface if the absorber is homogeneous and far enough from any wall and not installed with a metallic backplane. Introducing $S_0 = E^2/Z_0$ as the scalar power density in the RC, the total absorbed power associated to the flux of energy toward the planar surface A_f may be written as

$$P_{d-obj}^{A_f} = S_0 \frac{A_f}{2} \langle T \rangle \tag{6.38}$$

The average transmission coefficient depends on the intrinsic properties of the absorbing material. More precisely, it is calculated from the standard evaluation of plane wave Fresnel reflection coefficients depending on elevation angle and wave polarization [24]

$$\langle T \rangle = 2 \int\limits_0^{\pi/2} \left[1 - \frac{|\Gamma_{TM}(\theta)|^2 + |\Gamma_{TE}(\theta)|^2}{2} \right] \cos(\theta) \sin(\theta) d\theta \tag{6.39}$$

where $\Gamma_{TM}(\theta)$ and $\Gamma_{TE}(\theta)$ are the transverse magnetic and electric plane wave Fresnel reflection coefficients [24]

$$\Gamma_{TM}(\theta) = \frac{\mu_{abs} k_0 \cos(\theta) - \mu_0 \sqrt{k_{abs}^2 - k_0^2 \sin^2(\theta)}}{\mu_{abs} k_0 \cos(\theta) + \mu_0 \sqrt{k_{abs}^2 - k_0^2 \sin^2(\theta)}} \tag{6.40}$$

$$\Gamma_{TE}(\theta) = \frac{\mu_0 k_{abs}^2 \cos(\theta) - \mu_{abs} k_0 \sqrt{k_{abs}^2 - k_0^2 \sin^2(\theta)}}{\mu_0 k_{abs}^2 \cos(\theta) + \mu_{abs} k_0 \sqrt{k_{abs}^2 - k_0^2 \sin^2(\theta)}} \tag{6.41}$$

In these equations, $k_{abs} = \sqrt{\mu_{abs}(\varepsilon' - j\tau_{abs}/\omega)}$ and $k_0 = \omega/c$ are, respectively, the complex wave number of the absorbing material and the real wave number in free space (air). The terms μ_{abs} and ε' are the magnetic permeability and dielectric permittivity of the absorber, assuming homogeneous material. The parameter τ_{abs} is the equivalent conductivity of the absorber accounting for dielectric losses.

The overall AACS σ_{abs} of the parallelepiped object comes from the addition of the partial AACS $\sigma_{abs}^{A_f}$ of each face. Therefore, the AACS is expressed as a function of the total surface of the piece of material A_{tot} in a very simple way

$$\sigma_{abs} = \langle T \rangle \frac{A_{tot}}{2} \tag{6.42}$$

This expression is established in the hypothesis of a negligible skin depth ($\delta \to 0$). A correction may be introduced to take into account the edge of the absorber at grazing incidence of impinging waves. Traveling waves in the material with a path length smaller than the skin depth may be approximated as non-absorbed ones. Such a correction was proposed in [23]

$$\sigma_{abs} \approx \langle T \rangle \left[\frac{A_{tot}}{2} - \sum_{\substack{i,j=1,3 \\ i \neq j}} \left(\frac{2\delta(L_i + L_j)}{\pi} + \frac{\delta^2}{\pi} \right) \right] \tag{6.43}$$

6.2.3 Measurement examples

We illustrate the above theory by measuring different arrangements of one to four pieces of parallelepiped absorbing materials in a RC. Intrinsic properties of the absorbing material are supposed to be unknown, and the purpose of these measurements is to check the consistency of the above theory according to simple arrangements of these pieces of material. Figure 6.4 presents a photograph of the test setup in the case of four pieces of absorbing materials stacked along their largest side.

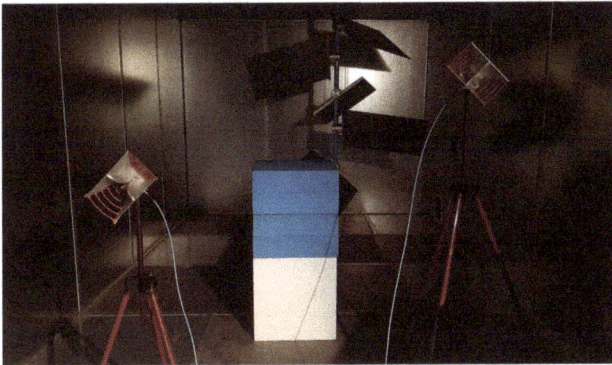

Figure 6.4   Test setup for the measurement of the average absorbing cross section of four pieces of absorbers in the 2–3-GHz frequency band. Two horn antennas are used in various transmitting and receiving configuration

Measurements are performed in the 2–3 GHz frequency range with a frequency resolution of 50 kHz using a VNA connected to two horn antennas. Q-Factors are determined as described in Section 6.1.4.5, using the estimations from the set of both antennas (6.13) and those available from each antenna separately, estimated with (6.20). The final estimation of the Q-factor comes from the average of those three estimations. A series of measurement was carried out with, successively, no piece of absorber in the cavity, a single piece (600 mm × 600 mm × 150 mm), two pieces and, finally, four pieces. The total external surface for these three last configurations are 1.08, 1.44 and 2.16 m^2, respectively.

Curves of Figure 6.5 show that the Q-factor estimated with a single absorbing piece of material is only one-third of that of the empty room. This is an obvious consequence of adding an efficient absorber in the original high-Q chamber. Despite its small volume compared to that of the RC, it has a huge impact on the quality factor and dissipates a large proportion of the transmitted power in the RC. The stack of additional pieces of absorbers continues to lower the Q-factor as expected.

According to the previous estimation of σ_{abs} in (6.43), a perfect parallelepiped absorber is an AACS that is equal to half of the total surface of the considered piece of absorbing material. Such an ideal absorber corresponds to a perfect average transmission coefficient $\langle T \rangle = 1$ and a null skin depth $\delta = 0$. We may, therefore, introduce a figure of merit that accounts for the efficiency of absorption under random field illumination. Thus, we define the normalized absorbing efficiency as

$$\eta_{abs} = \frac{2\sigma_{abs}}{S_{tot}} \tag{6.44}$$

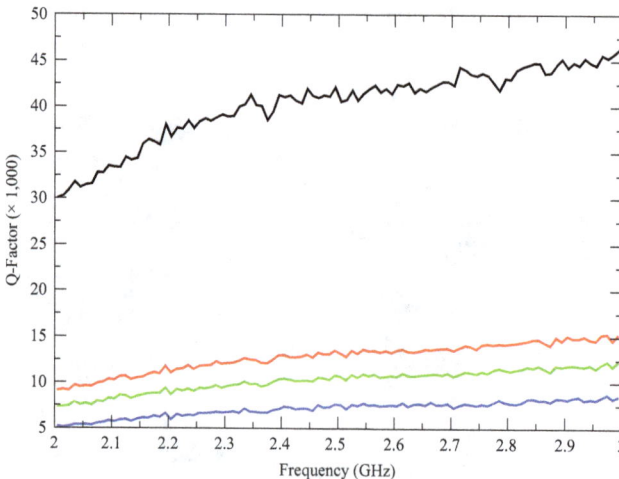

Figure 6.5   *Curves of Q-factor estimation for four different configurations of the RC: without absorber (black), with a single piece (red), two pieces (green) and four pieces of absorbing materials (blue)*

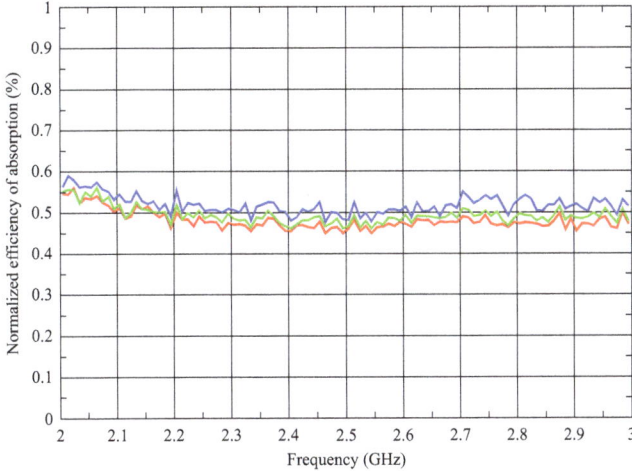

Figure 6.6   *Normalized absorbing efficiency with a single piece (red), two pieces (green) and four pieces of absorbing materials (blue)*

From the above Q-factor measurements we derive this normalized efficiency ($0 \leq \eta_{abs} \leq 1$) in Figure 6.6 for the three considered configurations, using (6.34) and (6.44). We may infer from these three curves that the efficiency of this absorbing material is roughly 50% in this frequency range. Fluctuations of these curves are linked to uncertainties in the determination of the respective Q-factors that could be enhanced from additional stirring positions and/or antenna positions. Figure 6.6 shows that the normalized efficiency of a single piece of absorber is slightly lower than that of two pieces, which is also a bit lower than that of four pieces. We assume that is due to the reduced impact of edge effects (6.43). These results confirm the consistency of the theory with respect to the absorbing external surface but does not confirm the theory itself. A further step of confirmation is shown in Section 6.3 where a piece of material with known intrinsic properties is used.

The control of the absorption factor in the RC is a key element to envision the use of RC for dosimetry. This is the subject of the next section.

6.3   Dosimetry for animals

6.3.1   Using RC for animal exposure

In the last decade, mode-stirred RCs have been investigated and implemented for in vivo animal exposures [25–29]. RC exposure systems have many advantages compared to other conventional whole-body systems: significant exposure levels can be generated, and the exposed animals, free of moving, are immersed in a statistically isotropic and uniform field regardless of their position in the volume of the chamber.

Thus, stressors in animals are reduced compared to other exposure systems where they are trapped in a small volume.

RCs were at the heart of in-depth investigations about long-term exposure to mobile phone radiation at 900 and 1,900 MHz. The US National Institute of Environmental Health Sciences (NIEHS) initiated a major study in 2006 to investigate the potential toxicity and carcinogenicity of long-term cell phone RF radiation in rodents (mice or rats). Partial results are available in [30]. Interested readers can find a detailed description in [29] of the whole exposure system, consisting of 21 RCs, to expose many mice and rats over a long-term period (months). A precise focus on the description of the calibration process is also provided in [31].

The calibration of such a test setup is indeed a challenge. On the one hand, animals are in interaction with a complex time-varying field (according to the rotation speed of the stirrer and their own movements in the RC). On the other hand, the animals themselves have complex intrinsic EM properties due to their heterogeneous nature of tissues and organs. In essence, the calibration in [31] is performed with numerical simulations performed with virtual but detailed numerical models of mice or rats. It consists of calculating the spatial distribution of EM fields in their body under random wave illumination as provided in a RC. It also accounts for uncertainties.

Beyond the EM exposure to mobile phone radiation of early generations, the advent of the 5G of mobile communication and other mm-wave devices (e.g. automotive radars) gives rise to similar questions with regard to mm-waves. A few research teams around the world focused on the specific nature of this interaction with living bodies and specific test setup for dosimetry [32]. Regarding mm-waves, RCs may be considered even more advantageous as, at such frequencies, RCs may be much smaller. According to achievable Q-factors in an aluminum cavity, an almost cubic RC of size lower than 0.5 m provides an ideal random field at frequencies as low as a few GHz only. With reasonably high Q-factors and small volume, the power density may be high enough to reach the required specific absorption rate (SAR) value for exposure. This was evidenced in early works by a group of researchers at IETR after building up a first prototype of a mm-wave RC of 0.423 m × 0.412 m × 0.383 m [33].

At such mm-wave frequencies, EM fields penetrate only within the skin of animals or human beings. Therefore, we proposed in [34] to calibrate the test setup by observing the temperature rise on the surface of a phantom that mimics the dielectric properties of the mouse skin.

The photography of the setup for calibration operation is shown in Figure 6.7. The mm-wave RC prototype is equipped with an infrared camera on the left handside. The camera lens is directed toward the interior of the cavity through an infrared transparent window having shielding properties for mm-waves. On the right hand side, a mm-wave source in the 58–62 GHz frequency range supply the RC through a transmitting WR-15 waveguide which is left open in the RC to radiate the EM energy. A sliding cylindrical brass tube is used to move the receiving WR-15 waveguide. This open-ended waveguide acts as a receiving antenna that is moved along the axis of the tube to perform measurements at different locations in the cavity.

The calibration procedure presented in this section uses the setup of Figure 6.7. This procedure involves the use of a dedicated piece of material properties of which

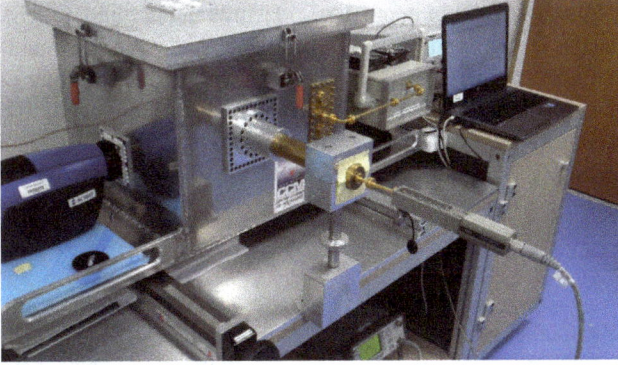

Figure 6.7 A prototype of mm-wave RC cavity equipped with DC voltage controlled mm-wave source (right) and an infrared camera (left) for imaging temperature through an optically transparent window

are close to the intrinsic properties of animals at such mm-wave frequencies. As a consequence, such a piece of material is called a "phantom" in the following. In this section, we highlight that it is possible to predict the temperature rise on the surface of the phantom installed in the RC as measured by the infrared camera. This calibration process is performed without resorting to numerical simulations.

6.3.2 Theory about heat transfer in a RC

The heat transfer on the surface of the phantom is closely related to the modification of the Q-factor of an object in the RC and, therefore, to its AACS as presented in Section 6.2. The knowledge of the Q-factor or its modification gives access to the power density within the RC, which, in turn, is proportional to the heat deposition at the phantom's surface. This modification can be predicted as far as the intrinsic properties of the inserted object are perfectly known. This is the case of the phantom described hereafter.

6.3.2.1 The phantom

A specific phantom of size $0.099\,\text{m} \times 0.099\,\text{m} \times 0.014\,\text{m}$ constituted of distilled water with a concentration of 4% of agar is used. The intrinsic properties of this phantom are represented by its complex permittivity [34]

$$\varepsilon^* = \varepsilon' - j\varepsilon'' \tag{6.45}$$

At 60 GHz, the agar phantom is equivalent to a lossy dielectric material with $\varepsilon^*/\varepsilon_0 = 11.9 - j19.5$. The skin depth, δ, is given by:

$$\delta = \left[\frac{2\pi f}{c} \sqrt{\frac{(\varepsilon'/\varepsilon_0)(\sqrt{1 + (\varepsilon''/\varepsilon')^2} - 1)}{2}} \right]^{-1} \tag{6.46}$$

Therefore, at 60 GHz, the skin depth $\delta \approx 250 \, \mu\text{m}$ is much thinner than the phantom thickness in each considered direction. From those data and the initial measurement of the empty RC Q-factor (Q_g^0), it is possible to deduce the resulting Q-factor (Q_g) once the phantom is installed in the RC. The skin depth being negligible with regard to the phantom size, we easily deduce from (6.32), (6.34) and (6.42) the modified Q-factor such as

$$Q_g = \left[\frac{1}{Q_g^0} + \frac{\lambda S_{ph} \langle T \rangle}{4\pi V} \right]^{-1} \tag{6.47}$$

where S_{ph} is the total external surface of the phantom.

Moreover, the thermal properties of this material are also determined and provided in Table 6.1.

The phantom is shown in Figure 6.8. It is shown together with the receiving open-ended waveguide at the end of the brass tube. The largest face of the phantom is oriented toward the camera lens that record the temperature rise at its surface as a function of time. From the knowledge of the power density in the RC, of the intrinsic EM and thermal properties of the phantom, the temperature rise is the solution of the electrothermal coupling analysis.

Table 6.1   Physical properties of the agar phantom

Thermal conductivity k_t (W/m K)	0.66
Specific heat C (J/kg K)	3,770
Mass density ρ(kg/m^3)	1,000

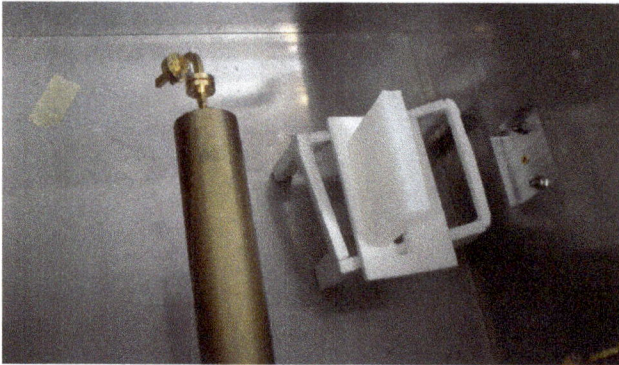

Figure 6.8   The agar phantom installed in a vertical position in the RC over a polystyrene support. The open WR-15 waveguide is shown at the end of the brass tube

Applying the procedure described in Section 6.2, the Q-factors are estimated before and after insertion of the phantom in the RC. The Q-factor of the loaded chamber is calculated from (6.47). These measurements are performed in the 58.8–61.2 GHz range at 61 receiving antenna positions by moving the brass tube in the RC. Moreover, both the Q-factors are calculated with a frequency step of 10 MHz and averaged over a 300-MHz frequency range. Each point of Figure 6.9 is therefore the average of 31 Q-factor estimations performed over 61 receiving antenna positions. The results are totally consistent. In other words, they tend to prove that the fabricated phantom possesses intrinsic characteristics in agreement with that expected from theory.

6.3.2.2 The heat equation

The purpose of this section is to evaluate theoretically the temperature rise on the surface of the above agar phantom sample exposed in a RC.

Heating at any location in the Cartesian coordinate system (O, x, y, z) and at any time t inside or on the surface of the phantom is governed by the 3-D heat equation [35]

$$\frac{\rho C}{k_t} \frac{\partial T}{\partial t} = \frac{\partial^2 T}{\partial x^2} + \frac{\partial^2 T}{\partial y^2} + \frac{\partial^2 T}{\partial z^2} + \frac{q(x, y, z, t)}{k_t} \tag{6.48}$$

where $T(x, y, z, t)$ is the phantom temperature (°C), ρ its mass density (kg/m^3), C its specific heat (J/kg K) and k_t its thermal conductivity (W/m K). The source term of this equation $q(x, y, z, t)$ is the heat deposition induced by the EM exposure (W/m^3). Heat

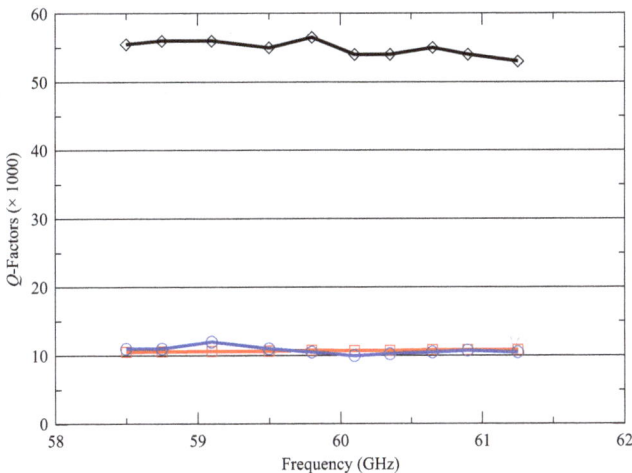

Figure 6.9 Q-Factors of the RC estimated from measurements before (black curve) and after (red curve) inserting the phantom in the RC. The blue curve represents the theoretical estimation from (6.47)

exchange between the phantom and the surrounding air is described by the following boundary conditions for the heat equation

$$k_t \cdot \frac{\partial T}{\partial x} = h \cdot (T - T_{ext}) \quad \forall x \in \partial \Omega_x,$$

$$k_t \cdot \frac{\partial T}{\partial y} = h \cdot (T - T_{ext}) \quad \forall y \in \partial \Omega_y,$$

$$k_t \cdot \frac{\partial T}{\partial z} = h \cdot (T - T_{ext}) \quad \forall z \in \partial \Omega_z \tag{6.49}$$

where $T_{ext}(x,y,z,t)$ is the ambient temperature (°C), $\partial \Omega_x \cup \partial \Omega_y \cup \partial \Omega_z$ defines the boundary surface of the phantom and h is the phantom heat transfer coefficient (W/m² K). Assuming a limited temperature rise, the vaporization effect is neglected. Therefore, the heat transfer only includes radiation (h_r) and convective (h_c) mechanisms. The radiative heat transfer coefficient, h_r, is expressed as

$$h_r = \sigma_{SB}(T + T_{ext})(T^2 + T_{ext}^2) \tag{6.50}$$

where σ_{SB} is the Stefan–Boltzmann constant ($\sigma_{SB} = 5.67 \times 10^{-8}$ W/m² K⁴). As far as convection is concerned, the corresponding heat transfer coefficient h_c may be approximated as (see [36] for a detailed analysis):

$$\frac{h_c L}{k_f} = 0.59 \left[\frac{L^3 \rho_f^2 \beta_f g \Delta T C_f}{\mu_f k_f} \right]^{1/4} \tag{6.51}$$

where L is the height of the phantom (0.099 m), ρ_f is the air density (1.184 kg/m³), β_f the volumetric coefficient of expansion of the air (3.4×10^{-3}K⁻¹), C_f its specific heat capacity (1.005 J/Kg K), μ_f its viscosity (1.51×10^{-5} m²/s), k_f its thermal conductivity (0.027 W/m K), g is the gravity acceleration (9.8 m s⁻²) and ΔT is the temperature difference between the air and the phantom surface.

6.3.2.3   The electrothermal coupling

The abovementioned heat equation (6.48) governs the temperature rise with time, depending on the temporal heat deposition $q(x,y,z,t)$. This heat deposition is related to the interaction between the EM field in the RC and the phantom. This EM field couples to thermal energy through EM energy losses in the material itself. Therefore, the total heat deposition q_v is equal to the total dissipated power in the volume V of the material

$$q_v = -\Re \left(\oint \int_{\partial V} \vec{\nabla} \cdot \vec{\Pi}_p \cdot d\vec{S} \right) \tag{6.52}$$

In this expression $\Re$ stands for the real part of the expression within the parentheses and $\vec{\Pi}_p$ represents the vectorial power density of the incident wave. The spatial dispersion of the heat quantity inside the phantom body may be retrieved from the Gauss–Ostrogradsky theorem [24]

$$q_{x,y,z} = -\Re \left(\int \int \int_V \vec{\nabla} \cdot \vec{\Pi}_p dV \right) \tag{6.53}$$

6.3.2.4 The 1-D approximation

With regard to the phantom geometry and its intrinsic properties at 60 GHz, two conclusions may be drawn. First, due to the small skin depth, the heat deposition is significant only at a pellicular surface of the phantom. Second, most of its surface is composed of two largest faces of the phantom in the x–y plane according to the schematic in Figure 6.10. Then we consider the only EM wave flux in the z direction on one face (A+) on the left and in the $-z$ direction on the right on the opposite side (A−). Therefore, the total heat deposition may be approximated as

$$q_v = \frac{L_x L_y}{2} \langle T \rangle S_0 \tag{6.54}$$

and

$$q_{x,y,z} \approx -\frac{\langle T \rangle S_0}{2} \frac{d}{dz} \left(\exp\left(\frac{-2z}{\delta}\right) - \exp\left(\frac{-2(L_z - z)}{\delta}\right) \right) \qquad z \in [0, L_z]$$
$$\forall x \in [0, L_x] \qquad \forall y \in [0, L_y] \tag{6.55}$$

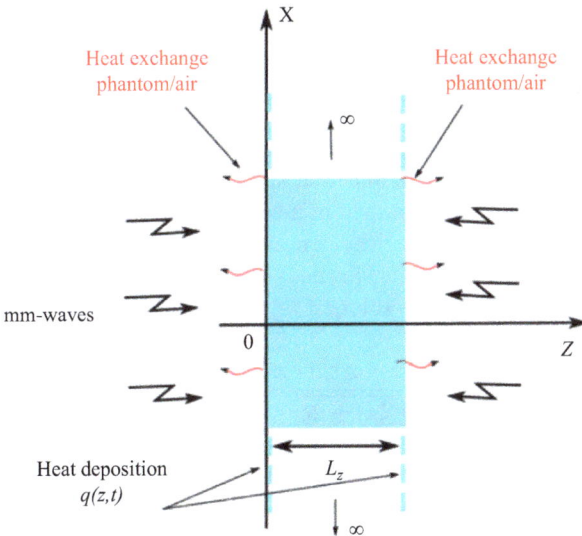

Figure 6.10 *1-D model of heat deposition on the phantom. Heat deposition comes from both sides of the phantom, virtually extended to infinity in the x-direction*

$$q_{x,y,z} \approx -\frac{\langle T \rangle S_0}{\delta}\left(\exp\left(\frac{-2z}{\delta}\right) - \exp\left(\frac{-2(L_z - z)}{\delta}\right)\right) \qquad z \in [0, L_z]$$

$$\forall x \in [0, L_x] \qquad \forall y \in [0, L_y] \qquad (6.56)$$

Moreover, if we are mainly interested in the temperature rise at the center of one of these two major faces, we may consider the case of the infinite extension of the phantom in the Oxy plane. Therefore, we solve the problem illustrated in Figure 6.10.

6.3.2.5   Solution of the 1-D heat equation

The solution of the 1-D heat equation is detailed in [34]. We provide only the key steps to obtain it. By introducing the diffusivity $\alpha_t = k_t/(\rho C)$ and the effusivity $\zeta = h/k$, the 1-D equivalent system of equations of the problem writes:

$$\frac{1}{\alpha_t}\frac{\partial T_r}{\partial t} = \frac{\partial^2 T_r}{\partial z^2} + \frac{q(z)}{k_t}, \qquad \text{for } 0 \leq z \leq L_z$$

$$-\frac{\partial T_r}{\partial z} + \zeta T_r = 0, \qquad \text{at } z = 0$$

$$\frac{\partial T_r}{\partial z} + \zeta T_r = 0, \qquad \text{at } z = L_z$$

$$T_r = 0, \qquad \text{at } t = 0 \qquad (6.57)$$

where $T_r = T - T_{ext}$ is the relative temperature rise and $q(z)$ is the heat deposited in the phantom:

$$q(z) \approx \langle T \rangle \frac{S_0}{\delta}\left[\exp\left(\frac{-2z}{\delta}\right) - \exp\left(\frac{-2(L_z - z)}{\delta}\right)\right] \qquad z \in [0, L_z] \qquad (6.58)$$

This equation is solved through integral transformation techniques using the Fourier transform as originally developed by Joseph Fourier for solving the heat equation. The Fourier transform aimed at separating space and time variables. The propagation of heat is then described in a series of spatial harmonics as a solution of an eigenvalue problem. Once the eigenvalue problem is solved, the heat equation is solved, in turn, in the dual (transformed) domain. Eventually, the inverse Fourier transform leads to the following solution

$$T_r(z) = \sum_{n=1}^{\infty} \frac{\sqrt{2}(v_n \cos v_n z + \zeta \sin v_n z)}{\sqrt{L_z(v_n^2 + \zeta^2) + 2\zeta}}$$

$$\times \frac{\phi}{\alpha_t v_n^2} \times \left[1 - \exp\left(-\alpha_t v_n^2 t\right)\right]$$

with

$$\phi = \frac{\alpha_t}{k_t}\frac{\sqrt{2}}{\sqrt{L_z(v_n^2 + \zeta^2) + 2\zeta}} \times \int_0^{L_z} (q(z)(v_n \cos v_n z + \zeta \sin v_n z))dz \qquad (6.59)$$

where the eigenvalues v_n are positive roots of

$$\tan v_n L_z = \frac{2v_n \zeta}{v_n^2 - \zeta^2} \tag{6.60}$$

6.3.3 Experimental validation

Measurements are performed at the central frequency $f = 59.3$ GHz with a frequency stirring bandwidth of 300 MHz and without any mechanical stirring or displacement of the sliding brass tube placed at the most extracted position in the RC. The operating frequency of the mm-wave source is adjusted to the desired value by application of a DC voltage signal generated by an external voltage source. The sweep time of the stirring bandwidth is 5 s. The input power in the RC is 1.81 W. The phantom is located at a distance of 35 cm away from the infrared camera. Figure 6.11 represents the infrared image of the temperature distribution on the phantom surface after 300 s (i.e. a total of 60 frequency sweeps) of exposure in the RC. The temperature rise is homogeneous in the central area of the phantom only. It is higher at phantom edges (particularly in corners) than in the central area. The solution of the 1-D heat-equation is obviously no longer valid for these edge positions since a correct solution must account for heat deposition on different sides of the phantom surface. The 3-D solution was implemented in [34] to account for these edge effects.

Therefore, the solution in (6.59) is approximately valid in the blue line window of Figure 6.11. In this zone, the temperature rise is reasonably uniform. In this example,

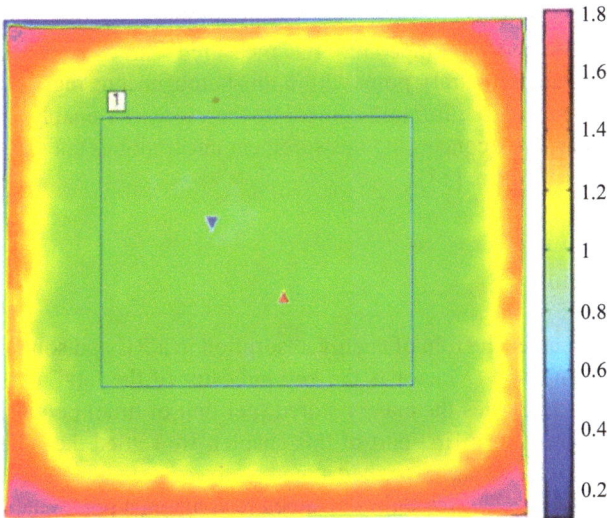

Figure 6.11   Temperature rise at the phantom surface after 300 s of exposure in the mm-wave prototype RC with an input power of 1.81 W. This image is transposed from the infrared camera measurement. The color scale indicates the temperature rise in the range 0°C–1.8°C

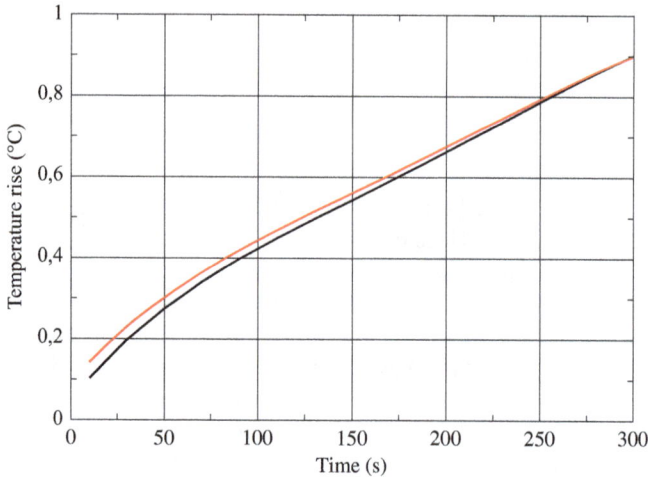

Figure 6.12   Temperature rise in °C at the surface of the phantom as a function of the time of exposure, measured in the homogeneous region of the infrared camera image (black curve), and predicted from the solution of the 1-D heat equation (red curve)

after 300 s of exposure the average temperature rise is 0.90°C with a standard deviation of 0.03°C. Then, we perform a comparison in Figure 6.12 between the solution provided by (6.59) and the temperature rise measured with the infrared camera as a function of time at an arbitrary point within this homogeneous region. The temperature rise observed with the infrared camera is, therefore, reasonably predicted from theory. Further analysis concerning measurement uncertainties and the estimation of the SAR were provided in [34].

6.4   Discussion

This chapter discusses the role of energy absorption in a RC and some related applications. The composite Q-factor is the key indicator of the total amount of losses in a RC since it quantifies the energy stored (per unit of time) per unit of transmitted/dissipated power. In a first part of this chapter, we settled the definition of this composite quality factor and described a method to estimate it in different ways with a set of two antennas. The scattering parameters at antenna ports enable to estimate the composite Q-factor in three different ways and enables to check for the enhanced backscattered coefficient. Obviously, an accurate enough estimation requires a high number of RC states. We suggest that averaging over small frequency bands is a solution to minimize intrinsic statistical variation of Q-factors estimated from a limited set of stirrer positions.

Beyond the determination of this important feature, any variation of losses due to the installation of objects in the chamber gives access to many different applications. We dealt with two of these that are strongly related to each other. The first application consists of evaluating the average absorption cross section of any piece of material within the RC. This absorption coefficient is retrieved from the contrast of quality factors measurement before and after putting the material in the chamber. For simple geometries of the inserted absorber, it is possible to estimate its efficiency. Moreover, from the knowledge of the intrinsic EM properties of homogeneous and thick (with regard to the skin depth) materials, we showed that the measured AACS computed from the modification of the Q-factor is consistent with the theoretical one.

This last fact was at the origin of a specific calibration procedure for mm-wave RC dedicated to animal exposure. This is the second application of this chapter. In order to calibrate the EM dose during exposure, the power density in the RC must be controlled and proved to be consistent with the expected temperature rise. Using a rectangular homogeneous phantom, mimicking the properties of animal's skin, we show that the temperature rise at its surface and measured with an infrared camera was indeed predictable. On a theoretical point of view, the measurement of the Q-factor, the knowledge of intrinsic parameters of the phantom and the properties of the diffuse field in the RC fully determine the heat deposition on the phantom surface. Solving the heat-equation gives then the correct temperature rise prediction with a reasonable accuracy.

The results presented in this chapter came from an early prototype of the RC. This gave birth in 2017 to the fabrication of a new mm-wave RC that is fully compliant with stringent requirements for animal exposure. This was further calibrated with not only rectangular phantoms but with phantoms of murine [37].

References

[1] Cozza A. The role of losses in the definition of the overmoded condition for reverberation chambers and their statistics. IEEE Trans Electromagn Compat. 2011;53(Pt 1):296–307.

[2] Gros JB, Legrand O, Mortessagne F, *et al*. Universal behaviour of a wave chaos based electromagnetic reverberation chamber. Wave Motion. 2014;51(Pt 4):664–672.

[3] Hill D. Plane wave integral equations for fields in reverberation chambers. IEEE Trans Electromagn Compat. 1998;40(Pt 3):209–217.

[4] Hill D. Electromagnetic fields in cavities: deterministic and statistical theories. IEEE Press, Wiley; 2009.

[5] Besnier P and Demoulin B. Electromagnetic reverberation chambers. ISTE Wiley & Sons; 2011.

[6] Gros JB, Kuhl U, Legrand O, *et al*. Lossy chaotic electromagnetic reverberation chambers: universal statistical behavior of the vectorial field. Phys Rev E. 2016;93:032108.

[7]   Lemoine C, Besnier P, and Drissi M. Investigation of reverberation chamber through high-power goodness-of-fit tests. IEEE Trans Electromagn Compat. 2007;49(Pt 4):745–755.

[8]   Adardour A, Andrieu G, and Reineix A. On the low-frequency optimization of reverberation chambers. IEEE Trans Electromagn Compat. 2014;56(Pt 2): 266–276.

[9]   Fall AK, Besnier P, Lemoine C, *et al.* Determining the lowest usable frequency of a frequency-stirred reverberation chamber using modal density. In: International Symposium on Electromagnetic Compatibility, EMC EUROPE; 2014 Sep; Göteborg, Sweden. IEEE; 2014. p. 263–267.

[10] Andrieu G. Calibration of reverberation chambers from S21 measurements. In: 2017 IEEE International Symposium on Electromagnetic Compatibility and Signal/Power Integrity; 2017 Aug; Washington, USA; 2017. p. 675–680.

[11] Andrieu G, Ticaud N, Lescoat F, *et al.* Fast and accurate assessment of the well stirred condition of a reverberation chamber from S11 measurements. IEEE Trans Electromagn Compat. 2019;61(Pt 4):874–982.

[12] Arnaut LR. Statistics of the quality factor of a rectangular reverberation chamber. IEEE Trans Electromagn Compat. 2003;45(Pt 1): 61–76.

[13] Arnaut LR and Gradoni G. Probability distribution of the quality factor of a mode-stirred reverberation chamber. IEEE Trans Electromagn Compat. 2013;55(Pt 1):35–44.

[14] Arnaut LR, Andries MI, Sol J, *et al.* Evaluation method for the probability distribution of the quality factor of mode-stirred reverberation chambers. IEEE Trans Antennas Propag. 2014;62(Pt 8):4199–4208.

[15] Arnaut LR, Besnier P, Sol J, *et al.* On the uncertainty quantification of the quality factor of reverberation chambers. IEEE Trans Electromagn Compat. 2019;61(Pt 3):823–832.

[16] Holloway CL, Shah HA, Pirkl RJ, *et al.* A three-antenna technique for determining the total and radiation efficiencies of antennas in reverberation chambers. IEEE Trans Electromagn Compat. 2012;54(Pt 1):235–241.

[17] Junqua I, Degauque P, and Lienard M. On the power dissipated by an antenna in transmit mode or in receive mode in a reverberation chamber. IEEE Trans Electromagn Compat. 2012;54(Pt 1):174–180.

[18] Cozza A. Power loss in reverberation chambers by antennas and receivers. IEEE Trans Electromagn Compat. 2018;60(Pt 6):2041–2044.

[19] Hallbjorner P, Carlberg U, Madsen K, *et al.* Extracting electrical material parameters of electrically large dielectric objects from reverberation chamber measurements of absorption cross section. IEEE Trans Electromagn Compat. 2005;47(Pt 2):291–303.

[20] Gifuni A. On the measurement of the absorption cross section and material reflectivity in a reverberation chamber. IEEE Trans Electromagn Compat. 2009;51(Pt 4):1047–1050.

[21] Dortmans JNH, Remley KA, Senic D, *et al.* Use of absorption cross section to predict coherence bandwidth and other characteristics of a reverberation

chamber setup for wireless-system tests. IEEE Trans Electromagn Compat. 2016;58(Pt 5):1653–1661.

[22]   Amador E, Andries MI, Lemoine C, *et al.* Absorbing material characterization in a reverberation chamber. In: International Symposium on Electromagnetic Compatibility, EMC EUROPE; 2011 Sep; York, United Kingdom. IEEE; 2011. p. 117–122.

[23]   Andries MI, Lemoine C, and Besnier P. On the prediction of the average absorbing cross section of materials from coherence bandwidth measurements in reverberation chamber. In: International Symposium on Electromagnetic Compatibility, EMC EUROPE; 2012 Sep; Rome, Italy. IEEE; 2012. p. 1–6.

[24]   Stratton JA. Electromagnetic theory. Wiley; 2007.

[25]   Jung KB, Kim TH, Kim JL, *et al.* Development and validation of reverberation-chamber type whole-body exposure system for mobile-phone frequency. Electromagn Biol Med. 2008;27(Pt 1):73–82.

[26]   Wu M, Hadjem A, Wong MF, *et al.* Whole-body new-born and young rats' exposure assessment in a reverberating chamber operating at 2.4 GHz. Phys Med Biol. 2010;55(Pt 6):1619.

[27]   De Gannes FP, Billaudel B, Haro E, *et al.* Rat fertility and embryo fetal development: influence of exposure to the Wi-Fi signal. Reprod Toxicol. 2013;36:1–5.

[28]   Jin YB, Choi HD, Kim BC, *et al.* Effects of simultaneous combined exposure to CDMA and WCDMA electromagnetic fields on serum hormone levels in rats. J Radiat Res. 2013;54(3):430–437.

[29]   Capstick M, Kuehn S, Berdinas-Torres V, *et al.* A radio frequency radiation exposure system for rodents based on reverberation chambers. IEEE Trans Electromagn Compat. 2017;59(Pt 4):1041–1052.

[30]   Wyde M, Cesta M, Blystone C, *et al.* Report of partial findings from the national toxicology program carcinogenesis studies of cell phone radiofrequency radiation in Hsd: Sprague Dawley SD rats (Whole body exposures). BioRxiv; 2018. https://doi.org/10.1101/055699(1).

[31]   Gong Y, Capstick M, Kuehn S, *et al.* Life-time dosimetric assessment for mice and rats exposed in reverberation chambers for the two-year NTP cancer bioassay study in cell phone radiation. IEEE Trans Electromagn Compat. 2017;59(Pt 6):1798–1808.

[32]   Zhadobov M, Chahat N, Sauleau R, *et al.* Millimeter-wave interactions with the human body: state of knowledge and recent advances. Int J Microwaves Wireless Technol. 2011;3(Pt 2):237–247.

[33]   Fall AK, Besnier P, Lemoine C, *et al.* Design and experimental validation of a mode-stirred reverberation chamber at millimeter-waves. IEEE Trans Electromagn Compat. 2015;57(Pt 1):12–22.

[34]   Fall AK, Besnier P, Lemoine C, *et al.* Experimental dosimetry in a mode-stirred reverberation chamber in the 60-GHz band. IEEE Trans Electromagn Compat. 2016;58(Pt 4):981–993.

[35]   Foster K, Kritikos H, and Schwan H. Effect of surface cooling and blood flow on the microwave heating of tissue. IEEE Trans Biomed Eng. 1978; 25(Pt 3):313–216.

[36]   Wissler EH. Steady-state temperature distribution in man. J Appl Physiol. 1961;16(Pt 4):734–740.

[37]   Fall AK, Lemoine C, Besnier P, *et al.* Exposure assessment in millimeter-wave reverberation chamber using murine phantoms. Bioelectromagnetics. 2020;41(2):121–135.

Chapter 7

Characterization of antenna efficiency in reverberation chambers

Wei Xue[1] and Xiaoming Chen[1]

The radiation efficiency (η_{rad}) is an important characteristic parameter of an antenna. It reflects the ability of an antenna to convert the input power into a radiated power. It is defined as the ratio of the total power radiated by the antenna (P_{rad}) to the total power accepted by the antenna (P_{in}) at its port [1]

$$\eta_{rad} = \frac{P_{rad}}{P_{in}}. \tag{7.1}$$

The antenna radiation efficiency is 100% for an ideal antenna and is usually less than 100% for a practical antenna due to conduction and dielectric losses. Higher antenna radiation efficiency means lower power loss and higher power utilization. Generally, there is a mismatch between the transmission line and the antenna, which results in reflection at the antenna port. Taking this mismatch into consideration, the total antenna efficiency (η_{tot}) is defined as

$$\eta_{tot} = \eta_{rad}(1 - |\Gamma|^2), \tag{7.2}$$

where Γ is the reflection coefficient at the antenna port and can be easily measured using a vector network analyzer (VNA) (i.e., S_{11}). It is obvious that the total antenna efficiency is lower than the antenna radiation efficiency.

In the past decades, many methods have been proposed to measure the antenna radiation efficiency accurately and efficiently, such as directivity/gain [1], pattern integration [2], Wheeler cap [3], radiometric [4], and reverberation chamber (RC). The anechoic chamber (AC) methods (i.e., directivity/gain method and pattern integration method) require three-dimensional (3D) measurement. Each complete 3D measurement can only determine the antenna efficiency at a single frequency. Obviously, the measurement process is very complex and time-consuming, especially for ultra-wideband (UWB) antennas under test. Moreover, ACs used to measure the efficiency of electrically large antennas are rather expensive in reason of the Faraday cage and the electromagnetic absorbers. In comparison, the RC methods are cost-effective and time efficient.

[1]School of Information and Communications Engineering, Xi'an Jiaotong University, Xi'an, China

In this chapter, we focus on the RC methods. Many RC methods have been proposed in the literature, for example, the standard reference antenna method [5], the one-antenna method [6], the two-antenna method [6,7], the three-antenna method [6], the time reversal method [8,9], the reflective character method [10], the quality factor method [11], the open-ended waveguide method [12], and the nested and contiguous RCs method [13]. Among those methods, the standard reference antenna method is probably the most widely used in practice, while the one-, two-, and three-antenna methods are efficient because they do not require the use of a reference antenna of known efficiency, and the time reversal method is the newest and the most relevant in the time domain. This is why, due to page limitation, only these representative methods are presented in detail in this chapter.

7.1 Antenna radiation efficiency measurement methods

7.1.1 Standard reference antenna method

A common setup for antenna radiation efficiency measurement using the standard reference antenna method [5] is shown in Figure 7.1.

The VNA and the motor controller are connected to a computer that controls the collection and storage of S-parameters and the rotation of the stirrers An antenna under test (AUT), a reference antenna (REF) with known radiation efficiency, and a fixed transmitting antenna (TX Ant) are inserted in the RC. The radiation efficiency of the reference antenna is provided by the manufacturer or measured previously in an AC. The TX Ant is connected to the port 1 of the VNA, and the reference antenna (AUT) is connected to the port 2 of the VNA for the reference (AUT) measurement. To ensure

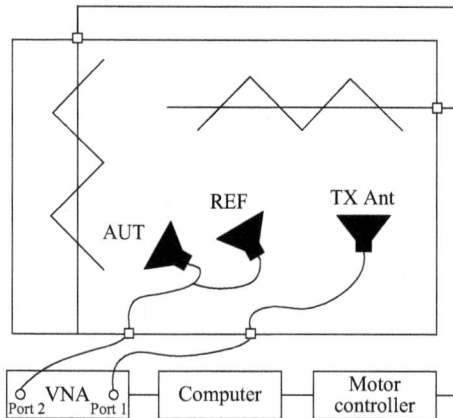

Figure 7.1   *Illustration of the measurement setup for antenna radiation efficiency measurement in a reverberation chamber using the standard reference antenna method*

that the RC has the same loading condition for both reference and AUT measurements, all the antennas are placed in the RC during the whole measurement procedure. Strictly speaking, the unused antenna should be loaded with a 50 Ω termination. However, practical measurements show little differences whether the unused antenna is loaded with a 50 Ω termination or not, especially when the antennas are sufficiently far from each other.

The theoretical derivation of the antenna radiation efficiency is as follows. According to the famous David Hill's equations [14], the average received power by the reference antenna ($\langle P_{REF} \rangle$) and the AUT ($\langle P_{AUT} \rangle$) can be, respectively, written as

$$\langle P_{REF} \rangle = \frac{E_{REF}^2}{Z_0} \frac{\lambda^2}{8\pi} \eta_{REF\_tot}, \tag{7.3}$$

$$\langle P_{AUT} \rangle = \frac{E_{AUT}^2}{Z_0} \frac{\lambda^2}{8\pi} \eta_{AUT\_tot}, \tag{7.4}$$

where $\langle \cdot \rangle$ represents an ensemble average over all the stirring positions (including mechanical stirring [15], frequency stirring [16], and source stirring [17]). E_{REF}^2/Z_0 and E_{AUT}^2/Z_0 (in which E^2 represents the mean-square value of the total electric field and Z represents the free-space wave impedance) can be interpreted as scalar power densities of the reference antenna and the AUT, respectively, both metrics are independent of the position in a well-stirred RC. $\lambda^2/8\pi$ (in which λ represents the free-space wavelength) is the average effective area containing a polarization mismatch factor $1/2$.

Note that the TX Ant is fixed during the reference and AUT measurements, thus both the reference antenna and AUT work in the same statistical environment, which results in $E_{REF}^2/Z_0 = E_{AUT}^2/Z_0 = E_0^2/Z_0$.

Assuming the power accepted by the TX Ant is P_{in} we can obtain the average power transmission coefficients of both the reference antenna and the AUT as

$$\left\langle \left| S_{21,REF} \right|^2 \right\rangle = \frac{\langle P_{REF} \rangle}{P_{in}} = \frac{1}{P_{in}} \frac{E_0^2}{Z_0} \frac{\lambda^2}{8\pi} \eta_{REF\_tot}. \tag{7.5}$$

$$\left\langle \left| S_{21,AUT} \right|^2 \right\rangle = \frac{\langle P_{AUT} \rangle}{P_{in}} = \frac{1}{P_{in}} \frac{E_0^2}{Z_0} \frac{\lambda^2}{8\pi} \eta_{AUT\_tot}. \tag{7.6}$$

Dividing (7.6)-(7.5), the common items therein can be eliminated and the total antenna efficiency is obtained as

$$\eta_{AUT\_tot} = \frac{\left\langle \left| S_{21,AUT} \right|^2 \right\rangle}{\left\langle \left| S_{21,REF} \right|^2 \right\rangle} \eta_{REF\_tot}. \tag{7.7}$$

The antenna radiation efficiency can be calculated as

$$\eta_{AUT\_rad} = \frac{\left\langle \left| S_{21,AUT} \right|^2 \right\rangle \left(1 - \left| \langle S_{22,REF} \rangle \right|^2 \right)}{\left\langle \left| S_{21,REF} \right|^2 \right\rangle \left(1 - \left| \langle S_{22,AUT} \rangle \right|^2 \right)} \eta_{REF\_rad}, \tag{7.8}$$

where the mean value of S_{22}, $\langle S_{22} \rangle$, measured in RC should be equal to S_{22} measured in free space [18]. This free-space mismatch factor is needed to be calibrated out of the channel transfer function (G_{ch}) [5]:

$$G_{ch} = \frac{\langle |S_{21}|^2 \rangle}{\eta_{rad1}\eta_{rad2}e_{mismatch1}e_{mismatch2}} = \frac{\langle |S_{21}|^2 \rangle}{\eta_{rad1}\eta_{rad2}\left(1 - |\langle S_{11} \rangle|^2\right)\left(1 - |\langle S_{22} \rangle|^2\right)}.$$

(7.9)

The measured S_{22} of an antenna at one mode stirrer state in an RC can be modeled as $S_{22} = S_{22}^{fs} + S_{22}^{sti} + S_{22}^{uns}$ [19], where S_{22} is the total reflection coefficient of the antenna in the RC, S_{22}^{fs} is the free-space reflection coefficient, and S_{22}^{sti} and S_{22}^{uns} are stirred and unstirred components in the RC, respectively If an effective mode stirring is used, for instance, a turn-table platform in addition with mechanical stirring [17], S_{22}^{uns} may be negligible. Since $\langle S_{22}^{sti} \rangle$ is close to zero $\langle S_{22} \rangle \approx S_{22}^{fs}$.

If the reference antenna has an efficiency close to 100%, we can omit the efficiency of the reference antenna for a rough efficiency measurement [12,14].

The standard reference antenna method is the most widely used among RC methods. An important advantage lies in the fact that the total radiation efficiency of the fixed TX Ant and the chamber loss are canceled out in (7.7). However, it requires a reference antenna of known radiation efficiency. This may lead inconvenience for measuring the efficiency of a broadband antenna on a large frequency range. In addition, the efficiency of the reference antenna provided by the manufacturers or measured in an AC may be inaccurate which can lead to systematic errors.

7.1.2   One-antenna method

In order to avoid using a reference antenna of known radiation efficiency, Holloway *et al.* proposed three new methods to measure the antenna radiation efficiency [6], i.e., one-antenna, two-antenna and three-antenna methods. These methods are derived based on the knowledge of the quality factor (Q) of the RC in time and frequency domains. All these methods have slightly different prerequisites. A common setup for antenna radiation efficiency measurement using the one-antenna method is shown in Figure 7.2.

We assume that there are two antennas in the RC, i.e., antenna 1 and antenna 2 (not shown in Figure 7.2), the following derivation showing that antenna 2 is not required in fact. The frequency-domain quality factor (Q_{FD}) and the time-domain quality factor (Q_{TD}) can be calculated as follows [14]

$$Q_{FD} = C_{RC}\langle |S_{21,s}|^2 \rangle,$$

(7.10)

$$Q_{TD} = \omega\tau_{RC},$$

(7.11)

where $C_{RC} = 16\pi^2 V/\lambda^3$ with V being the volume of the RC $S_{21,s} = S_{21} - \langle S_{21} \rangle$ the stirred part of S_{21}, and τ_{RC} the chamber decay time, which reflects the decrease of the energy within the RC as a function of the time for a given loading condition.

It should be noted that (7.10) is only applicable to impedance-matched, lossless transmitting and receiving antennas. Therefore, Q_{FD} offers a lower estimation of the

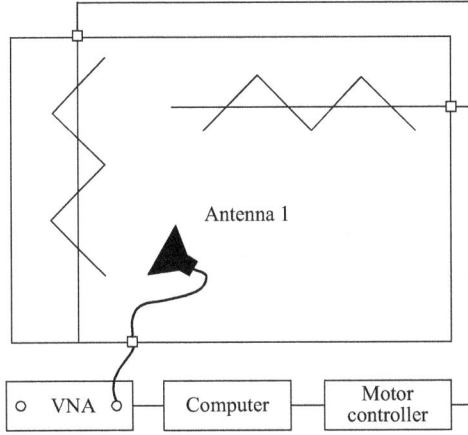

Figure 7.2   Setup for antenna radiation efficiency measurement in a reverberation chamber using the one-antenna method

quality factor than the true value due to the antenna losses in practice. However, Q_{TD} offers a better estimation of the quality factor because it is less affected by the antenna efficiency and impedance mismatch [6,14].

In order to obtain an accurate estimation of the quality factor using the frequency-domain method, Q_{FD} can be corrected as [6]

$$Q_{FD}^{cor} = C_{RC} \frac{\left\langle |S_{21,s}|^2 \right\rangle}{\eta_{1,tot}\eta_{2,tot}} = \frac{Q_{FD}}{\eta_{1,tot}\eta_{2,tot}}, \tag{7.12}$$

where Q_{FD}^{cor} is the corrected frequency-domain quality factor. Actually, the correction of Q_{FD} using (7.12) is the same as the correction of G_{ch} using (7.9). Obviously, the effects of the antenna losses on Q_{FD} are removed using (7.12). There are still discrepancies between the estimators (i.e., Q_{FD}^{cor} and Q_{TD}) and the true value due to the inherent estimation uncertainties, but Q_{FD}^{cor} and Q_{TD} offer basically consistent estimations of the quality factor, that is $Q_{FD}^{cor} = Q_{TD}$ [14,20]. Substituting $Q_{FD}^{cor} = Q_{TD}$ into (7.12), the relationship between total antenna efficiencies and Q values can be expressed as [6]

$$\eta_{1,tot}\eta_{2,tot} = \frac{Q_{FD}}{Q_{TD}}. \tag{7.13}$$

Assuming the two antennas are identical, which means that the two antennas have the same efficiency, i.e., $\eta_{1,tot} = \eta_{2,tot} = \eta_{tot}$, (7.13) can be rewritten as

$$\eta_{tot} = \sqrt{\frac{C_{RC}\left\langle |S_{21,s}|^2 \right\rangle}{\omega \tau_{RC}}}. \tag{7.14}$$

If there is only one antenna in the RC $\left\langle \left| S_{21,s} \right|^2 \right\rangle$ in (7.14) is not available. To avoid this problem, we introduce the enhanced backscatter coefficient (e_b) [14,21], which is defined as

$$e_b = \frac{\left\langle \left| S_{11,s} \right|^2 \right\rangle}{\left\langle \left| S_{21,s} \right|^2 \right\rangle}, \tag{7.15}$$

$e_b = 2$ for a well-stirred RC. For the continuity of the description of the one-antenna method, e_b will be discussed in the end of this section.

Combining (7.14) with (7.15), we can express the total radiation efficiency as

$$\eta_{1,tot} = \sqrt{\frac{C_{RC}\left\langle \left| S_{11,s} \right|^2 \right\rangle}{2\omega\tau_{RC}}}. \tag{7.16}$$

The antenna radiation efficiency estimated using the one antenna method can then be expressed as

$$\eta_{1,rad} = \frac{1}{\left(1 - |\langle S_{11} \rangle|^2\right)}\sqrt{\frac{C_{RC}\left\langle \left| S_{11,s} \right|^2 \right\rangle}{2\omega\tau_{RC}}}. \tag{7.17}$$

Using the one-antenna method, it is therefore sufficient to measure S_{11} of the AUT (provided that τ_{RC} is determined *a priori*). However, a prerequisite of the method is that the enhanced backscatter coefficient is equal to two, which is only an approximation of a well-stirred RC. Moreover, precise calibration of the RC should be conducted and proper sampling parameters of the VNA (ensuring the application of the inverse Fourier transform) should be chosen to obtain accurate estimation of the chamber decay time.

From the point of view of the definition of the enhanced backscatter coefficient, a ray model is established to explain the relationship between $\left\langle \left| S_{11,s} \right|^2 \right\rangle$ and $\left\langle \left| S_{21,s} \right|^2 \right\rangle$ via geometrical optics [14,21], and an illustration of the model is shown in Figure 7.3.

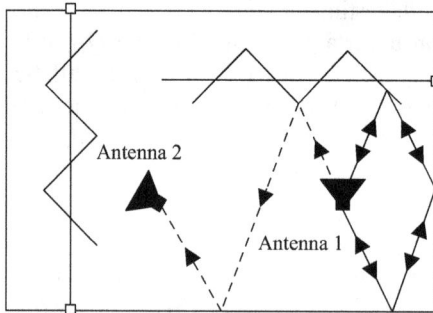

Figure 7.3   Illustration of a ray travelling from antenna 1 to antenna 2 contributing to $S_{21,s}$ (dashed line), and a ray emitted by antenna 1 travelling back to antenna 1 contributing to $S_{11,s}$ (solid line)

Taking the chamber loss (imperfect conductivity of the chamber walls and stirrers) into consideration, $S_{21,s}$ can be approximated by the accumulation of a large (but finite) number of rays:

$$S_{21,s} = \sum_{n=1}^{N} A_n \frac{\exp(ikr_n)}{r_n}, \tag{7.18}$$

where r_n is the length of the nth ray and A_n is a complex coefficient depending of the radiation pattern and of the characteristics of the chamber. The square of the absolute value of $S_{21,s}$ can be calculated based on (7.18)

$$\left|S_{21,s}\right|^2 = S_{21,s}S_{21,s}^* = \sum_{n=1}^{N} A_n \frac{\exp(ikr_n)}{r_n} \sum_{m=1}^{N} A_m \frac{\exp(ikr_m)}{r_m}, \tag{7.19}$$

where $*$ denotes the complex conjugate operator. Since the rays are assumed to be not correlated with any other, $\left\langle \left|S_{21,s}\right|^2 \right\rangle$ can be calculated and simplified as

$$\left\langle \left|S_{21,s}\right|^2 \right\rangle = \left\langle \sum_{n=1}^{N} \frac{|A_n|^2}{r_n^2} \right\rangle = N \left\langle \frac{|A_n|^2}{r_n^2} \right\rangle. \tag{7.20}$$

The second average $\langle \cdot \rangle$ denotes the ensemble of both stirrer positions and ray paths.

Different when dealing with $S_{21,s}$, the locations of the TX Ant and receiving antenna are identical to $S_{11,s}$. For each ray emitted by the TX Ant and travelling back to the TX Ant, there must be a companion ray travelling along the same path but in the reverse direction (as depicted in Figure 7.3). Hence, there are still N rays contributing to $S_{11,s}$ but every two reciprocal rays that possess the same phase can be merged as one ray that contributes double to the $S_{11,s}$:

$$S_{11,s} = \sum_{n=1}^{N/2} 2A_n \frac{\exp(ikr_n)}{r_n}, \tag{7.21}$$

$\left\langle \left|S_{11,s}\right|^2 \right\rangle$ can then be calculated in a similar way as (7.20)

$$\left\langle \left|S_{11,s}\right|^2 \right\rangle = 4 \left\langle \sum_{n=1}^{N/2} \frac{|A_n|^2}{r_n^2} \right\rangle = 2N \left\langle \frac{|A_n|^2}{r_n^2} \right\rangle. \tag{7.22}$$

Combining (7.20) with (7.22), we achieve $\left\langle \left|S_{11,s}\right|^2 \right\rangle = 2\left\langle \left|S_{21,s}\right|^2 \right\rangle$, that is, $e_b{=}2$. This value is supposed to be obtained under ideal conditions, i.e. when the number of stirring samples tending toward the infinite.

To have a better understanding of e_b, measurements are performed from 1 to 4 GHz in an RC. The RC has a size of $1.50 \times 1.44 \times 0.92$ m^3 and the lowest usable frequency around 0.87 GHz. The measured e_b with different stirring sample numbers (i.e., $N_S = 20$ and 100) is shown in Figure 7.4. For $N_S = 100$, the measured e_b oscillates in a small range around 2 over the whole test frequency. However, the measured e_b oscillates seriously around a value that is slightly higher than 2 for $N_S = 20$.

It can be concluded that $e_b=2$ is only satisfied in an ideal situation, while there exist oscillations in practical measurements. In order to reduce the oscillations around 2 and make e_b closer to 2, the increase of the independent sample number is an effective way. However, the oscillations are definitely obtained regardless of the independent sample number (even large), which results in the inherent measurement uncertainty of the one-antenna method.

7.1.3   Two-antenna method

A common setup for antenna radiation efficiency measurement using the two-antenna method is shown in Figure 7.5.

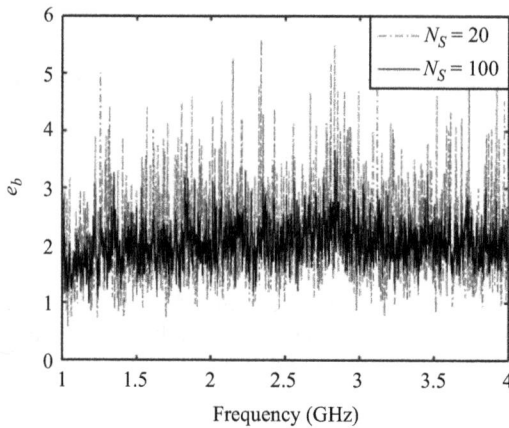

Figure 7.4   *Measured e_b for different stirring sample numbers from 1 to 4 GHz*

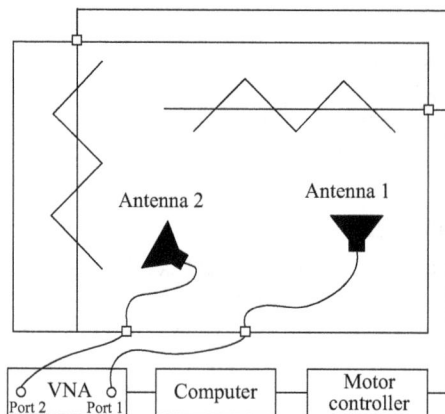

Figure 7.5   *Setup for antenna radiation efficiency measurement in a reverberation chamber using the two-antenna method*

The measurement setup is similar to the one-antenna method, while the difference is that another antenna (i.e., antenna 2) is connected to the other port of a VNA. The derivation of the two-antenna method is based on the one-antenna method. From previous discussions, we have:

$$\eta_{1,tot}^2 = \frac{C_{RC}\left\langle\left|S_{11,s}\right|^2\right\rangle}{e_{b1}\omega\tau_{RC}}.$$

(7.23)

$$\eta_{2,tot}^2 = \frac{C_{RC}\left\langle\left|S_{22,s}\right|^2\right\rangle}{e_{b2}\omega\tau_{RC}}.$$

(7.24)

Combining (7.23) with (7.24) and (7.13), we can obtain:

$$\frac{\eta_{1,tot}}{\eta_{2,tot}}e_{b1} = \frac{\left\langle\left|S_{11,s}\right|^2\right\rangle}{\left\langle\left|S_{21,s}\right|^2\right\rangle}.$$

(7.25)

$$\frac{\eta_{2,tot}}{\eta_{1,tot}}e_{b2} = \frac{\left\langle\left|S_{22,s}\right|^2\right\rangle}{\left\langle\left|S_{21,s}\right|^2\right\rangle}.$$

(7.26)

Assuming $e_{b1} = e_{b2}$ (an assumption discussed later):

$$e_b = e_{b1} = e_{b2} = \frac{\sqrt{\left\langle\left|S_{11,s}\right|^2\right\rangle\left\langle\left|S_{22,s}\right|^2\right\rangle}}{\left\langle\left|S_{21,s}\right|^2\right\rangle}.$$

(7.27)

Thus, (7.23) and (7.24) can be rewritten as

$$\eta_{1,tot} = \sqrt{\frac{C_{RC}\left\langle\left|S_{11,s}\right|^2\right\rangle}{e_b\omega\tau_{RC}}},$$

(7.28)

$$\eta_{2,tot} = \sqrt{\frac{C_{RC}\left\langle\left|S_{22,s}\right|^2\right\rangle}{e_b\omega\tau_{RC}}}$$

(7.29)

respectively.

The radiation efficiency can be expressed in a similar way as (7.17).

The advantage of the two-antenna method over the one-antenna method is that the former does not require $e_b = 2$ (which is not always very accurate in a practical RC). Instead, it only assumes that $e_{b1} = e_{b2}$, i.e., e_b should be identical to different antenna positions under a certain loading condition, which indicates that a good spatial uniformity is required in the chamber.

The theoretical assumption of $e_{b1} = e_{b2}$ means that e_{b1} and e_{b2} can be a reasonable value close to 2, that is, e_{b1} and e_{b2} are not required to be strictly equal to 2. The assumption of the two-antenna method is therefore more relaxed than that of the one-antenna method. However, the measured e_b oscillates around a certain value (the theoretical value) that depends on the independent sample number as discussed in Section 7.1.2. In other words, higher measured e_b means smaller independent sample number, further larger oscillation range, and worse spatial uniformity. Therefore, the assumption is relaxed, but a larger independent sample number is still needed to ensure a satisfying measurement accuracy. Furthermore, it should be noted that there definitely exists discrepancy between e_{b1} and e_{b2} due to the oscillation of the measured e_b (as shown in Figure 7.4), which means $e_{b1} \neq e_{b2}$ in practice. Similar to the one-antenna method, the uncertainty related to the new assumption on e_b (i.e., $e_{b1} = e_{b2}$) still exists (even reduced) regardless of the independent sample number, which results in the inherent measurement uncertainty of the two-antenna method.

It can be seen that the measured antenna efficiency is determined mainly by $S_{ii,s}(i = 1, 2)$, and τ_{RC} can also be extracted from $S_{ii,s}$. If we suppose the antenna radiation efficiency to be very low, it makes $S_{ii,s}$ indistinguishable with the VNA noise. Xu *et al.* proposed a modified two-antenna method described hereafter to overcome this problem [7].

Assuming $e_b = e_{b1} = e_{b2} = 2$ in (7.27), we have:

$$2\left\langle |S_{21,s}|^2 \right\rangle = \sqrt{\left\langle |S_{11,s}|^2 \right\rangle \left\langle |S_{22,s}|^2 \right\rangle}. \tag{7.30}$$

If antenna 1 is very lossy while antenna 2 has a high efficiency, (7.28) can be rewritten as

$$\eta_{1,tot} = \left\langle |S_{21,s}|^2 \right\rangle \sqrt{\frac{2C_{RC}}{\omega \tau_{RC} \left\langle |S_{22,s}|^2 \right\rangle}}. \tag{7.31}$$

In the meantime, if antenna 2 is very lossy while antenna 1 has a high efficiency, (7.29) can be rewritten as

$$\eta_{2,tot} = \left\langle |S_{21,s}|^2 \right\rangle \sqrt{\frac{2C_{RC}}{\omega \tau_{RC} \left\langle |S_{11,s}|^2 \right\rangle}}. \tag{7.32}$$

The modified two-antenna method can therefore be used to measure a highly lossy antenna, the key is to utilize the enhanced backscatter coefficient to remove the small $S_{ii,s}$ parameter from the formula of the two-antenna method. This modified method has two prerequisites. First, the RC should be well stirred to ensure the enhanced backscatter coefficient is equal to two, which implies that this method suffers from the same limitation as the one-antenna method. Second, one of the antennas under test must have a high efficiency. In other words, this method cannot measure two highly lossy antennas simultaneously.

7.1.4　Three-antenna method

A common setup for antenna radiation efficiency measurement using the three-antenna method is shown in Figure 7.6.

Three antennas are inserted in the RC: antenna 1, antenna 2, and antenna 3. Measurements are conducted between each pair of two antennas. The derivation of

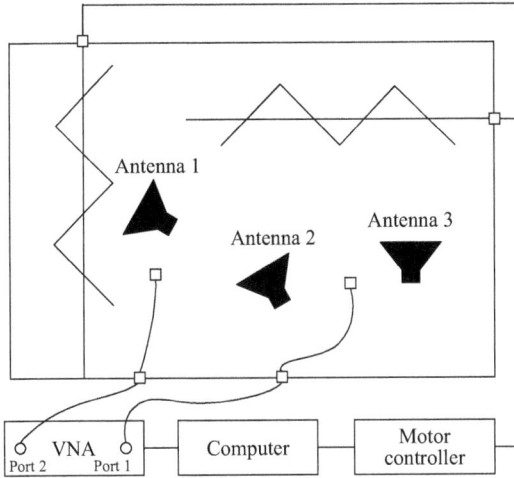

Figure 7.6 Setup for antenna radiation efficiency measurement in a reverberation chamber using the three-antenna method

the three-antenna method is based on the two-antenna method. According to (7.13), we can write:

$$\eta_{1,tot}\eta_{2,tot} = \frac{C_{RC}\langle|S_{21,s}|^2\rangle}{\omega\tau_{RC}}. \tag{7.33}$$

$$\eta_{1,tot}\eta_{3,tot} = \frac{C_{RC}\langle|S_{31,s}|^2\rangle}{\omega\tau_{RC}}. \tag{7.34}$$

$$\eta_{2,tot}\eta_{3,tot} = \frac{C_{RC}\langle|S_{32,s}|^2\rangle}{\omega\tau_{RC}}. \tag{7.35}$$

The antenna efficiencies can be easily solved from the following three equations:

$$\eta_{1,tot} = \sqrt{\frac{C_{RC}\langle|S_{21,s}|^2\rangle\langle|S_{31,s}|^2\rangle}{\omega\tau_{RC}\langle|S_{32,s}|^2\rangle}}. \tag{7.36}$$

$$\eta_{2,tot} = \sqrt{\frac{C_{RC}\langle|S_{21,s}|^2\rangle\langle|S_{32,s}|^2\rangle}{\omega\tau_{RC}\langle|S_{31,s}|^2\rangle}}. \tag{7.37}$$

$$\eta_{3,tot} = \sqrt{\frac{C_{RC}\langle|S_{31,s}|^2\rangle\langle|S_{32,s}|^2\rangle}{\omega\tau_{RC}\langle|S_{21,s}|^2\rangle}}. \tag{7.38}$$

Compared with both the one-antenna and two-antenna methods, this method has nearly no limitations, i.e., prerequisites on e_b. Therefore, there is no inherent measurement uncertainty on e_b of this method. However, it should be stressed that this method requires three antennas under test.

In practical measurements, except for the uncertainty of e_b, the uncertainty of S-parameters and τ_{RC} also contribute to the total uncertainty of all the non-reference antenna methods (as can be seen from (7.16), (7.28), and (7.36)). Thus, it is hard to determine the relationship between the uncertainties of these three non-reference antenna methods.

7.1.5   Time reversal method

Most researches on antenna efficiency measurements in the RC are conducted in the frequency domain. These methods have two requirements: a continuous wave excitation and a large number of stirring samples. This may lead to a time-consuming measurement, especially when the requirement on the measurement accuracy is high or if the AUT is a UWB antenna. Le Fur *et al.* proposed a time-domain measurement method for UWB antennas utilizing a pulse excitation [8]. The method is time efficient compared with the frequency-domain method. Measurement setup for this method is shown in Figure 7.7.

This method requires three antennas, i.e., a reference antenna with known efficiency (REF), the UWB AUT, and a fixed TX Ant. An arbitrary waveform generator (AWG) and a digital storage oscilloscope (DSO) with an adequate bandwidth and

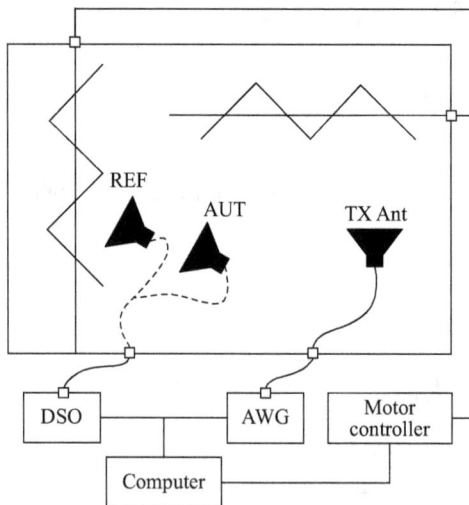

Figure 7.7   Setup for the antenna radiation efficiency measurement in a reverberation chamber using the time reversal method

sampling rate are needed instead of a VNA, as in the frequency-domain method. The measurement procedure is as follows:

1. Connect the fixed TX Ant to the AWG and the reference antenna to the DSO. The UWB AUT is loaded with a 50 Ω termination. Ensure that all the antennas are placed in the RC during the whole measurement procedure.
2. A pulse (referred to as "signal 1") is generated by the AWG and transmitted by the fixed antenna
3. The signal received by the reference antenna is sampled, averaged on a large number of acquisitions (e.g., 512 is adopted in [8]), and recorded by the DSO. The averaged signal is the truncated pulse response of the RC, which is referred to as "signal 2."
4. Signal 2 is time reversed and amplified, ensuring that the amplitude of the processed signal is the same as signal 1. The time-reversed and amplified signal is referred to as "signal 3," which is transmitted by the reference antenna.
5. The signals are received by the fixed antenna, but only the initial received pulse is recorded due to the multipath propagation of the RC. The received signal is referred to as "signal 4," and the amplitude of signal 4 referred to as $A_{REF}(t)$
6. Connect the UWB AUT to the DSO and load the reference antenna with a 50 Ω termination, repeat steps 2–5 to obtain the amplitude $A_{AUT}(t)$ of the "signal 4" received by the fixed antenna.

In order to have a better understanding of the time-reversed signal, a model of two antennas is established. One antenna transmits the impulse signal, and the other antenna works as receiving antenna, and the received impulse response and the time-reversed impulse response are shown in Figure 7.8.

Note that the pulse response of signal 1 is averaged over a large number of acquisitions, to improve the signal-to-noise ratio (SNR). More signals can be recorded to further improve the SNR by repeating steps 4 and 5. Different stirring positions can be also used to improve the signals at the cost of an increase of the measurement time.

Once all the signals are recorded, the antenna efficiency could be derived as described in the following subsections.

7.1.5.1 Averaged antenna efficiency

The pulse energy of signal 4 is computed as

$$E_4 = c \int_0^T |A(t)|^2 dt, \tag{7.39}$$

where c is a constant and T is the duration of signal 4 after filtering.

From the point of view of the generation process of signal 4, it can also be written as

$$E_4 = (\eta_{avg}^t \eta_{avg}^r)^2 KE_1 IL_{RC}^2, \tag{7.40}$$

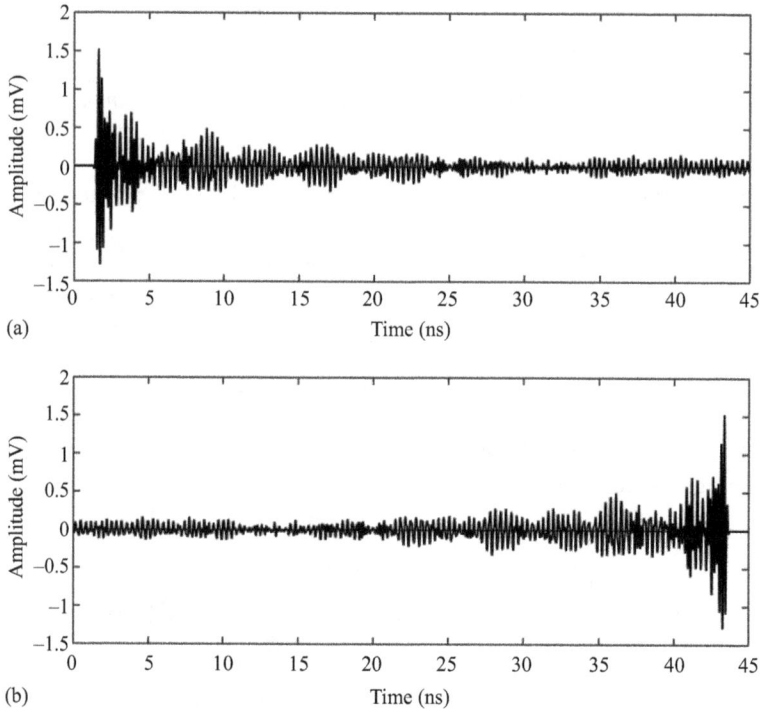

Figure 7.8   Example of (a) an impulse response and (b) a time-reversed impulse response from inverse Fourier transform of S_{21} measurement

where η_{avg}^{t} is the averaged antenna efficiency of the fixed TX Ant, η_{avg}^{r} is the averaged antenna efficiency of the receiving antenna (the UWB AUT or the reference antenna), K is the amplification coefficient of signal 2, E_1 is the pulse energy of signal 1, and IL_{RC} is the chamber insertion loss. The two squared items in (7.40) are the consequence of both transmission procedure made to obtain "signal 1" and "signal 3."

Since the UWB AUT and the reference antenna are measured separately, we have two equations based on (7.40), and the frequency averaged antenna efficiency of the UWB AUT can be calculated as

$$\eta_{avg}^{AUT} = \eta_{avg}^{REF} \sqrt{\frac{E_{4,AUT} K_{REF}}{E_{4,REF} K_{AUT}}}. \tag{7.41}$$

If we have two reference antennas and two antennas under test, we can use two reference antennas as the transmitting and receiving ones in the first time and substitute the reference antennas with both antennas under test in the second time. We have in this case:

$$\eta_{avg}^{AUT} = \eta_{avg}^{REF} \left(\frac{E_{4,AUT} K_{REF}}{E_{4,REF} K_{AUT}} \right)^{1/4}. \tag{7.42}$$

The averaged antenna efficiency can also be obtained in a onetime measurement using a modified version of the method. We still need three antennas as the original method, but the reference antenna and the AUT are used as receiving antennas simultaneously. The measurement procedure is the same, while the only difference is that signal 3 is obtained by processing the sum of signal 2 received by both antennas. Obviously, half of the measurement time is saved in this modified time reversal method.

Assuming there are three antennas under test, measurements are conducted between each pair of two antennas. According to (7.40), we have:

$$E_{4,12} = (\eta_{avg}^1 \eta_{avg}^2)^2 K_{12} E_1 IL_{RC}^2. \tag{7.43}$$

$$E_{4,13} = (\eta_{avg}^1 \eta_{avg}^3)^2 K_{13} E_1 IL_{RC}^2. \tag{7.44}$$

$$E_{4,23} = (\eta_{avg}^2 \eta_{avg}^3)^2 K_{23} E_1 IL_{RC}^2. \tag{7.45}$$

The averaged antenna efficiency is obtained from (7.43)–(7.45) as

$$\eta_{avg}^1 = \left(\frac{E_{4,12} E_{4,13} K_{23}}{E_{4,23} E_1 K_{12} K_{13} IL_{RC}^2} \right)^{1/2}. \tag{7.46}$$

The averaged antenna efficiency can also be obtained without the need of a reference antenna by using time-domain three-antenna method (in order to distinguish with the three-antenna method proposed by Holloway). This method shows better measurement accuracy. However, it requires three antennas working in the same bandwidth, and the measurement time is increased. The chamber insertion loss (IL_{RC}) has also to be measured to obtain the averaged antenna efficiency.

7.1.5.2   Total frequency-dependent antenna efficiency

It is worth noting that the measurement procedure given in Section 7.1.5.1 to obtain the averaged antenna efficiency only provides a single value for the entire bandwidth, which makes it not really useful in practice. Indeed, it is important to obtain the antenna efficiency as a function of the frequency.

When Fourier transformations are applied, all the time-domain signals in the measurement procedure can be transformed to the frequency domain. Similar to (7.40), we have

$$S_{4,REF}(f) = \eta_{tot}^{Tx^2}(f)\,\eta_{tot}^{REF^2}(f)\,K_{REF}S_1(f)\,IL_{RC}^2(f), \tag{7.47}$$

$$S_{4,AUT}(f) = \eta_{tot}^{Tx^2}(f)\,\eta_{tot}^{AUT^2}(f)\,K_{AUT}S_1(f)\,IL_{RC}^2(f), \tag{7.48}$$

where $S_{4,REF}(f)$ and $S_{4,AUT}(f)$ are the spectrum of signal 4 for the reference antenna and for the AUT, respectively, and $S_1(f)$ is the spectrum of the initial pulse. Thus, the antenna efficiency in the frequency domain is obtained as

$$\eta_{tot}^{AUT}(f) = \eta_{tot}^{REF}(f)\sqrt{\frac{S_{4,AUT}(f)\,K_{REF}}{S_{4,REF}(f)\,K_{AUT}}}. \tag{7.49}$$

The antenna efficiency in the frequency domain measured using other methods discussed in Section 7.1.5.1 can also be obtained in a similar way as (7.49).

The relationship between the averaged antenna efficiency and the total efficiency can be obtained through the following expression:

$$\eta_{avg} = \frac{\int_{f_1}^{f_2} S_e(f)\eta_{tot}(f)df}{\int_{f_1}^{f_2} S_e(f)df}, \tag{7.50}$$

where $S_e(f)$ is the Fourier transformation of the initial transmitted pulse power, f_1 and f_2 correspond to the bandwidth where the antenna is well matched.

Compared with the frequency-domain methods, the time-domain pulse method does not deal with continuous wave excitations. Due to the substantial reduction of stirrer positions (see detailed discussion in Section 7.1.6), this method can reduce the measurement time significantly when a UWB antenna is measured. However, this method has some drawbacks, i.e., the measured efficiency can be influenced by the characteristics of the pulse. In other words, if the pulse is changed, the measured efficiency may also change. The signals need to be preprocessed and post-processed (e.g. sampling, truncation, averaging, and time reversal), which complicates the measurement procedure.

7.1.5.3   Sub-band time reversal method

The time reversal method offers a good estimation of the antenna efficiency. However, the power spectral density of the time-reversed signal (signal 3) varies within the bandwidth, which leads to a higher measurement uncertainty. In order to improve the measurement uncertainty of the time reversal method, Naqvi *et al.* proposed a sub-band time reversal method [9].

The sub-band time reversal method aims at obtaining a flat power spectral density of the time-reversed signal over the bandwidth, ensuring that the antenna efficiency over the whole bandwidth contributes equally to the power related of the received signal. Therefore, the sub-band filtering scheme (cf. Figure 7.9) is adopted, in which $S(t)$ represents the input time-reversed signal, and $S'(t)$ represents the output time-reversed signal after filtering.

The time-reversed signal in the frequency domain is obtained using the fast Fourier transform (FFT). Then the signal is filtered by N band-pass filters and transformed back into time domain via the inverse FFT. The power of each sub-band is normalized using equal power control (EPC), and, thus, each sub-band contributes

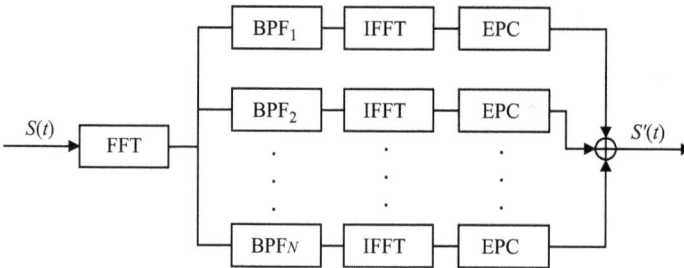

Figure 7.9   Block diagram of the sub-band filtering scheme

equally to the power spectral density. The new time-reversed signal is generated by adding up all the normalized signals. Note that the total power of the signals before and after filtering should be the same.

The sub-band time reversal method has been proven to reduce the measurement uncertainty of time reversal method [8,9].

7.1.6 Discussion on antenna efficiency measurement methods

7.1.6.1 Non-reference antenna method

For an intuitive comparison of the one-, two-, and three-antenna methods proposed by Holloway, measurements of three antennas were conducted and the corresponding results are shown in [6]. It can be seen that the measurement results are close to each other, especially for the two-antenna and three-antenna methods. However, the one-antenna method estimates a higher efficiency than the other two methods. This is due to the prerequisite $e_b = 2$ which is not satisfied in a practical RC.

As discussed in previous sections, the one-antenna method has a prerequisite that $e_b = 2$. As shown in Section 7.1.2, e_b oscillates around a certain value (which may be higher than 2) due to imperfect stirring in practical RC measurements. Moreover, e_b is also measured in [6]. The results show that e_b is higher than 2 and varies as much as 4%. It should be stressed that the variation of e_b depends on the number of independent samples (as discussed in Section 7.1.2); thus, the obtained variation (i.e., 4%) corresponds to the specific measurement configurations in [6]. It is also shown that the difference of the measured efficiencies using the one-antenna and two-antenna methods is about 4%. This explains why the one-antenna method gives a higher estimation of the antenna efficiency.

Holloway *et al.* gave a rough estimation of measurement uncertainty of the three non-reference antenna methods in [6]. The measurement uncertainty is composed of two parts [22]. The first part is the uncertainty related to S-parameter measurement (i.e., S_{11}, S_{22}, S_{21}, and τ_{RC}). The second part is the uncertainty caused by the fluctuation of measurement equipment (i.e., VNA and cable), the inconsistence of the measurement environment, and the theoretical assumption on e_b. For the convenience of expression, we denote the first part of the uncertainty as u_1, and the second part as u_2. On the whole, the measurement uncertainties of the one-antenna, two-antenna, and three-antenna methods are 0.38, 0.35, and 0.49 dB, or 9%, 8%, and 12%, respectively. The measurement uncertainty in percentage form (σ_p) can be transformed to the dB form (σ_{dB}) using $\sigma_{dB} = 10 \times \log_{10}(1 + \sigma_p)$. These uncertainties are conservative estimations, and the uncertainties can be reduced if more stirring samples are adopted [6].

The results in [6] show that the measurement uncertainty of the three-antenna method is the highest among the non-reference antenna methods. As mentioned in Section 7.1.4, the uncertainty of e_b contributes only a portion of the total uncertainty. Meanwhile, the S-parameter-related measurements and the uncertainty of measurement equipment also contribute to the total uncertainty. For the three-antenna method, the uncertainty of e_b is removed; thus, u_2 is the lowest [6]. This is in accordance with the theoretical analysis. However, the three-antenna method requires more

S-parameter measurements than the other two methods (comparing (7.16), (7.28) with (7.36)), which leads to the highest u_1 [6]. Combining u_1 with u_2, the total uncertainty of the three-antenna method is the highest. It should be stressed that the results do not mean the relationships between the uncertainties of the three non-reference antenna methods are the same as [6], because the uncertainty depends on the specific measurement setup.

The RC should be calibrated carefully to obtain accurate value of the chamber decay time (τ_{RC}). If the RC loading is the same during all the measurements, only one calibration is needed. The non-reference antenna methods have also another assumption, i.e., the chamber losses (including wall losses, paddle losses, and absorbers losses) dominate in the RC. If the antenna losses dominate, the received signal level drops significantly, which makes τ_{RC} hard to determine [6,7].

Li *et al.* have compared antenna efficiency measurements in two different RCs using the one- and two-antenna methods [23]. The measurements were performed in two different RCs (located at UK National Physical Laboratory and the University of Liverpool) for two different antennas (i.e., a directional antenna and an omnidirectional antenna). The two RCs have different sizes, Q factors, stirrers, chamber decay time and enhanced backscatter coefficients, but the measured radiation efficiencies of the two antennas from both RCs show good agreements (cf. Figure 7.10). Since e_b of the University of Liverpool's RC is much closer to 2 than NPL's above 400 MHz [23], the differences between the one-antenna method and the two-antenna method corresponding to University of Liverpool's RC are smaller (cf. Figure 7.10). This is in accordance with the researches in [6]. The discrepancy of the measured radiation efficiencies between both RCs is below 5% for the omnidirectional antenna (Figure 7.10(b)) and 10% for the directional antenna (Figure 7.10(a)). This indicates the robustness of the one- and two-antenna methods.

Further measurements were performed with combinations of two polarizations (i.e., vertical polarization and horizontal polarization) for two antennas, and the results are shown in [23]. It can be seen that the discrepancy for the omnidirectional antenna is within 8% for both methods, while for the directional antenna, the discrepancy is mostly within 10%. The slightly larger discrepancy obtained for the directional antenna can be explained as this antenna is more sensitive to the polarization mismatch in the RC. This implies that polarization stirring can be used to improve the measurement accuracy, especially for directional antennas.

7.1.6.2   Time reversal method and sub-band time reversal method

Several points have to be noted on the measurement procedure related to the time reversal method. A measurement example and related parameters are given in [8]. The first received signal (signal 2) has to be averaged over a certain amount of acquisitions. As mentioned earlier, the SNR is improved because the noise is averaged out, especially for the signals, the power of which is lower than the noise level. Unexpected signals are sometimes recorded due to the DSO trigger defaults or synchronization issues between the DSO and the AWG [8]. Thus, the hidden "wrong signals" should be identified and removed in the post-processing. Signal 4 has to be obtained through several acquisitions to reduce the measurement uncertainty.

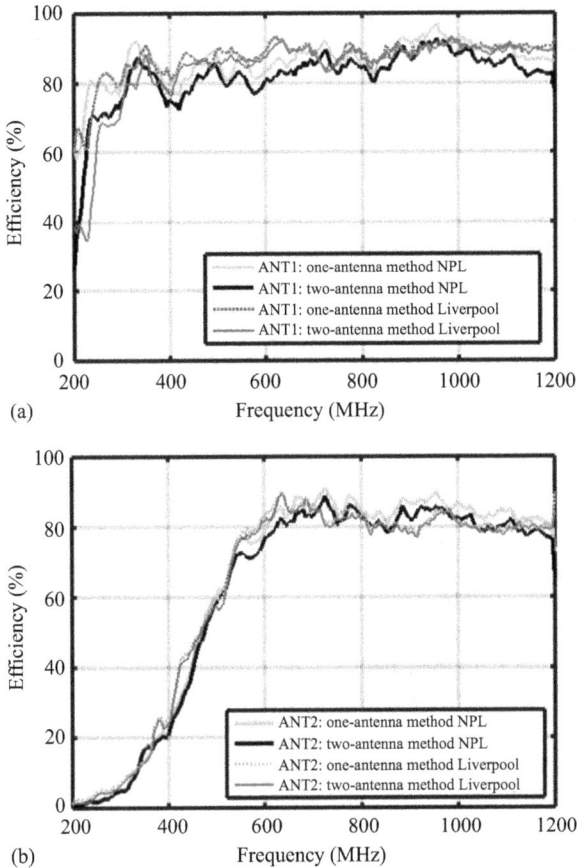

Figure 7.10   *Measured antenna efficiency using the one-antenna and two-antenna*
                methods in two RCs: (a) log-periodic antenna (directional) and
                (b) biconical antenna (omnidirectional) (extracted from [23])

The measurement results in [8] show that increasing the number of both stirring positions and acquisitions can reduce the measurement uncertainty. The total efficiency converges to a particular value as the number of acquisitions increases. However, increasing the number of stirring positions does not improve the performance significantly. This may be due to the correlations between the samples.

The time needed for the measurement example (including four stirring positions and ten acquisitions) in [8] is about half an hour.

The UWB antenna efficiency measured using the time reversal method and sub-band time reversal method for 12 stirring positions is shown in [9]. It can be seen that the general trend of the results is similar, but there is an obvious improvement in measurement uncertainty using the sub-band time reversal method.

For a better comparison of the time reversal method and the sub-band time reversal method, the standard deviation (STD) of the antenna efficiency with both methods is shown in [9]. The STD corresponding to the time reversal method fluctuates widely over the interested bandwidth and reaches 3.91% at the peak, whereas the STD corresponding to the sub-band time reversal method remains below 1.72%.

7.1.6.3   Comparison

As can be seen from the discussion in previous sections, all the methods can determine the antenna total efficiency. For the frequency-domain methods, the radiation efficiency can be straightforwardly obtained without additional measurements. However, the reflection coefficient of the antenna is unknown with the time-domain method, which means that additional measurements are needed to obtain the radiation efficiency.

The antenna efficiency can be measured with or without a reference antenna, in time or frequency domain. Each method has its own characteristics (e.g., measurement setup, measurement time, and complexity of the pre- or post-processing, inherent uncertainty). To facilitate the comparison of these methods, the characteristics of each method are summarized in Table 7.1. The measurement accuracies and uncertainties of these different methods (which are shown in the published papers) are calculated following different schemes (some are not even provided) and therefore not provided in the table. Instead, the corresponding inherent uncertainties of these methods are stressed. Without special statements in the table, all the methods are suitable for electrically small and large antennas, and stirring methods (i.e., mechanical stirring, frequency stirring, and source stirring) and VNA are needed for frequency-domain methods.

7.2   Statistical analysis of antenna efficiency measurement uncertainty

Since these methods are based on statistics, they are incomplete without an analysis of the measurement uncertainty. In this section, we introduce a statistical method to analyze the measurement uncertainty of measured antenna efficiency [24]. The statistical method presented here computes the theoretical measurement uncertainty as a function of the used method. The procedures are mainly composed of three steps:

1.   Extracting the relationship between the measured antenna efficiency and the true antenna efficiency, that is, representing the measured antenna efficiency using an expression containing the true antenna efficiency;
2.   Deriving and verifying the probability density function (pdf) of the expression obtained in step 1;
3.   Based on the pdf obtained in step 2, further deriving the related statistics, e.g., mean, variance.

Obviously, the key point of the statistical method is to derive the pdf of the specified expression. Theoretically, the statistical method can be used to analyze the measurement uncertainty of different RC methods. However, the pdf of the specified

Table 7.1 *Comparison of different RC methods to determine the antenna efficiency*

Measurement method	Characteristics
Reference antenna method	1. A reference antenna with known radiation efficiency is required. 2. Two measurements (reference and AUT measurements) over a complete mode stirring process are needed
One-antenna method	1. No more antennas are needed except the AUT. 2. One measurement over a complete mode stirring process is needed. 3. Precise calibration and proper sampling parameter are needed to ensure an accurate estimation of the chamber decay time τ_{RC} 4. Inherent uncertainty caused by enhanced backscatter coefficient ($e_b = 2$)
Two-antenna method	1. Two AUTs and one measurement over a complete mode stirring process are needed. 2. Precise calibration and proper sampling parameter are needed to ensure an accurate estimation of the chamber decay time τ_{RC} 3. Inherent uncertainty caused by enhanced backscatter coefficient ($e_{b1} = e_{b2}$)
Modified two-antenna method	1. "Specified suitable" for highly lossy antennas and one measurement over a complete mode stirring process is needed, but cannot measure two highly lossy antennas simultaneously. 2. Inherent uncertainty caused by enhanced backscatter coefficient ($e_{b1} = e_{b2}$)
Three-antenna method	1. Three AUTs and measurements over a complete mode stirring process are needed. 2. Precise calibration and proper sampling parameter are needed to ensure an accurate estimation of the chamber decay time τ_{RC}
Time reversal method	1. "Specified suitable" for UWB antennas, measurement time can be reduced significantly. 2. Time-domain pulse is used instead of the continuous wave excitation, AWG and DSO are needed instead of VNA 3. Efficiency of reference antenna is needed for reference antenna method, and chamber insertion loss is needed for the time-domain three-antenna method. 4. Inherent uncertainty caused by characteristics of the pulse. 5. Preprocessing and post-processing of the signals are needed. 6. Mechanical stirring is optional to further improve uncertainty
Sub-band time reversal method	1. Measurement uncertainty can be further reduced compared with time reversal method. 2. Mechanical stirring is no more needed. 3–7 are the same as 1–5 of time reversal method

expression and the related statistics are usually difficult to derive, which makes the theoretical measurement uncertainty not an easy work.

7.2.1 Statistics of measured antenna efficiency

Since the reference antenna method is widely used, we analyze the measurement uncertainty of this method. Certainly, the measurement uncertainty of other antenna efficiency methods can also be analyzed using the statistical method by following the procedures described in the previous sections. Nevertheless, we focus on the statistics of the popular reference antenna method here.

Based on the concept of average power transfer function, (7.7) can be rewritten as,

$$\hat{\eta}_{AUT\_tot} = \frac{\hat{P}_{AUT}}{\hat{P}_{REF}} \eta_{REF\_tot}, \tag{7.51}$$

where the hat $\hat{}$ denotes the estimate of the true value, and P_{AUT} and P_{REF} are the average power transfer functions for the AUT and reference antenna, respectively.

Note that (7.51) provides an estimate of the total antenna efficiency of the AUT. Thus, there is a deviation between the measured efficiency and the actual efficiency.

To facilitate the derivation of the distribution of the measured efficiency, the net average power transfer function is defined as

$$\hat{G} = \frac{\hat{P}}{\eta_{tot}}. \tag{7.52}$$

Thus, (7.51) can be rewritten as

$$\hat{\eta}_{AUT\_tot} = \frac{\hat{G}_{AUT}}{\hat{G}_{REF}} \eta_{AUT\_tot}. \tag{7.53}$$

Since η_{AUT_tot} is the true value, we are interested in the distribution of the random variable $T = \hat{G}_{AUT}/\hat{G}_{REF}$ only. Here, we assume that the net average power transfer function (i.e., $\hat{G}_{AUT}$ and $\hat{G}_{REF}$) is obtained by N-independent samples: $\hat{G} = \sum_{i=1}^{N} G_i/N$. In a well-stirred RC, G_i is an independent and identically distributed random variable that follows the exponential distribution [25]. Thus, $\hat{G}$ follows the Gamma distribution [26], where G is the mean of G_i. The pdf of Gamma is

$$f(x) = \frac{x^{N-1} \exp(-x/G_0)}{G_0^N \Gamma(N)}, \tag{7.54}$$

where Γ is the Gamma function. For an integer N, $\Gamma(N) = (N-1)!$, where ! denotes the factorial operator.

Since $\hat{G}_{AUT}$ and $\hat{G}_{REF}$ follow the Gamma distribution, the pdf of T can be derived [24] as

$$f(t) = \frac{\Gamma(2N)}{(\Gamma(N))^2} \frac{t^{N-1}}{(1+t)^{2N}}. \tag{7.55}$$

The mean, variance, and other statistics of the random variable T can be derived easily based on the distribution:

$$E(T) = \frac{N}{N-1}, \tag{7.56}$$

$$Var(T) = \frac{N(2N-1)}{(N-2)(N-1)^2}, \tag{7.57}$$

where *Var* denotes the variance. As T is related to the measured antenna efficiency by $\hat{\eta}_{AUT\_tot} = \eta_{AUT\_tot} T$, the mean and the variance of $\hat{\eta}_{AUT\_tot}$ are

$$E(\hat{\eta}_{AUT\_tot}) = \frac{N}{N-1} \eta_{AUT\_tot}. \tag{7.58}$$

$$Var(\hat{\eta}_{AUT\_tot}) = \frac{N(2N-1)}{(N-2)(N-1)^2} \eta_{AUT\_tot}^2. \tag{7.59}$$

It can be seen from (7.58)–(7.59) that the standard method (referred hereafter to the estimator (7.51)) is asymptotically unbiased. As N tends toward the infinity, the mean goes to η_{AUT_tot} and the variance goes to zero.

The bias of an estimator is the discrepancy between the mean of the estimator and the true value of the parameter being estimated. An estimator with zero bias is unbiased, otherwise, the estimator is biased. Obviously, an unbiased estimator is usually preferable to a biased estimator. Estimator (7.51) is a biased estimator obviously ($E(\hat{\eta}_{AUT\_tot}) \neq \eta_{AUT\_tot}$); therefore, it is straightforward to propose an unbiased estimator (referred to estimator (7.60) hereafter) based on (7.58) and (7.59):

$$\hat{\eta}_{AUT\_tot}^{unbiased} = \frac{N-1}{N} \frac{\hat{P}_{AUT}}{\hat{P}_{REF}} \eta_{REF\_tot}. \tag{7.60}$$

The mean and the variance of the unbiased estimator (7.60) are

$$E(\hat{\eta}_{AUT\_tot}^{unbiased}) = \eta_{AUT\_tot}. \tag{7.61}$$

$$Var(\hat{\eta}_{AUT\_tot}^{unbiased}) = \frac{2N-1}{N(N-2)} \eta_{AUT\_tot}^2. \tag{7.62}$$

The root mean squares (RMS) (defined as the square root of the mean square, i.e., $rms(X) = \sqrt{E(X^2)}$) of $\hat{\eta}_{AUT\_tot}$ and $\hat{\eta}_{AUT\_tot}^{unbiased}$ can be easily derived from (7.58)–(7.59) and (7.61)–(7.62) as

$$rms(\hat{\eta}_{AUT\_tot}) = \eta_{AUT\_tot} \sqrt{\frac{N(N+1)}{(N-2)(N-1)}}. \tag{7.63}$$

$$rms(\hat{\eta}_{AUT\_tot}^{unbiased}) = \eta_{AUT\_tot} \sqrt{\frac{N^2-1}{N(N-2)}}. \tag{7.64}$$

Note that for a large N, (7.59) can be approximated by

$$Var(\hat{\eta}_{AUT\_tot}) \approx \frac{2}{N} \eta_{AUT\_tot}^2. \tag{7.65}$$

Since (7.51) is a biased estimator, the mean square error (MSE) (defined as the average squared difference between the estimated value and the true value, i.e., $mse(\hat{X}) = E[(\hat{X} - X)^2]$) is the most appropriate metric to evaluate its performance, which can be easily derived from (7.58)–(7.59):

$$mse(\hat{\eta}_{AUT\_tot}) = \frac{2(N+1)}{(N-2)(N-1)} \eta_{AUT\_tot}^2. \tag{7.66}$$

For an unbiased estimator, its MSE is equal to its variance.

It should be noted that N represents the independent sample numbers and is usually not equal to the practical measured sample numbers due to the correlation of the samples. In order to evaluate the performance of the two estimators through the related statistics, it is necessary to estimate the independent sample numbers N from the measurements. The estimation of N is discussed in Section 7.2.3.

7.2.2 Simulations

In order to verify the derived pdf and related statistics of the antenna efficiency estimators, simulations utilizing Monte Carlo method are adopted. The Monte Carlo method is a mean of statistical evaluation of mathematical functions by generating suitable and massive random samples. The larger the number of random samples is, the more accurate the result is. For simplicity and without loss of generality, the antenna efficiency is assumed to be unity (i.e., $\eta_{AUT\_tot} = 1$) hereafter.

Three cases are chosen for simulations, i.e., $N = 10$, 30, and 50. Since the real and imaginary parts of each rectangular component of the electric field (e.g., E_{xr} and E_{xi}) follow the normal distribution [6], we randomly generate $1{,}000 \times 2$ sets of data that follow the standard normal distribution ($N(0, 1)$) to simulate E_{xr} and E_{xi}. Each set of data is composed of N stirring samples. Thus, we have $1{,}000 \times N$ samples for E_{xr} and E_{xi}, respectively. By using $\hat{G}_{AUT} = \langle |E_x|^2 \rangle = \langle E_{xr}^2 + E_{xi}^2 \rangle_N$, we have $1{,}000\ \hat{G}_{AUT}$, and $\hat{G}_{REF}$ can be calculated in a similar way. Finally, we obtain 1,000 realizations of $\hat{\eta}_{AUT\_tot}$ using estimator (7.51) for each case, the empirical pdf, mean, variance, and other statistics can be easily obtained further.

Figure 7.11 shows the analytical pdf (7.55) and the corresponding empirical pdf for three cases. It can be seen that the analytical pdf is in accordance with the

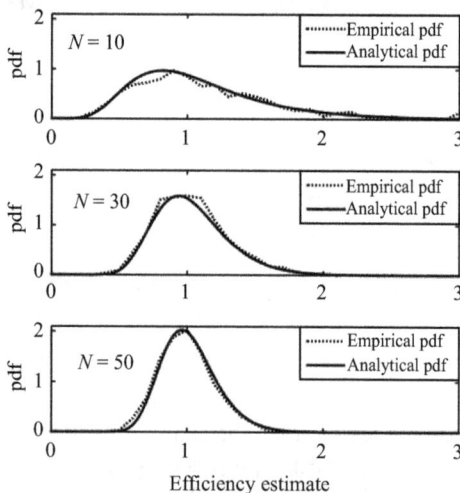

Figure 7.11   The analytical pdf (7.55) and the corresponding empirical pdf for $N = 10$, 30, and 50, respectively

empirical pdf, which proves the correctness of the derived pdf. As N becomes larger, the first and second terms in the right-hand side of (7.55) become extremely large and extremely small, respectively. This makes (7.55) infeasible to calculate the pdf for a large N from the point of view of numerical analysis. However, the statistics are always available regardless of the value of N.

Figure 7.12 shows the analytical and empirical means, variances, and RMS of estimators (7.51) and (7.60) as a function of N, respectively. It can be seen that the analytical mean and variance are in accordance with the empirical ones for both two estimators. The unbiased estimator (7.60) performs better than the biased estimator (7.51) in estimation mean, variance, and bias for a small N. However, the improvement becomes indistinguishable as N becomes larger.

7.2.3 Measurements

Measurements are conducted from 700 to 3,000 MHz to further verify the derived statistics of the antenna efficiency estimators. The RC used has a size of 1.80 m × 1.75 m × 1.25 m (as depicted in Figure 7.13). It contains a turn-table platform, two mechanical stirrers, and three antennas located on three orthogonal walls (referred to as wall antennas hereafter). A trestle with adjustable height which supports the reference antenna and the AUT is fixed on the turn-table platform. The turn-table platform moves stepwise to 20 angles that are evenly distributed over one complete rotation (i.e., 18° between two successive positions). At each platform rotating angle, the two stirrers move simultaneously and stepwise to 50 angles that are evenly distributed over one complete rotation (i.e., 7.2° between two successive positions). At each stirrer position and for each wall antenna, the S-parameters are sampled and recorded by the VNA. Hence, for each measurement, we have three wall antennas, 50 stirrer positions, and 20 platform positions, corresponding to 3,000 samples.

To facilitate the characterization of the antenna efficiency estimators, we repeat the measurement 12 times by adjusting the trestle of the reference antennas to four heights and placing the reference antenna with three orthogonal orientations (i.e., one vertical and two horizontal orientations) at each height. Arbitrary combinations of antenna heights and orientations are chosen to estimate η_{AUT_tot}. In addition, we introduce three attenuators (i.e., 0, −3, and −6 dB) to change the loss character of the AUT.

In order to confirm the RC loading effect on the estimation performance of the antenna efficiency, we repeat the measurement in two loading conditions, i.e., load 1 (empty RC) and load 2 (a head phantom that is equivalent to a human head in terms of microwave absorption). Hereafter, measured data from these two different loading conditions are simply referred to as load 1 and load 2.

The biased estimator (7.51) results in biased estimate of related statistics (i.e., (7.58) and (7.59)). As mentioned in previous sections, the MSE is adopted as the performance metric to evaluate the characteristics of the statistics. Note that the MSE of $\hat{\eta}_{AUT\_tot}$ can be calculated using (7.66), but a prerequisite is to know the number N of independent samples. This is done here with the DoF method [27,28]. For the sake of completeness, we briefly introduce the DoF method as follows.

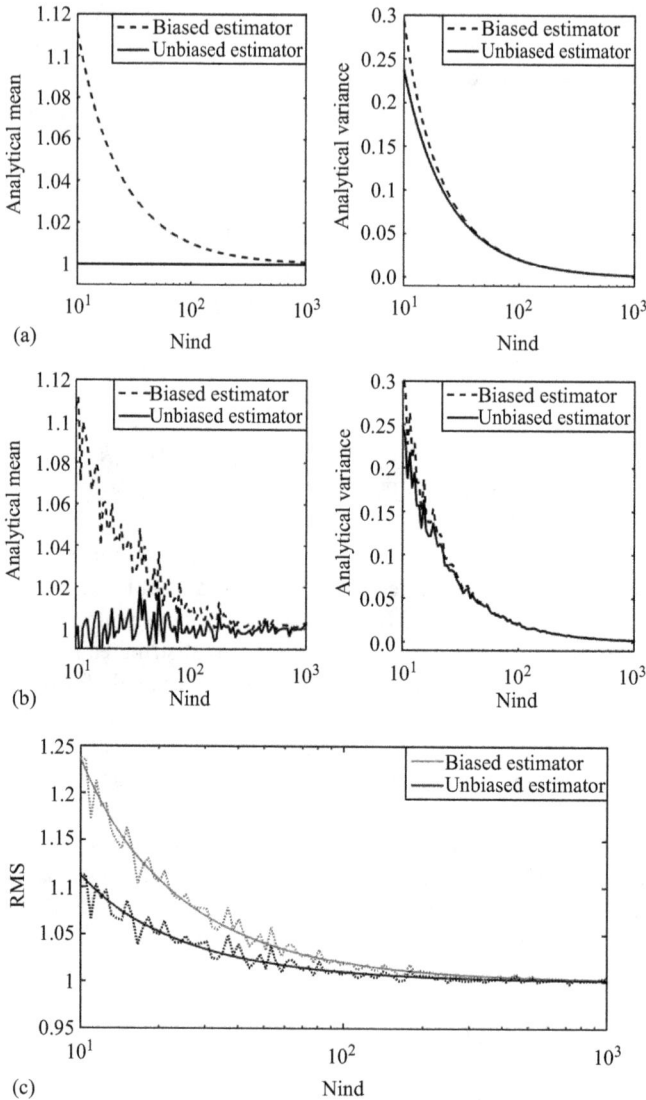

Figure 7.12   Means and variances of estimators (7.51) and (7.60) for the case
$\eta_{AUT\_tot} = 1$ (a) analytical mean and variance, (b) empirical mean and
variance, and (c) analytical (solid) and empirical (dotted) root mean
squares

The key thought of the DoF method is to divide the samples into several sub-sets according to each stirring process, estimating the independent sample number of each subset, and integrating the independent sample number of all subsets to be the total independent sample number. Here we have three independent stirring

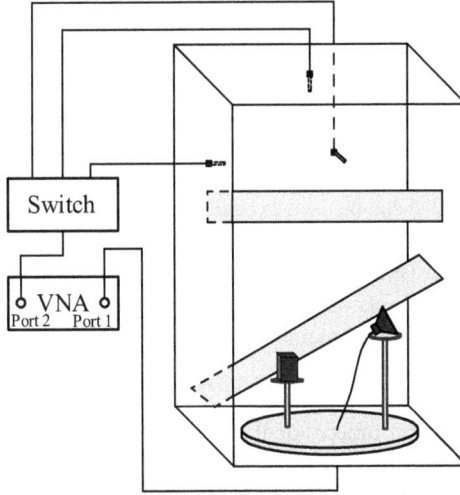

Figure 7.13   Drawing of the RC with two mechanical stirrers, a platform, three wall antennas, and a head phantom

mechanisms (i.e., wall antennas, mechanical stirrers, and turn-table platform), thus the total independent sample number can be obtained as

$$N = N_{ind,ant} N_{ind,st} N_{ind,pf}, \tag{7.67}$$

where $N_{ind,l}$ (l represents *ant*, *st*, or *pf*) is the independent sample number of the specified stirring method. $N_{ind,l}$ can be obtained following these procedures:

1. Denote the complex channel transfer function sampled at the mth stirring position (i.e., wall antenna position ($l = ant$), stirrer position ($l = st$), and platform position ($l = pf$)) as a column vector $\mathbf{x}_m$, $m = 1 \ldots N_l$, the dimension of $\mathbf{x}_m$ being $N_{tot}/N_l \times 1$, where $N_{tot} = N_{ant} \times N_{st} \times N_{pf}$
2. Concatenate $\mathbf{x}_m$ into a matrix $\mathbf{X} = [\mathbf{x}_1 \ldots \mathbf{x}_{N_l}]$, the dimension of $\mathbf{X}$ being $N_{tot}/N_l \times N_l$
3. Estimate the correlation matrix of $\mathbf{X}$ as

$$\mathbf{R} = \mathbf{X}^H \mathbf{X}, \tag{7.68}$$

    where $\mathbf{X}^H$ denotes the conjugate transpose of $\mathbf{X}$, the dimension of $\mathbf{R}$ being $N_l \times N_l$
4. Estimate the independent antenna ($l = ant$), stirrer ($l = st$), or platform ($l = pf$) sample number as

$$N_{ind,l} = \frac{tr(\mathbf{R})^2}{tr(\mathbf{R}^2)} = \frac{\left(\sum_i \lambda_i\right)^2}{\sum_i \lambda_i^2}, \tag{7.69}$$

    where tr represents the trace operator, and λ_i represents the ith eigenvalue of $\mathbf{R}$.

The estimated independent sample numbers (i.e., $N_{ind,ant}$, $N_{ind,st}$, $N_{ind,pf}$, and N) using the DoF method are shown in Figure 7.14. It can be seen that the estimated independent sample numbers increase as frequency increases and loading decreases. Since the DoF method processes the measured data regardless of the measurement procedures, all the connotative influence factors (e.g., the stirring technique, the loading condition and the unstirred component) are considered inherently in the method. In addition, the DoF method can be used to evaluate the performance of different stirring mechanism separately. It makes the evaluation convenient and efficient, especially for multiple stirring mechanisms measurement. However, if the measured data generated by multiple stirring mechanisms cannot be separated, or there is only one stirring mechanism, there are still other methods [5,28–30] that can be used to determine the independent sample number.

Once the independent sample number is confirmed, the empirical MSE of $\hat{\eta}_{AUT\_tot}$ can be estimated based on independent measurements, and the analytical MSE of $\hat{\eta}_{AUT\_tot}$ can be calculated using (7.66). The empirical MSE (solid) and the analytical MSE (dotted) of $\hat{\eta}_{AUT\_tot}$ (for $\eta_{AUT\_tot} = 1$, 0.5, and 0.25) are shown in Figure 7.15. In order to facilitate the comparison of different MSEs, the dB value of MSE is calculated by averaging the dB values of $(1+mse)$ and $(1 - mse)$, that is, the following dB-transformation [31]:

$$mse_{dB} = \frac{10[\log_{10}(1+mse) - \log_{10}(1-mse)]}{2} = 10\log_{10}\sqrt{\frac{(1+mse)}{(1-mse)}}, \quad (7.70)$$

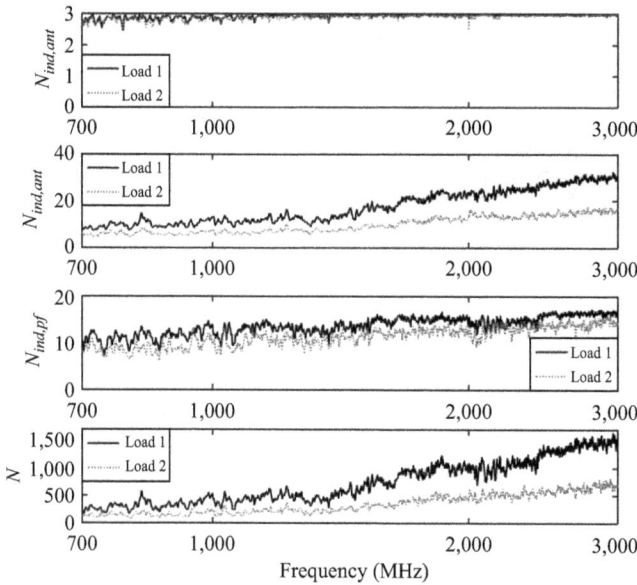

Figure 7.14   Estimated independent sample numbers $N_{ind,ant}$, $N_{ind,st}$, $N_{ind,pf}$, and N using the DoF method

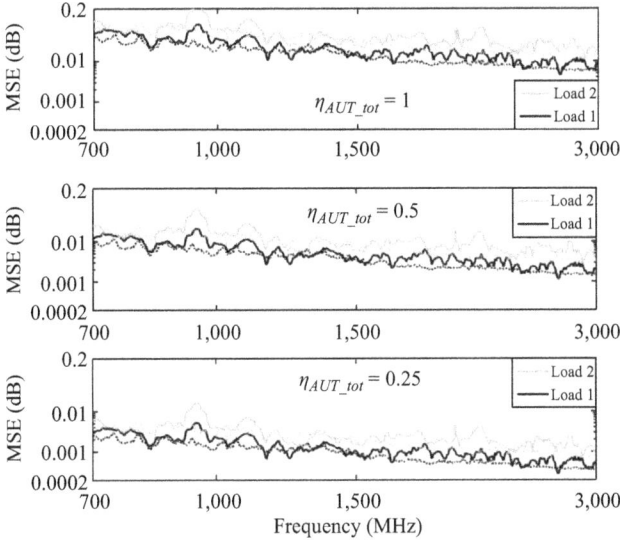

Figure 7.15 *Empirical MSE of* $\hat{\eta}_{AUT\_tot}$ *based on independent measurements (solid) and the analytical MSE model (7.66) (dotted). The three graphs from top to bottom correspond to* $\eta_{AUT\_tot} = 1$, *0.5, and 0.25, respectively*

where *mse* is the MSE of the normalized antenna efficiency of all the 12 times measurements.

As can be seen, the empirical MSE (estimated based on independent measurements) and the analytical MSE (calculated using (7.66)) are in accordance with each other. It can be also seen that the MSE decreases as η_{AUT_tot} decreases and N increases (in reason of the loading decreases and the frequency increases). This can be easily explained by the expression (7.66).

Examining the expression (7.66) again, we can conclude that the MSE decreases as N increases, and the convergence rate is about $1/N$. This indicates that the uncertainty of the measured antenna efficiency can be reduced by increasing the independent sample numbers. From Figure 7.14, we can see that the independent wall antenna number tends toward the number of wall antennas as frequency increases (the three antennas being orthogonally polarized and located sufficiently away from each other). However, the independent stirrer and platform numbers are smaller than the actual number of stirrer and platform positions. Thus, we can add more wall antennas (located sufficiently away from each other) to reduce the measurement uncertainty while increasing the number of stirrer, and platform positions would probably have a lower effect on the improvement. To increase the independent sample numbers, of course, one can increase the radius of the turn-table platform, optimize the design of the stirrers, and also introduce more stirring mechanisms.

As discussed in previous section, the performances of estimator (7.51) and (7.60) are similar for a large N. In order to show the advantage of the unbiased estimator

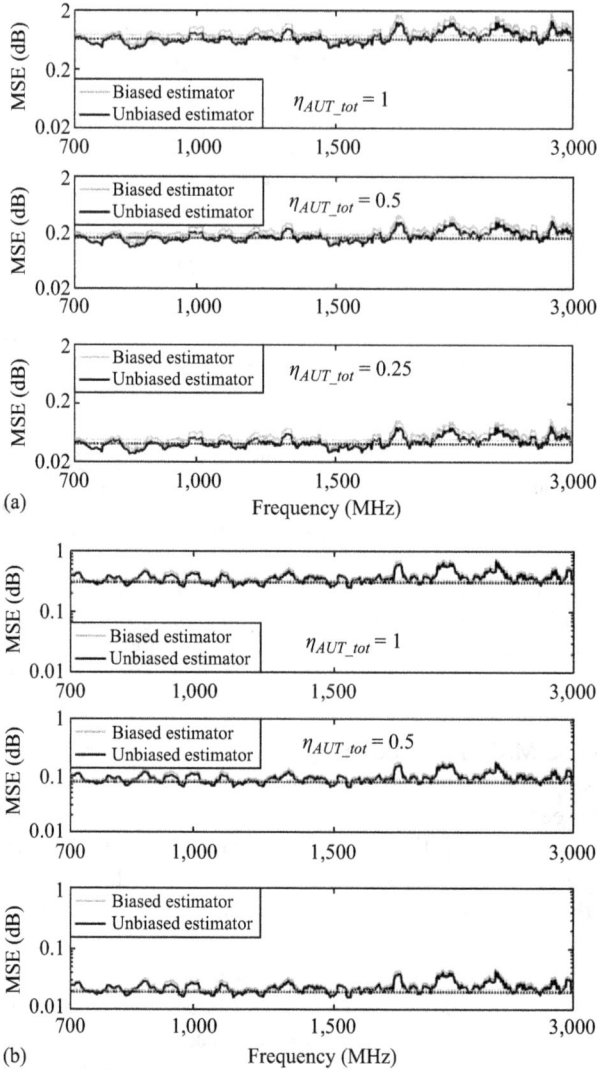

Figure 7.16   Empirical MSEs of $\hat{\eta}_{AUT\_tot}$ (7.51) and $\hat{\eta}_{AUT\_tot}^{unbiased}$ (7.60) based on independent measurements (solid) and the analytical MSE expressions (dotted) for (a) $N = 15$ and (b) $N = 30$ under the loading condition of load 1

(7.60) over the biased estimator (7.51), additional measurements of different independent sample numbers are conducted. Figure 7.14 shows the independent wall antenna number is three, and the independent platform position number for two loading conditions is about 10 at 700 MHz (varying slightly for both loading conditions). Therefore, we choose one stirrer position, three wall antennas, five and ten platform positions, in order to have $N = 15$ and 30, respectively.

The empirical MSE (estimated based on independent measurements) and the analytical MSE (calculated using (7.66) and (7.62)) of estimator (7.51) and (7.60) for $N = 15$ and 30 are shown in Figure 7.16. Note that the MSE of the unbiased estimator (7.60) is its variance (7.62). Since the loading effect on the MSEs has been discussed previously, $N=15$ and 30 are the independent sample numbers for load 1 and load 2, only the results of load 1 are shown in Figure 7.16. It can be seen that there is a small but noticeable improvement of estimator (7.60) over (7.51), and the improvement becomes indistinguishable for a larger N (i.e., $N = 30$). This indicates that the independent sample number should be much larger than 30 in practical antenna efficiency measurements.

In summary, the results show that the unbiased estimator performs slightly better than the standard estimator for $N < 30$, which indicates that the standard antenna efficiency estimator (7.51) is as good as the unbiased estimator (7.60), since N is usually much larger than 30 in practical RC measurements.

Note that it is time-consuming to directly determine the MSE of the measured antenna efficiency by performing many independent measurements. Once the analytical expressions of the derived statistics (e.g., mean, variance, RMS, and MSE) with estimated N are available, single measurement used to determine N is enough to characterize the measurement performance.

7.3   Conclusion

In this chapter, five typical methods to measure the antenna radiation efficiency in an RC have been discussed. The standard reference antenna method is general and performs well, but a reference antenna is needed. Three non-reference antenna methods overcome the problem to use a reference antenna of known efficiency. However, they require additional prerequisites or antennas. The time-domain methods are time efficient, especially for UWB antennas. All these methods have different advantages and disadvantages, thus there is not a "best" method. A suitable method can be chosen depending on the measurement conditions, requirements, and type of antennas under test. Statistical analyses of the measured antenna efficiency are also given in this chapter. In particular, the presented method offering an accurate theoretical analysis on the measurement uncertainty has been illustrated on the standard reference method.

References

[1]   Balanis C.A. *Antenna Theory: Analysis and Design*. New York, NY, USA: Wiley-Interscience; 2005.

[2]   Kabacik P., Byndas A., Hossa R., and Bialkowski M. 'A measurement system for determining radiation efficiency of a small antenna'. *1st European*

Conference on Antennas and Propagation (EuCAP)*; Nice, France, Nov. 2006, pp. 1–6.

[3]   Wheeler H.A. 'The radiansphere around a small antenna'. *Proceedings of the IRE*. 1959;**47**(8):1325–31.

[4]   Ashkenazy J., Levine E., and Treves D. 'Radiometric measurement of antenna efficiency'. *Electronics Letters*. 1985;**21**(3):111–12.

[5]   IEC 61000-4-21. 'Electromagnetic Compatibility (EMC) – Part 4-21: Testing and Measurement Techniques – Reverberation Chamber Test Methods', 2011.

[6]   Holloway C.L., Shah H.A., Pirkl R.J., Young W.F., Hill D.A., and Ladbury J. 'Reverberation chamber techniques for determining the radiation and total efficiency of antennas'. *IEEE Transactions on Antennas and Propagation*. 2012;**60**(4):1758–70.

[7]   Xu Q., Huang Y., Zhu X., Xing L., Tian Z., and Song C. 'A modified two-antenna method to measure the radiation efficiency of antennas in a reverberation chamber'. *IEEE Antennas and Wireless Propagation Letters*. 2016;**15**:336–39.

[8]   Le Fur G., Besnier P., and Sharaiha A. 'Time reversal efficiency measurement in reverberation chamber'. *IEEE Transactions on Antennas and Propagation*. 2012;**60**(6):2921–28.

[9]   Naqvi I.H., Le Fur G., Sol J., Besnier P., and Sharaiha A. 'Sub-band time reversal efficiency measurement: an enhanced method for efficiency characterization of UWB antennas'. *IEEE Transactions on Antennas and Propagation*. 2012;**60**(3):1657–60.

[10] Hallbjörner P. 'Reflective antenna efficiency measurements in reverberation chambers'. *Microwave and Optical Technology Letters*. 2001;**30**(5): 332–35.

[11] Besnier P., Sol J., Presse A., Lemoine C., and Tarot A.C. 'Antenna efficiency measurement from quality factor estimation in reverberation chamber'. *46th European Microwave Conference (EuMC)*; London, UK, 2016, pp. 715–18.

[12] Senic D., Williams D.F., Remley K.A., *et al.* 'Improved antenna efficiency measurement uncertainty in a reverberation chamber at millimeter-wave frequencies'. *IEEE Transactions on Antennas and Propagation*. 2017;**65**(8): 4209–19.

[13] Gifuni A., Flintoft I.D., Bale S.J., Melia G.C.R., and Marvin A.C. 'A theory of alternative methods for measurements of absorption cross section and antenna radiation efficiency using nested and contiguous reverberation chambers'. *IEEE Transactions on Electromagnetic Compatibility*. 2016;**58**(3): 678–85.

[14] Hill D.A. *Electromagnetic Fields in Cavities: Deterministic and Statistical Theories*. Piscataway, NJ, USA: Wiley-IEEE; 2009.

[15] Moglie F. and Primiani V.M. 'Analysis of the independent positions of reverberation chamber stirrers as a function of their operating conditions'. *IEEE Transactions on Electromagnetic Compatibility*. 2011;**53**(2):288–95.

[16] Hill D.A. 'Electronic mode stirring for reverberation chamber'. *IEEE Transactions on Electromagnetic Compatibility*. 1994;**36**(4):294–99.

[17] Chen X. 'Scaling factor for turn-table platform stirring in reverberation chamber'. *IEEE Antennas and Wireless Propagation Letters*. 2017;**16**: 2799–802.

[18] Yang J., Carlsson J., Kildal P-S., and Carlsson C. 'Calculation of self-impedance and radiation efficiency of a dipole near a lossy cylinder with arbitrary cross section by using the moment method and a spectrum of two-dimensional solutions'. *Microwave and Optical Technology Letters*. 2002; **32**(2):108–12.

[19] Andrieu G., Ticaud N., Lescoat F., and Trougnou L. 'Fast and accurate assessment of the "well stirred condition" of a reverberation chamber from S11 measurements'. *IEEE Transactions on Electromagnetic Compatibility*. 2018:1–9.

[20] West J.C., Dixon J.N., Nourshamsi N., Das D.K., and Bunting C.F. 'Best practices in measuring the quality factor of a reverberation chamber'. *IEEE Transactions on Electromagnetic Compatibility*. 2018;**60**(3):564–71.

[21] Ladbury J. and Hill D.A. 'Enhanced backscatter in a reverberation chamber: inside every complex problem is a simple solution struggling to get out'. *IEEE International Symposium on Electromagnetic Compatibility*; Honolulu, HI, USA, Jul. 2007, pp. 1–5.

[22] Taylor B.N. and Kuyatt C.E. 'Guidelines for Evaluating and Expressing the Uncertainty of NIST Measurement Results'. NIST Tech. Note 1297, Sep. 1994.

[23] Li C., Loh T.H., Tian Z.H., Xu Q., and Huang Y. 'Evaluation of chamber effects on antenna efficiency measurements using non-reference antenna methods in two reverberation chambers'. *IET Microwaves, Antennas & Propagation*. 2017;**11**(11):1536–41.

[24] Chen X. 'On statistics of the measured antenna efficiency in a reverberation chamber'. *IEEE Transactions on Antennas and Propagation*. 2013;**61**(11):5417–24.

[25] Kostas J.G. and Boverie B. 'Statistical model for a mode-stirred chamber'. *IEEE Transactions on Antennas and Propagation*. 1991;**33**(4):366–70.

[26] Grimmett G. and Stirzaker D. *Probability and Random Processes*. New York, NY, USA: Oxford University Press; 2001.

[27] Chen X. 'Experimental investigation of the number of independent samples and the measurement uncertainty in a reverberation chamber'. *IEEE Transactions on Antennas and Propagation*. 2013;**55**(5):816–24.

[28] Pirkl R.J., Remley K.A., and Patané C.S.L. 'Reverberation chamber measurement correlation'. *IEEE Transactions on Electromagnetic Compatibility*. 2012;**54**(3):533–44.

[29] Lemoine C., Besnier P., and Drissi M. 'Estimating the effective sample size to select independent measurements in a reverberation chamber'. *IEEE Transactions on Electromagnetic Compatibility*. 2008;**50**(2):227–36.

[30] Gradoni G., Mariani Primiani V., and Moglie F. 'Reverberation chamber as a multivariate process: FDTD evaluation of correlation matrix and independent positions'. *Progress in Electromagnetics Research*. 2013;**133**:217–34.

[31] Kildal P.-S., Chen X., Orlenius C., Franzén M., and Lötbäck Patané C. 'Characterization of reverberation chambers for OTA measurements of wireless devices: physical formulations of channel matrix and new uncertainty formula'. *IEEE Transactions on Antennas and Propagation*. 2012;**60**(8):3875–91.

Chapter 8

Characterization of antenna radiation pattern in reverberating enclosures

Guillaume Andrieu[1]

An increasing interest appears recently in order to characterize in an oversized (with respect to the wavelength) metallic enclosure (i.e., a reverberation chamber (RC) if a mode stirring process is necessary or a reverberating enclosure (RE) if not) the radiation pattern of antennas, a measurement traditionally done in an anechoic chamber. This measurement that may appear counter-intuitive to perform in such a highly multipath environment (see also the case of radar cross-section measurement in such enclosures discussed in Chapter 9) presents some non-negligible advantages. Indeed, such metallic enclosures are not disturbed by the external electromagnetic environment while being substantially less expensive than an anechoic chamber, especially in reason of the absence of absorbers within the facility. Moreover, the efficiency of absorbers used in anechoic chambers is not constant with the frequency and may be insufficient in the lowest frequency range (i.e., below a few hundred of MHz, a value depending on the absorber dimensions).

In a multipath environment such the one depicted in Figure 8.1, the challenge to measure the radiation pattern of the antenna under test (i.e., the AUT) at a frequency of interest f_0 is to extract the line-of-sight (LOS) path (in red) between the AUT and

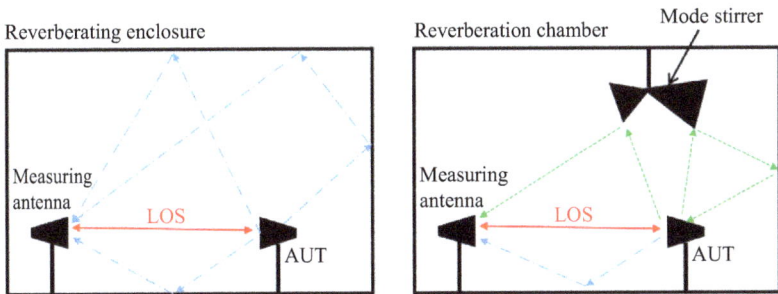

Figure 8.1   Schematic illustration of LOS (in red, solid line), NLOS unstirred paths (in blue, dash-dot line) and stirred paths (in green, dashed line) between two antennas inserted in an RE (on the left) and an RC (on the right)

[1]XLIM Laboratory, UMR7252, University of Limoges, Limoges, France

a measuring antenna from all the other indirect or non-LOS (NLOS) paths reflected by the enclosure walls (in blue and green).

Three methods (experimentally validated) have already been proposed to perform such measurements in RE, stirred or not. Their basic principles are briefly presented in this chapter as well as their main strengths and weaknesses (if applicable). However, for the sake of completeness, it is worth noting that another method has been proposed recently [1] in an RC. This method is based on a spherical wave decomposition computed from scattering parameter measurements between the AUT and the measuring antenna (i.e., S_{21}) for two AUT positions. The subsequent computation of self-correlation coefficients allows its radiation pattern to be computed. This method, not tested by the author of this chapter, is therefore not presented here. Among the three other methods presented, an emphasis is made on the time-gating technique published recently by the author which presents probably the best performance when considering together the simplicity of the experimental setup, the measurement time and the ability to filter the unwanted indirect NLOS paths.

8.1   *K*-factor method

Chronologically, the first method proposed in the literature (by the IETR Laboratory, Rennes, France) requires the determination of the K-factor of an RC for each angle of the radiation pattern [2,3]. The experimental setup to install in a mechanically stirred RC, depicted in Figure 8.2, requires to locate the AUT (i.e., Port 1) in front of a measuring antenna (i.e., Port 2), both antennas being separated by a distance d. The aim is to measure the S_{21} scattering parameter with a vector network analyzer (VNA) for a sufficient number of mode stirrer positions (and if necessary on a narrowband

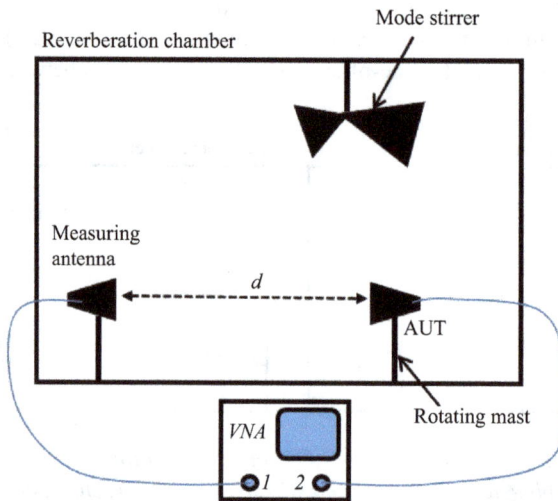

Figure 8.2   Experimental setup related to the radiation pattern measurement of an AUT using the K-factor method

around f_0 in order to add frequency stepping samples and therefore reinforce the measurement accuracy). Finally, the rotation of the AUT on itself leads, using the post-processing process described hereafter, to the AUT radiation pattern.

In a multipath environment such as a mechanically stirred RC, the K-factor is defined as the ratio of the unstirred power P_{uns} (related to the paths having no interaction with the mode stirrer) and the average stirred power P_{sti} (related to the paths interacting with the mode stirrer)

$$K = \frac{P_{uns}}{\langle P_{sti} \rangle}. \tag{8.1}$$

When using S_{21} measurements between both antennas, the K-factor can be assessed as [4]

$$K = \frac{|\langle S_{21} \rangle|^2}{\langle |S_{21} - \langle S_{21} \rangle|^2 \rangle}. \tag{8.2}$$

where the operator $\langle \cdot \rangle$ denotes averaging on all the stirring conditions (i.e., for instance the number of mode stirrer positions for a mechanically stirred RC).

In [4], a relation between the RC K-factor and the directivity D of a transmitting antenna (here, the AUT) pointing toward a receiving antenna is given as a function of the distance d between both the antennas, the volume V, the quality factor Q of the RC and the wavelength λ

$$K = \frac{3VD}{2\lambda Q d^2}. \tag{8.3}$$

The principle of the method requires first to determine the K-factor of the RC, i.e., K_{REF}, for each angle of the desired radiation pattern using a reference antenna of known directivity D_{REF} at f_0. Then, the same measurements are repeated using the AUT (of unknown directivity D_{AUT}) instead of the reference antenna for the same distance d between both antennas. The K-factor measured in these conditions is denoted as K_{AUT}.

If we assume that the quality factor Q of the RC is not modified when the reference antenna is substituted by the AUT, the directivity of the AUT can be deduced

$$D_{AUT} = \frac{D_{REF} K_{AUT}}{K_{REF}}. \tag{8.4}$$

This approach seems promising considering its simplicity but suffers of two important drawbacks discussed hereafter.

First, equation (8.10) assumes that the only unstirred path between both antennas is the LOS one. This directly implies that all the other paths interact with the mode stirrer. As illustrated in Figure 8.1, this is not true if we consider the effect of the NLOS-unstirred paths. Indeed, the validity of this approach is therefore limited to the main directions of the AUT radiation pattern when the power related to the NLOS unstirred paths is negligible with respect to the power related to the LOS path.

Second, the method requires the RC K-factor to be measured accurately for each desired angle of the radiation pattern. As the uncertainty related to K is directly related to the number of independent samples considered during the measurements, the measurement process becomes rapidly time consuming and not comparable to the time typically required in an anechoic chamber for such measurement.

8.2   Doppler spectrum method

The second method presented in this chapter has been developed at the XLIM Laboratory, Limoges, France. Unlike the K-factor method, the method can be used in an RE or in other terms does not require to use a mode-stirring process within the enclosure.

The principle of the method is to measure (at f_0) and for each angle of the desired radiation pattern the transmission coefficient as a function of time between the AUT and the measuring antenna when the AUT is moving (at a constant speed) in the direction of the second antenna. Thus, for each angle, these measurements allow the Doppler spectrum (reminded hereafter) to be computed through a simple Fourier transform (FT). Finally, the knowledge of the AUT velocity allows the frequency (in the Doppler domain) related to the LOS signal between both antennas, i.e., f_d^{LOS}, to be identified.

Before presenting with more details the method principle, it is important to summarize what is the Doppler spread spectrum and how it can be obtained within an RC.

The Doppler effect corresponds to the frequency offset of a signal observed on a receiver when the distance with the emitter varies over the time. In a highly multipath environment as the one obtained in RC, the movement of an object (the rotating mode stirrer [5] or the antenna [6,7] for instance) induces a Doppler spectrum around each frequency of interest f [8]. This corresponds to all the frequency offsets (generally a few Hertz in an RC if we consider the usual rotation speed of mode stirrers) observed by the receiver around f.

On a more theoretical point of view, the Doppler spectrum D is the FT of the time autocorrelation function of the transfer function between an emitter and a receiver. Therefore, D at a frequency f is calculated by the following expression [9]

$$D(f, f_d) = |FT\{H(f, t)\}|^2, \tag{8.5}$$

where f is the working frequency, f_d the Doppler frequency, $H(f, t)$ the transfer function between the transmitter and the receiver at f as a function of the time t. One understands that if the transfer function is evaluated from S_{21} scattering parameter measurements, $H(f, t)$ equals $S_{21}(f, t)$.

Two versions of the method have been proposed successively by the authors. The initial method [10] consists in measuring, for each angle of the desired radiation pattern, the S_{21} parameter between both antennas at different fixed positions p_n (i.e., $S_{21}^{p_n}$) with a constant distance δ_d between two successive positions of the AUT as sketched in Figure 8.3.

The Doppler spread spectrum is then calculated after associating each measurement to a particular time t_n (i.e., $S_{21}^{t_n}$) in order to reproduce "virtually" the movement of the AUT and to obtain a "fictitious" Doppler spectrum [11]. Obviously, the final

Reverberating enclosure

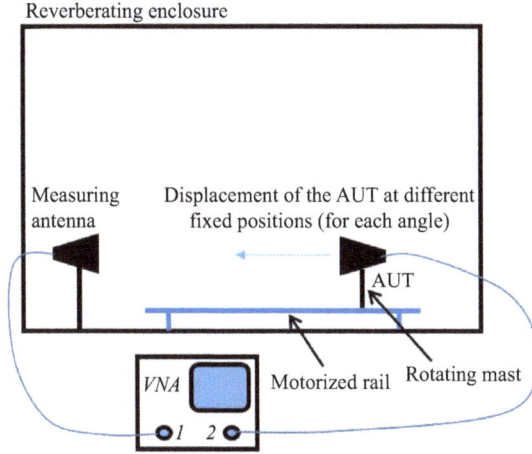

Figure 8.3 Experimental setup related to the radiation pattern measurement of an AUT using the step-wise Doppler spectrum method

result (i.e., the obtained radiation pattern) has to be independent of the chosen time step δ_t considered between two successive instants.

Thus, for each desired angle of the AUT radiation pattern (which can be measured in 3-dimensions), $S_{21}^{t_n}$ is a discrete function and consequently the Doppler spread spectrum $D(f_0, f_d)$ at f_0 writes as a sum

$$D(f_0, f_d) = \left| \sum_{n=1}^{N} S_{21}(f_0, t_n) e^{-j2\pi f_d t_n} \right|^2. \tag{8.6}$$

The next step consists in identifying the Doppler frequency f_d^{LOS} corresponding to the LOS path between both antennas. This is computed straightforwardly as a function of the velocity of the AUT movement, the working frequency f_0 and the celerity C

$$f_d = \frac{vf}{C}. \tag{8.7}$$

It is worth noting that f_d^{LOS} is also the maximum frequency of the Doppler spectrum D. Indeed, all the other indirect paths within the RC corresponds to a lower Doppler frequency as their relative velocity between the emitting and the receiving antenna are lower than v.

The normalized radiation pattern of the AUT is then computed by dividing the amplitude of f_d^{LOS} for each considered angle by the maximum value.

Figure 8.4 shows the Doppler spectrum obtained in the XLIM RC (the characteristics of this enclosure are given in Chapter 1) for two different values of the angle θ between two double-ridged horn antennas at the working frequency of 10 GHz when 163 fixed positions shifted of 1.11 cm are considered (the length of our motorized rail being 1.8 m, i.e., 60 λ at f_0).

When selecting a δ_t time step of 10 ms, the fictitious velocity v of the antenna equals 1.11 m s^{-1} leading in these conditions to $f_d^{LOS} = 37$ Hz.

This method has been improved in [12], the Doppler spectrum being computed from a continuous movement of the AUT along the motorized rail. This can be achieved as shown in Figure 8.5 by measuring $S_{21}(f_0, t)$ with a VNA working in constant wave (CW) mode at f_0. This allows the total measurement time to be severely decreased.

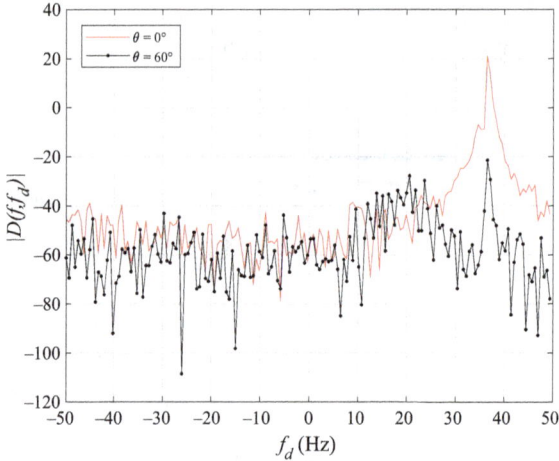

Figure 8.4   *Doppler spread spectrum obtained in the XLIM RC for two different angles θ of the AUT. As mentioned in the text, the Doppler frequency related to the LOS path f_d^{LOS} equals 37 Hz*

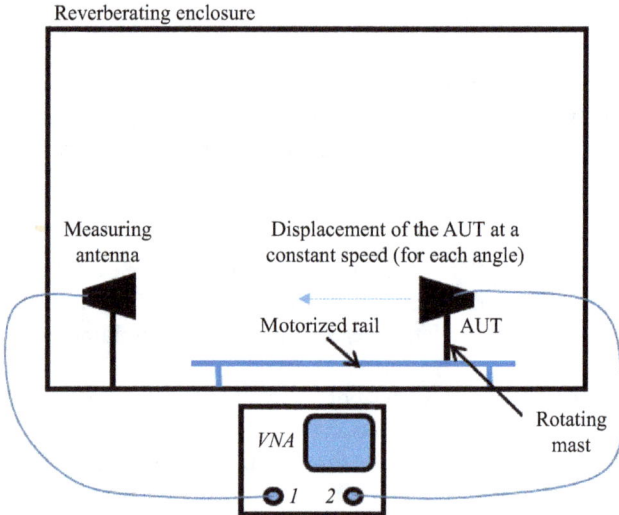

Figure 8.5   *Experimental setup related to the radiation pattern measurement of an AUT using the continuous Doppler spectrum method. The VNA is here working in the CW mode at f_0*

This method is theoretically perfectly able to extract the LOS path from the unstirred paths (especially the shortest ones). However, the accuracy of this method depends on the total distance swept by the AUT in terms of wavelength (i.e., the accuracy increases with the frequency). Moreover, the experimental setup is quite complex as a long motorized rail is required. In the variant of the method requiring a continuous movement of the AUT, it is fundamental to ensure that the AUT velocity is constant in order to compute accurately the amplitude of f_d^{LOS}. This is not so easy to achieve especially when considering the acceleration and deceleration phases.

The measurement time, even with the "continuous" method, is large with respect to the time typically required in an anechoic chamber. Indeed, each angle requires an entire movement of the AUT along the motorized rail. Thus, a complete 2D radiation pattern over 360° using an angular step of 1° requires 180 round trips of the antenna along the rail.

8.3   Time-gating method

8.3.1   Principle of the method

The third and last method presented in this chapter and using the time-gating technique has also been proposed recently at the XLIM Laboratory, Limoges, France [13] within an RE, no mode stirring process being required. The time-gating method was originally proposed for radiation pattern measurements in imperfect anechoic environments [14,15] as for instance in outdoor environments, in order to remove the effect of ground reflections.

The experimental setup depicted in Figure 8.6 consists in measuring the scattering parameter S_{21} between the AUT (and a measuring antenna located at the same height) within the RC for each desired angle of the radiation pattern. The aim is to extract the LOS signal traveling the distance d_{LOS} (i.e., arriving at the instant t_{LOS}) by discarding all the NLOS paths and especially the one traveling the shortest distance d_{NLOS}^{min} (arriving at t_{NLOS}^{min}). This is achieved in the time-domain by applying the time-gating technique described hereafter.

Each $S_{21}(f, \theta, \varphi)$ measurement is performed on a bandwidth B centered around f_0 (with $B < f_0$). Then, the impulse response of the system, including both antennas and the RE, is computed with an inverse FT (IFT) of $S_{21}(f, \theta, \varphi)$

$$h(t, \theta, \varphi) = IFT\left[S_{21}(f, \theta, \varphi) \cdot W(f)\right], \qquad (8.8)$$

with $W(f)$ the frequency-domain windowing function. Different windowing functions (Hann, Hamming, Blackman, etc.) can be used in addition to the rectangular one in order to reduce the influence of the secondary lobes of the impulsional response. However, it is worth noting that these windowing functions have also the effect to enlarge the main lobe (the influence on the obtained results of a rectangular and a Hann window is studied in the next subsection of this chapter).

The time-gating technique consists in truncating the signal $h(t)$ at a time t_{tg} included in the range between t_{LOS} and t_{NLOS}^{min}, i.e., after the arrival of the LOS path

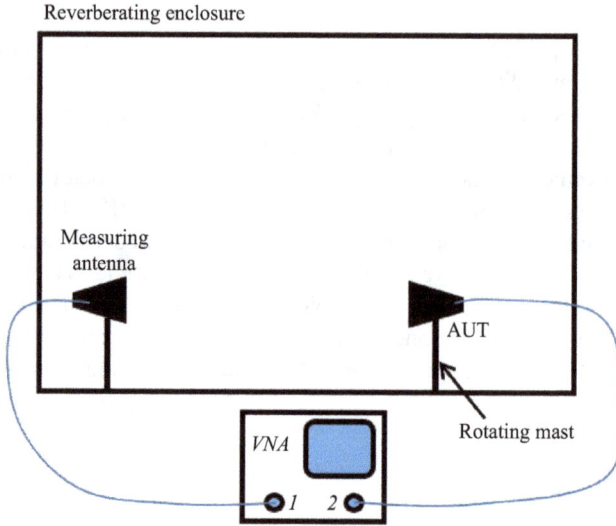

Figure 8.6   Experimental setup related to the radiation pattern measurement of an AUT using the time-gating method

on the measuring antenna but before the arrival of all the other paths. More exactly, we realize a zero-padding operation on all the points not included in the time-gating window in order to keep the same frequency-domain resolution at the end of the posttreatment process.

The final step consists in doing the FT of the truncated signal $h_{tg}(t)$ in order to examine its value at f_0 (or in theory at any frequency included in the bandwidth B, this being also analyzed in the next subsection). The AUT radiation pattern at f_0 is obtained after repeating this process for each desired angle.

As for the other methods presented previously, the measuring antenna has to be located in the far-field of the AUT. Respecting this condition, the distance d_{LOS} has to be chosen in order to increase (if possible) the relative difference between d_{LOS} and d_{NLOS}^{min}. This can help one to measure the radiation pattern of dispersive or resonating antennas. In such case, the measurement is possible if the spread signal related to the LOS path is entirely received on the measuring antenna before the arrival of the first NLOS path.

Different measurement parameters influence the results obtained by this approach. First, the choice of the bandwidth B is fundamental as B drives the ability to dissociate the LOS and the shortest NLOS path. Indeed, the minimum bandwidth B_{min} of the S_{21} measurement is directly related to the maximum time domain resolution δ_t^{max} (or to the maximum tolerable spatial resolution δ_d^{max})

$$B_{min} = \frac{1}{\delta_t^{max}} = \frac{c}{\delta_d^{max}},$$ (8.9)

with c the celerity.

Then, $B_{\min}$ can be easily calculated as a function of the difference between t_{LOS} and $t_{NLOS}^{\min}$ (or of the difference of distance between d_{LOS} and $d_{NLOS}^{\min}$)

$$B_{\min} = \frac{1}{(t_{NLOS}^{\min} - t_{LOS})} = \frac{c}{(d_{NLOS}^{\min} - d_{LOS})}. \tag{8.10}$$

Second, the frequency-domain resolution δ_f of the S_{21} measurement is also important in order to avoid any aliasing effect when performing the IFT. Indeed, a δ_f parameter insufficiently small has the effect to increase the level of the IFT in all the time domain and especially during the first instants in reason of the aliasing effect. This adds a nonphysical noise reducing the signal-to-noise ratio within the time-gating window. The last parameters influencing the results are the frequency-domain windowing function and the fact to insert absorbers within the enclosure [16,17]. The influence of all these parameters is discussed in the next subsection of this chapter.

8.3.2 Example of results

The results shown in this subsection have been obtained on a double-ridged horn AUT working in the frequency range from 0.75 to 18 GHz, the measuring antenna being an antenna of the same model pointing toward the AUT.

Reference results have been obtained in one anechoic chamber of the XLIM Laboratory where both antennas were located at a distance of 7 m from each other. Measurements have been performed on both *E*- and *H*-planes from 1.5 to 12 GHz over 22 frequencies (frequency step of 0.5 GHz), each with an angular step of 1° (*H*-plane results are not shown here but the conclusions are the same).

The experimental setup used in the RC of the XLIM Laboratory is described in Figures 8.7 and 8.8. The rotating mode stirrer being removed (the enclosure being no

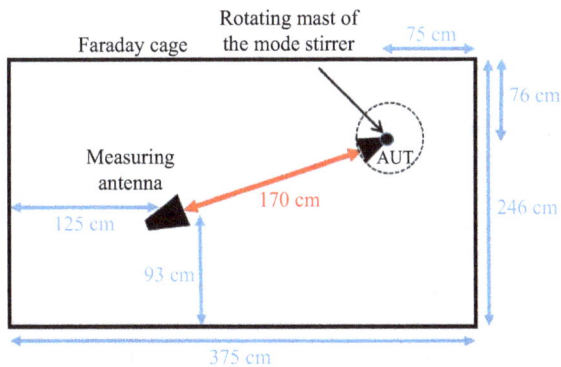

Figure 8.7 *Geometrical description (top view) of the experimental setup used in the RE of the XLIM Laboratory. According to the distance between both antennas, the LOS should theoretically arrived after 5.5 ns if we neglect the additional propagation time related to both antennas themselves*

*Figure 8.8   Picture of the experimental setup used in the RE of the XLIM
Laboratory for the time-gating technique. The measurement antenna is
in the foreground, while the AUT is in the background. The orientation
of both antennas on this picture corresponds to an H-plane radiation
pattern measurement for the angle $\theta = -180°$. The influence of the
movement of the blue coaxial cable connected to the AUT during its
rotation is assumed to be negligible*

longer an RC but an RE), the AUT has been fixed to its rotating mast. S_{21} measurements have been performed with the same angular step of 1° on the bandwidth from 1 to 13 GHz over 48,001 points corresponding to a frequency shift δ_f of 0.25 MHz between two successive frequencies. Measurements have been repeated for the RE left empty (i.e., the "unloaded" configuration) and when three blocks of absorbers have been inserted in (i.e., the "loaded" configuration).

Figure 8.9 presents the power-delay profile [18,19] (averaged over all the considered angles) obtained in the RE for both loading configurations as well as the average mode bandwidth Δ_f (in the legend) defined as the ratio between the frequency f and the average composite quality factor Q [6]–[20].

Each RE measurement lasts around 260 min. This long measurement time cannot be directly compared to the measurement time typically required in an anechoic chamber. Indeed, it is possible from these measurements to compute the radiation pattern using the time-gating technique on the whole frequency range (with the exception of the bounds) and not only at the 22 frequencies measured in the anechoic chamber. Moreover, it should be noted that 60 min (i.e., 10 s for each angle) are devoted to the stabilization of the AUT after each rotation of 1°.

Before applying the time-gating process on these measurements, it is useful to identify the instant of arrival of the LOS and of the first three NLOS paths in order to determine a suitable value for t_{tg}. To do this, we have computed the absolute value

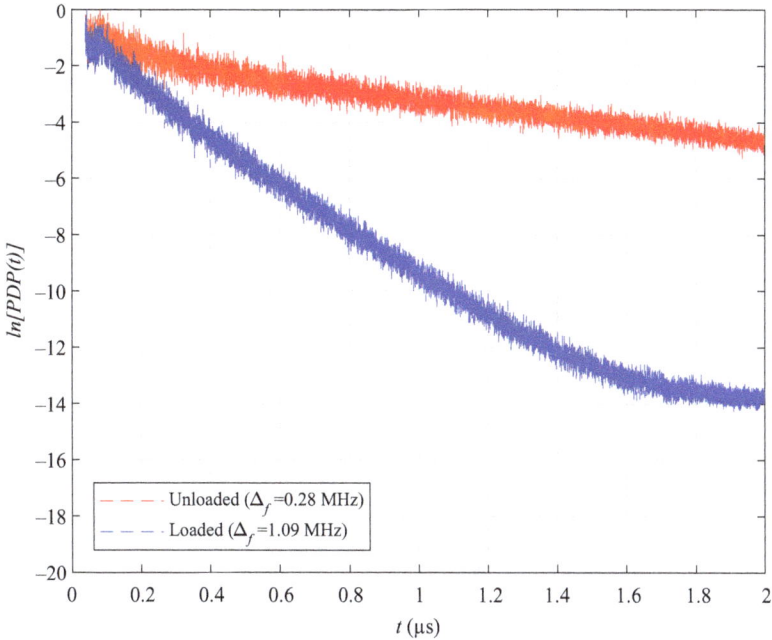

*Figure 8.9 Normalized power delay profile of the RE for both loading
configurations*

of the IFT of S_{21} for each angle and calculated its maximum value at any instant. It
is shown in Figure 8.10 that these paths arrive at, respectively, $t = 7.5$ ns for the LOS
and $t = 10.8$, 12.3 and 14.5 ns for the first three NLOS paths. Therefore, t_{tg} has to be
included in the range between 7.5 and 10.8 ns and the minimum bandwidth B equals
300 MHz using (8.9).

For each sub-band B of 1 GHz around each frequency f_0 where the radiation
pattern has been measured in the anechoic chamber, the truncated signal (using
a symmetrical time-gating window from $-t_{tg}$ to t_{tg}) is computed for each angle
of the radiation pattern (as shown in Figure 8.11 for two particular angles) and
subsequently the AUT radiation pattern is obtained from the FT of the truncated
signal.

Figures 8.12 and 8.13 present a comparison of the results obtained in the anechoic
chamber and in RE using the time-gating method at two extreme frequencies of our
frequency range, i.e., $f = 1.5$ and 12 GHz. It is also shown that the results obtained
without any post-processing are, as expected, totally erroneous.

To go further, Figure 8.14 presents the magnitude of the S_{21} parameter before
and after the time-gating process. The smoother aspect of the plot obtained after

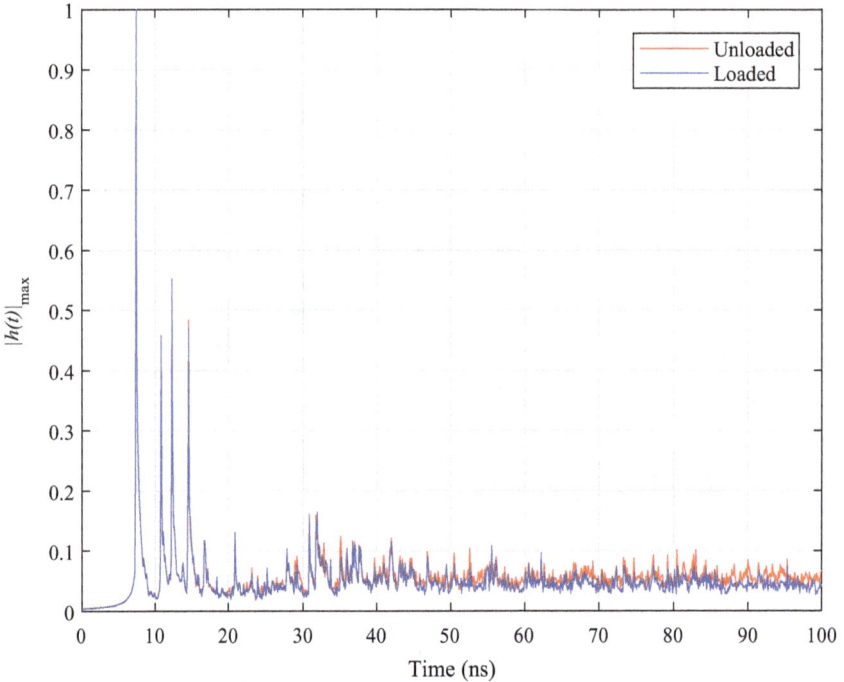

Figure 8.10   *Normalized maximum impulsional response* $|h(t)|_{max}$ *in the E-plane from* S_{21} *measurements performed from 1 to 13 GHz over 48,001 frequency points for both loading configurations of the RE. The LOS and the first three NLOS paths appear clearly. It is also shown that the insertion of three blocks of absorbers does not have any influence during the early instants but becomes gradually visible after*

time-gating makes sense and looks like what we could obtain during a free space measurement in an anechoic chamber.

In order to quantify on a rigorous basis the difference between the reference radiation pattern (D_{AC}) measured in the anechoic chamber and the one (D_{RE}) measured in our RE, the root-mean-square error (RMSE) is computed as

$$RMSE = \sqrt{\langle [D_{AC}(\alpha) - D_{RE}(\alpha)]^2 \rangle}, \tag{8.11}$$

for each entire radiation pattern, α being the angle θ for the E-plane and ϕ for the H-plane. However, it is important to emphasize here that the measurements performed in both facilities (and therefore also in the anechoic chamber) are subject to measurement uncertainties (in particular positioning errors) especially near the nulls of the beam pattern.

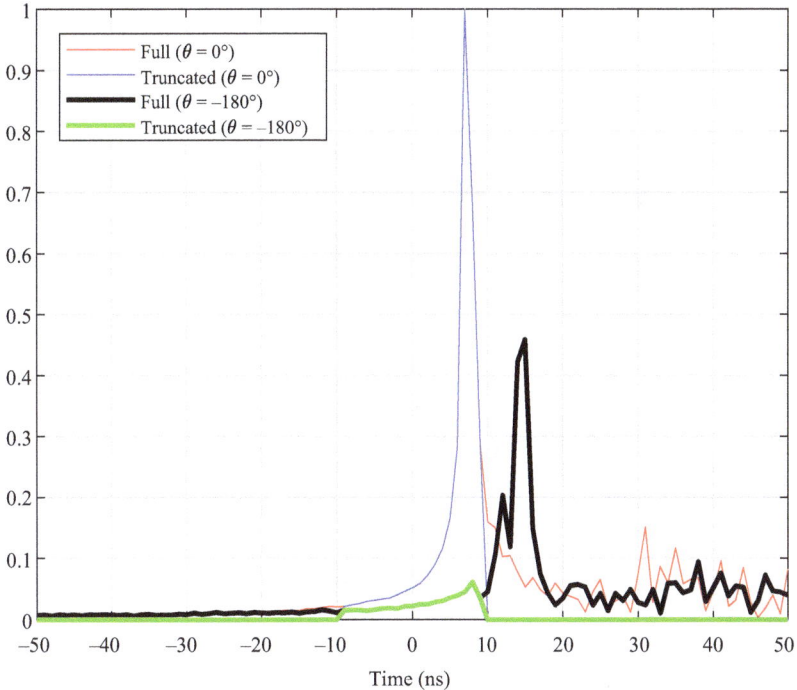

Figure 8.11   Full and truncated (with $t_{tg} = 9$ ns) magnitude of the impulsional response $|h(t)|$ around $f = 12$ GHz (with $B = 1$ GHz) in the E-plane for two different θ angles. These curves are normalized with respect to the $\theta = 0°$ plot. It appears clearly that the LOS has a lower amplitude with respect to the first two NLOS paths for $\theta = -180°$

The RMSE shown in Figure 8.15 as a function of the frequency is plotted for both loading configurations with and without a Hann filtering in the E-plane for $B = 1$ GHz. It is shown that, in such favorable conditions, there is no clear effect of loading the chamber and of using a Hann filter. More importantly, there is no dependence of the RMSE with the frequency, the error being satisfying (i.e., < -25 dB) and almost constant on the entire frequency range investigated.

This error computed for one particular frequency has been averaged over the 22 considered frequencies and converted in dB with $\langle RMSE_{dB} \rangle = 20\log_{10}(\langle RMSE(f)\rangle)$. The average RMSE presented in Table 8.1 for four different bandwidths B (including a value of 0.25 GHz lower than B_{min}) is presented for both loading configurations with and without a Hann filtering. A satisfying average RMSE lower than -27 dB is obtained for all the configurations when $B > B_{min}$, i.e., 0.3 GHz. The decrease of the accuracy for $B = 0.25$ GHz is clear. In these results, the benefit to use a Hann filter or to load the RE is once again not clear.

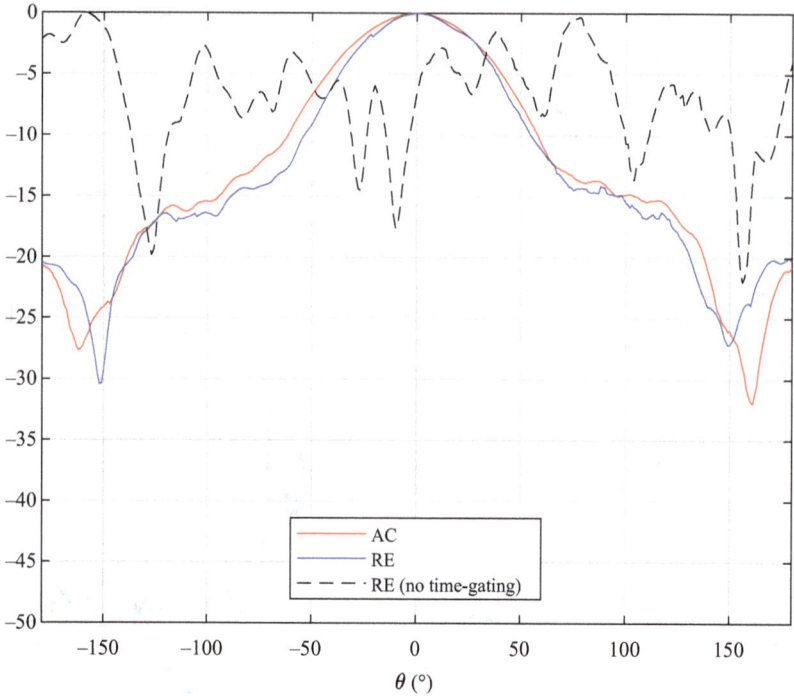

Figure 8.12   Normalized radiation pattern (in dB) in the E-plane at 1.5 GHz in the anechoic chamber and in the unloaded RE using optimal time-gating parameters (B = 1 GHz, δ_f = 0.25 MHz, symmetric window)

To reduce the measurement time, it can be of interest to decrease the number of collected frequencies during the S_{21} measurements. RC measurements have been repeated at only 4,001 frequencies over the same frequency range (i.e., δ_f = 3 MHz). The total measurement time reduces to 83 min (with 1 hour still related to the stabilization of the AUT). Table 8.2 containing the results obtained for the same configurations from this new measurement conditions clearly illustrates the degradation of the results' quality (the RMSE being higher than -20 dB), particularly in the case of the "unloaded" RE, the results obtained for the loaded configuration remaining satisfying.

These results illustrate the importance of the aliasing effect in the case of an insufficient frequency-domain resolution. Indeed, even if the time-gating window truncates the signal early, it is fundamental to have a sufficiently maximum time t_{max} of the IFT, which is obtained with a sufficiently low δ_f. Otherwise, aliasing effects have the effect to add a nonphysical offset during the early time of the impulsional response. This is illustrated in Figure 8.16 where the normalized early-time impulsional response $|h(t)|$ is shown for different number of frequencies collected from 1 to 13 GHz. It

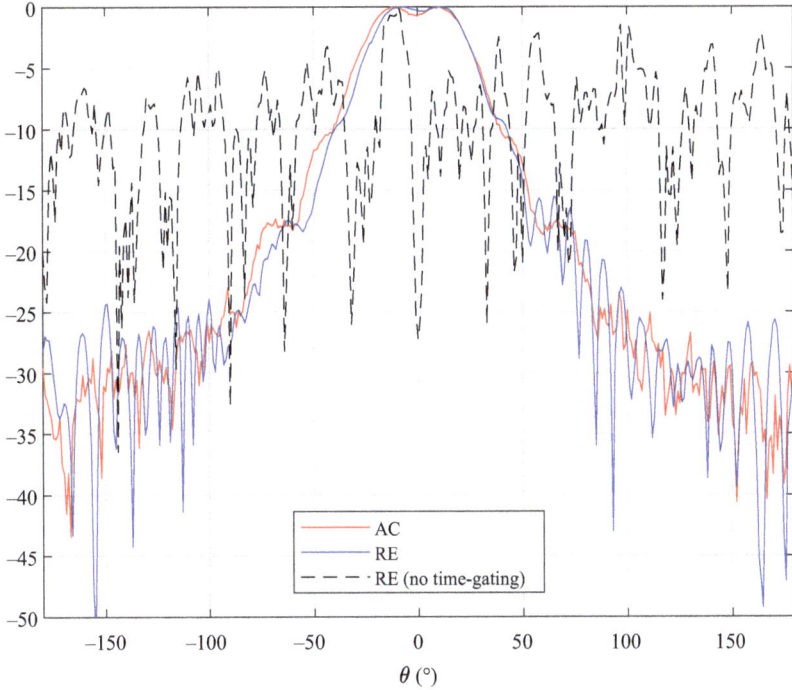

Figure 8.13   *Normalized radiation pattern (in dB) in the E-plane at 12 GHz in the anechoic chamber and in the unloaded RE using optimal time-gating parameters (B = 1 GHz, δ_f = 0.25 MHz, symmetric window)*

appears clearly that the aliasing increases artificially the level of $|h(t)|$ in the early instants even if the peak related to the LOS at $t = 7.5$ ns seems not to be disturbed. This phenomenon also explains why loading the chamber improves the results. Indeed, the loading accelerates the decrease of energy, which reduces the aliasing effect during the earlier instants.

In addition with the potential insertion of absorbers that can solve the problem, another solution consists in reducing the time-gating window around the peak corresponding to the LOS path in order to cancel the unwanted signal around it. Table 8.3 contains the average RMSE in the case of a "tight" time-gating window (from t_{tg1} to t_{tg2}, N being the number of points included in the time-gating window) for $\delta_f = 3$ MHz. These results show an enhancement of the time-gating performance for both loading configurations. However, there is still some differences between the unloaded and the loaded configurations as the aliasing effect still disturbs the LOS peak especially for the angles when the AUT radiation is small.

After analyzing the time-gating method accuracy and particularly what parameters can help to increase it, it is worth interesting to check if the method is able

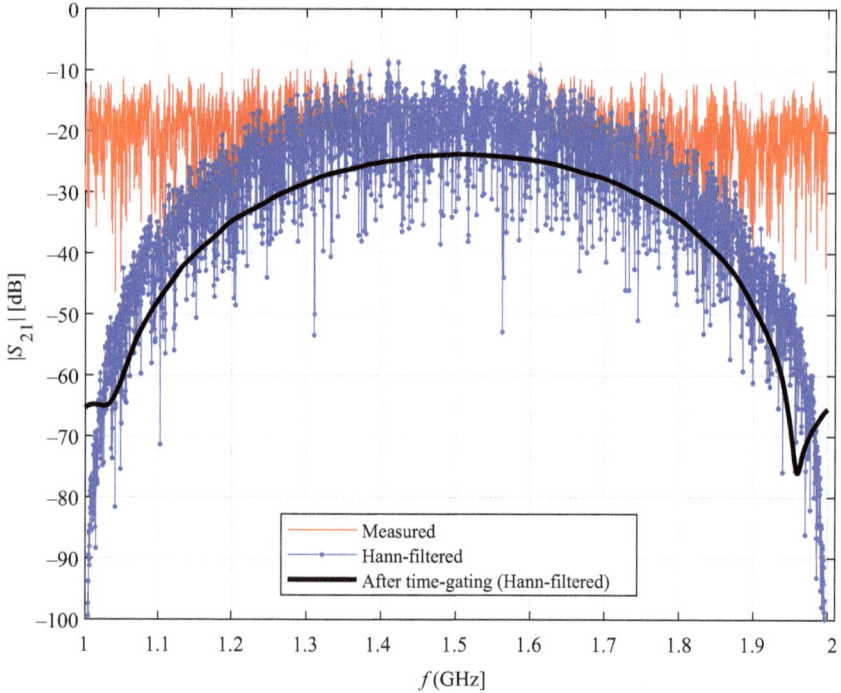

Figure 8.14   Magnitude of the S_{21} parameter measured with and without Hann-filtering around $f = 1.5$ GHz (with $B = 1$ GHz) in the E-plane (for $\theta = 0°$). The Hann-filtered result is shown before and after time-gating (with $t_{tg} = 9$ ns)

to compute accurately the radiation pattern of the AUT on the whole measurement bandwidth, i.e., from $f_0 - B/2$ to $f_0 + B/2$ or if there is a degradation for frequencies away from the center frequency of the measurement band.

A simple way to analyze it is to move the frequency window. For instance, to calculate the radiation pattern at $f = 9$ GHz with a bandwidth B of 1 GHz, the reference result is calculated from S_{21} measurements collected between 8.5 and 9.5 GHz. Then, the frequency window can be moved in both directions with a frequency shift δ_B. For instance, a δ_B value of 0.1 GHz leads to use the measurements collected in the 8.6–9.6 GHz bandwidth.

Figure 8.17 presents the results obtained at two different frequencies (3 and 9 GHz) for a rectangular and a Hann window as a function of δ_B.

For small shifts lower than $B/4$ in absolute value, there is no clear degradation of the result accuracy with respect to the reference case obtained for $\delta_B = 0$ Hz. There is also no particular difference obtained according to the windowing function used. As a rule-of-thumb, we can then consider that the radiation pattern of the AUT is correct

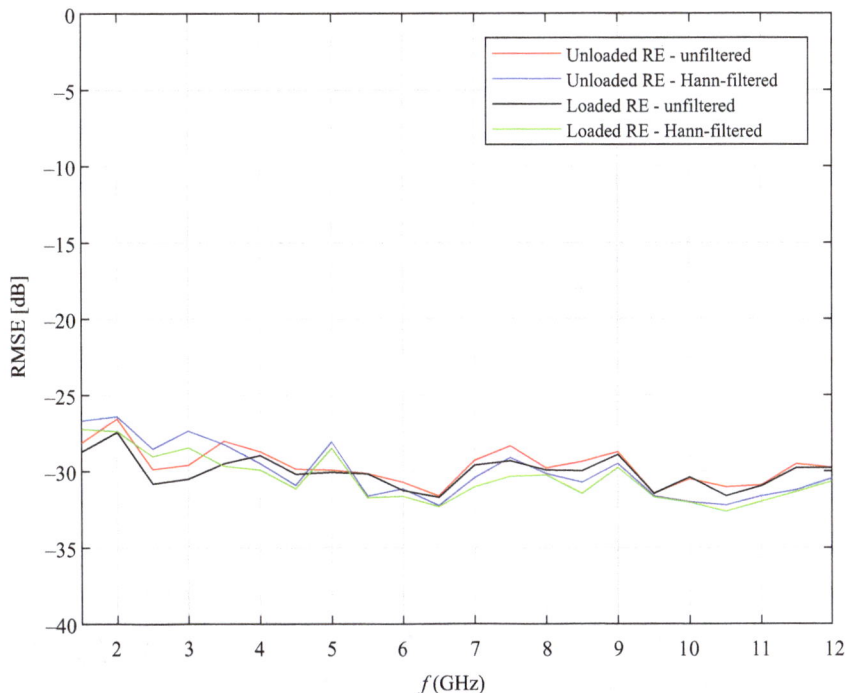

Figure 8.15 RMSE obtained as a function of the frequency in the E-plane for B = 1 GHz for both loading configurations with and without a Hann filtering

Table 8.1 Average RMSE (in dB) in the E-plane for $\delta_f = 250$ kHz and a symmetric time-gating window

B (GHz)	0.25	0.5	0.75	1
Time-domain resolution δ_t (ns)	4	2	1.33	1
t_{tg} (ns)	8	8	9.33	9
Unloaded RE and unfiltered	−26.5	−27.8	−29.8	−29.7
Unloaded RE and Hann-filtered	−23	−28	−28.1	−29.8
Loaded RE and unfiltered	−26.8	−28.1	−30.1	−30.2
Loaded RE and Hann-filtered	−23	−28.3	−28.5	−30.3

for any frequency included in the range from $f_0 - B/4$ to $f_0 + B/4$ obviously if B is sufficiently large and δ_f is sufficiently low.

For larger shifts, the result accuracy is degraded particularly for the Hann window. It makes sense as this corresponds to the frequencies where the Hann filtering is more important (as illustrated in Figure 8.14).

Table 8.2   *Average RMSE (in dB) in the E-plane for $\delta_f = 3$ MHz and a symmetric time-gating window*

B (GHz)	0.25	0.5	0.75	1
t_{tg} (ns)	10	9	9.33	9
Unloaded RE and unfiltered	−17.3	−18.6	−17.2	−19.2
Unloaded RE and Hann-filtered	−14.9	−16.8	−17.9	−17.8
Loaded RE and unfiltered	−26.3	−27.2	−28.1	−26.9
Loaded RE and Hann-filtered	−23.2	−27.2	−27.5	−26.5

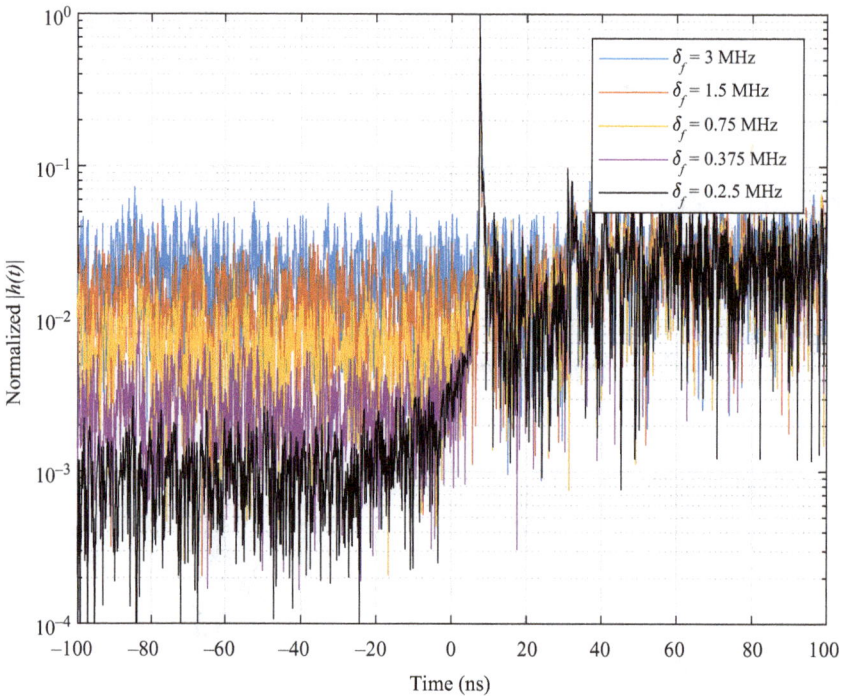

Figure 8.16   *Aliasing effect in the early-time as a function of the frequency-domain resolution f_d. This effect artificially increases the level of $|h(t)|$ in the early instants when f_d is increased*

8.4   Comparison of the methods

It is difficult to compare the three methods presented in this chapter on a rigorous basis. For this, it would be necessary to implement these methods in the same conditions, i.e., on the same AUT and in the same RE (or RC). Nevertheless, it is possible to

Table 8.3   *Average RMSE (in dB) in the E-plane for $\delta_f = 3$ MHz and a "tight" time-gating window*

B (GHz)	0.25	0.5	0.75	1
t_{tg1} (ns)	6	7	6.66	7.5
t_{tg2} (ns)	10	9	8	8.5
N	2	2	2	2
Unloaded RE and unfiltered	-21.6	-24.4	-25.8	-26.4
Unloaded RE and Hann-filtered	-19	-23.8	-25.2	-26.1
Loaded RE and unfiltered	-27.5	-27.2	-26.8	-26.5
Loaded RE and Hann-filtered	-23.5	-28.2	-28.3	-28.1

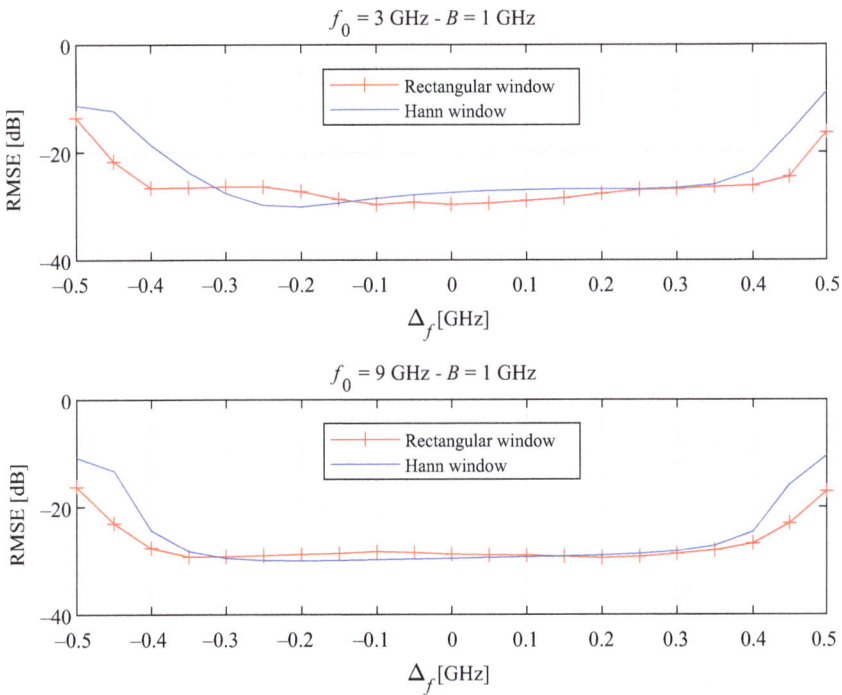

Figure 8.17   *RMSE (in dB) obtained as a function of the frequency shift δ_B in the E-plane for $B = 1$ GHz at $f = 3$ GHz (top subplot) and $f = 9$ GHz (bottom subplot)*

compare these methods according to their advantages and drawbacks (if applicable) as shown hereafter in Table 8.4.

As a conclusion, we can consider that, to date, the time-gating technique out-matches the other techniques proposed in REs when considering the complexity of

Table 8.4   Comparison of the three presented methods from different criteria

	K-factor method	**Doppler spectrum method**	**Time-gating method**
Setup complexity (excluding the rotation of the AUT on itself)	Mechanical mode stirrer required	No mode stirring process required but a rectilinear motion of the AUT	No mode stirring process and no other movement of the AUT required
Measurement time (with respect to an anechoic chamber measurement)	Greatly higher as K has to be determined accurately for each couple of angles of the radiation pattern	Greatly higher as one motion of the AUT is required for each couple of angles of the radiation pattern	Slightly higher as the S_{21} parameter has to be measured on the bandwidth B (and not only at f_0)
Possibility to distinguish the LOS and the NLOS unstirred paths	Impossible	Possible	Possible
Other comments	Valid only when K is high	Accuracy proportional to the length of the rail. Difficult to ensure a constant speed of the motorized rail all along its movement	Invalid if the reception of both the LOS and the first NLOS paths occurs at the same time (i.e., for ultrawide band antennas for instance)

the measurement setup, the duration of the measurements and the potential ability of the method to cancel the influence of the indirect paths.

References

[1] Xu Q, Huang Y, Xing L, *et al.* 3-D Antenna Radiation Pattern Reconstruction in a Reverberation Chamber Using Spherical Wave Decomposition. IEEE Transactions on Antennas and Propagation. 2017;65(4):1728–1739.

[2] Lemoine C, Amador E, Besnier P, *et al.* Antenna Directivity Measurement in Reverberation Chamber From Rician K-Factor Estimation. IEEE Transactions on Antennas and Propagation. 2013;61(10):5307–5310.

[3] Besnier P, Lemoine C, Sol J, *et al.* Radiation pattern measurements in reverberation chamber based on estimation of coherent and diffuse electromagnetic fields. In: 2014 IEEE Conference on Antenna Measurements Applications (CAMA), IEEE, Antibes, France; 2014. p. 1–4.

[4] Holloway CL, Hill DA, Ladbury JM, *et al.* On the Use of Reverberation Chambers to Simulate a Rician Radio Environment for the Testing of Wireless Devices. IEEE Transactions on Antennas and Propagation. 2006;54(11): 3167–3177.

[5] Soltane A, Andrieu G, and Reineix A. Analytical model for the assessment of Doppler spectrum of rotating objects. In: 2017 International Symposium on Electromagnetic Compatibility – EMC Europe, IEEE, Wroclaw, Poland; 2017. p. 1–4.

[6] Kildal PS, Chen X, Orlenius C, *et al.* Characterization of Reverberation Chambers for OTA Measurements of Wireless Devices: Physical Formulations of Channel Matrix and New Uncertainty Formula. IEEE Transactions on Antennas and Propagation. 2012;60(8):3875–3891.

[7] Cerri G, Primiani VM, Pennesi S, *et al.* Source Stirring Mode for Reverberation Chambers. IEEE Transactions on Electromagnetic Compatibility. 2005;47(4):815–823.

[8] Jeong MH, Park BY, Choi JH, *et al.* Doppler spread spectrum of antenna configurations in a reverberation chamber. In: 2013 Asia-Pacific Microwave Conference Proceedings (APMC), Seoul, South Korea; 2013. p. 675–677.

[9] Choi JH, Lee JH, and Park SO. Characterizing the Impact of Moving Mode-Stirrers on the Doppler Spread Spectrum in a Reverberation Chamber. IEEE Antennas and Wireless Propagation Letters. 2010;9:375–378.

[10] García-Fernández MA, Carsenat D, and Decroze C. Antenna Radiation Pattern Measurements in Reverberation Chamber Using Plane Wave Decomposition. IEEE Transactions on Antennas and Propagation. 2013;61(10):5000–5007.

[11] Karlsson K, Chen X, Kildal PS, *et al.* Doppler Spread in Reverberation Chamber Predicted From Measurements During Step-Wise Stationary Stirring. IEEE Antennas and Wireless Propagation Letters. 2010;9:497–500.

[12] García-Fernández MA, Carsenat D, and Decroze C. Antenna Gain and Radiation Pattern Measurements in Reverberation Chamber Using Doppler Effect. IEEE Transactions on Antennas and Propagation. 2014;62(10):5389–5394.

[13] Soltane A, Andrieu G, Perrin E, *et al.* Antenna Radiation Pattern Measurement in a Reverberating Enclosure Using the Time-Gating Technique. IEEE Antennas and Wireless Propagation Letters. 2019:1–5.

[14] Loredo S, Pino MR, Las-Heras F, *et al.* Echo Identification and Cancellation Techniques for Antenna Measurement in Non-Anechoic Test Sites. IEEE Antennas and Propagation Magazine. 2004;46(1).

[15] Piasecki P and Strycharz J. Measurement of an omnidirectional antenna pattern in an anechoic chamber and an office room with and without time domain signal processing. In: 2015 Signal Processing Symposium (SPSympo), IEEE, Debe Village, Poland; 2015. p. 1–4.

[16] Cozza A. The Role of Losses in the Definition of the Overmoded Condition for Reverberation Chambers and Their Statistics. IEEE Transactions on Electromagnetic Compatibility. 2011;53(2):296–307.

[17] Adardour A, Andrieu G, and Reineix A. On the Low-Frequency Optimization of Reverberation Chambers. IEEE Transactions on Electromagnetic Compatibility. 2014;56(2):266–275.

[18] Genender E, Holloway CL, Remley KA, *et al.* Simulating the Multipath Channel With a Reverberation Chamber: Application to Bit Error

Rate Measurements. IEEE Transactions on Electromagnetic Compatibility. 2010;52(4):766–777.

[19]   Holloway CL, Shah HA, Pirkl RJ, *et al.* Early Time Behavior in Reverberation Chambers and Its Effect on the Relationships Between Coherence Bandwidth, Chamber Decay Time, RMS Delay Spread, and the Chamber Buildup Time. IEEE Transactions on Electromagnetic Compatibility. 2012;54(4):714–725.

[20]   Andrieu G, Ticaud N, Lescoat F, *et al.* Fast and Accurate Assessment of the "Well Stirred Condition" of a Reverberation Chamber From S_{11} Measurements. IEEE Transactions on Electromagnetic Compatibility. 2019;61(4):974–982.

Chapter 9

Radar cross-section estimation in reverberation chambers

Philippe Besnier[1]

This chapter is dedicated to recent advances in a rather unexplored range of applications. It deals with the ability of performing radar cross-section (RCS) measurements, using a reverberation chamber (RC). If feasible, it would allow broadband measurements in a simple Faraday cage at a much lower cost, absorbers being no longer required. However, using a RC for RCS measurement seems inadequate at a first glance. Adding an object under test in the chamber consists in introducing another scatterer, indistinguishable in a rich multipath environment offered by the RC.

However, we may take advantage of several interesting properties which are relevant to investigate their ability to do so. Regarding celerity, a proper time filtering of RC response is obviously a possible strategy. Indeed, transmitting and receiving antennas (if different) may be oriented with their maximum gain in the line-of-sight of the illuminated target. We may use a short pulse at emission or equivalently a frequency scan on a bandwidth that is compatible with the requested time resolution. It enables to segregate the target response at early times of delayed and spread echoes from the RC. By doing so, RCS measurements using this principle may be performed in any medium of propagation providing sufficiently distinct times of arrival of the target and other scatterer responses.

Ideal RCs also provide perfect diffuse field patterns when modal density is high enough and modal overlap is small enough. In these conditions, the complex transfer function of the RC is described by centred Gaussian distributions. Therefore, any direct echo from the target appears as a signal modulated by a random interference. Fluctuations of this direct echo and of this random interference are very different from each other. This provides possible strategies to retrieve the direct echo alone. This underlies the proposed method presented hereafter.

This chapter is organized as follows. We first recall the basic definition of RCS. Then, we introduce the frame of RCS measurements before focusing on RCS estimation in RCs. From these introductory elements, we provide a theory for RCS pattern measurements in RC. Some measurement results on simple targets illustrate this measurement principle. Finally, the role of the mechanical stirrer is discussed.

---

[1]INSA Rennes, CNRS, IETR – UMR 6164, Rennes, France

9.1   Definition of radar cross section and the radar equation

The RCS of an object illuminated by an electromagnetic wave is related to the proportion of energy which is backscattered from it [1]. More precisely, the monostatic RCS quantifies the backscattered energy in the opposite direction to the incidence direction of the illuminating wave. The bistatic RCS quantifies the scattered energy in any other direction of observation. This is schematically represented in Figure 9.1. In a spherical coordinate system (O, r, θ, ϕ), an antenna A located at (R, θ_1, ϕ_1) transmits electromagnetic energy in the direction of the target, where R represents its distance from the target. If the same antenna A is used as a receiver of the scattered energy in the opposite direction, it defines a monostatic measurement. If a second antenna B located at (R, θ_2, ϕ_2) is used to measure the scattered energy in another (θ_2, ϕ_2) direction, it is called a bistatic measurement. Since the scattered energy is proportional to the energy density impinging on the object and related to the total power scattered by the target, RCS unity is m^2. Throughout this chapter, RCS is denoted by σ^T where T denotes the target. Assuming a locally incident plane wave at the scale of the target, i.e. a far-field illumination, σ^T is defined as

$$\sigma^T(\theta_x, \phi_x) = \lim_{R \to \infty} 4\pi R^2 \frac{\| W_s(\theta_x, \phi_x) \|}{\| W_i \|}. \tag{9.1}$$

According to Figure 9.1, the subscript x in (9.1) takes the values 1 and 2 for a monostatic and a bistatic configuration, respectively. The incident plane wave at target location transports an energy density magnitude denoted by W_i in (9.1) and scatters the electromagnetic energy W_s, which is observed at a theoretically infinite distance from the target. It is given by the vectorial product of the electric field strength vector $(\overrightarrow{E_i})$ and the magnetic excitation vector $(\overrightarrow{H_i})$. In the plane wave approximation, we get

$$\| W_i \| = \| \overrightarrow{E_i} \wedge \overrightarrow{H_i} \| = \frac{E_i^2}{Z_w}. \tag{9.2}$$

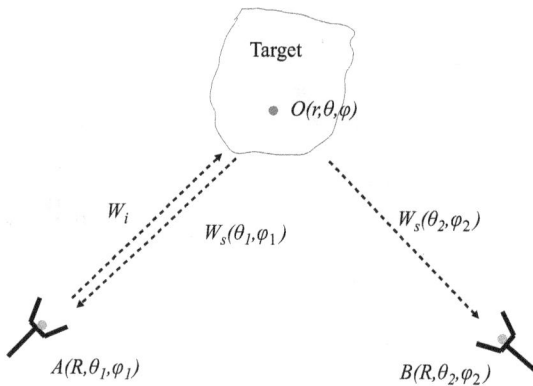

Figure 9.1   Radar cross section: a schematic representation

The energy density, W_i, of the incident plane wave is therefore determined from the electric field strength and the medium wave impedance, Z_w. The scattered energy from the illuminated target depends on the angles of incidence and observation and on its geometric and electromagnetic properties. In practice, this energy density is observed at a finite distance R, being at a far-field distance from the target, assuming again a locally plane wave. More precisely, with regard to (9.1), the scattered energy density W_s is therefore assumed to be inversely proportional to R^2.

The radar equation defines the power transfer function between the radar transmitting antenna and its receiving antenna. In the monostatic configuration, neglecting all losses in the measurement chain, it is easily derived from the Friis equation [2] and antenna radiation properties. By definition, the backscattered power from the target in the direction of observation (opposite direction to the incidence wave in the monostatic configuration) is that which would have been backscattered by an isotropic antenna supplied with the following transmitted power, denoted as P_{target}

$$P_{target} = \sigma^T \times \frac{E_i^2}{Z_w}. \tag{9.3}$$

The impinging electric field density, E_i^2/Z_w, in (9.3) is proportional to the transmitted power P_t and to the gain G_t of the transmitting antenna. We may therefore write

$$E_i^2 = \frac{Z_w G_t}{4\pi R^2} P_t. \tag{9.4}$$

The source of scattered energy radiates a scattered field back to the source such as

$$E_s^2 = \frac{Z_w}{4\pi R^2} P_{target}. \tag{9.5}$$

Eventually, the received power, P_r at the receiving/transmitting antenna, is given by

$$P_r = \frac{\lambda^2}{4\pi} G_t \frac{E_s^2}{Z_w}. \tag{9.6}$$

Using successively (9.6), (9.5), (9.3) and (9.4), we find the expression of the so-called radar equation

$$\frac{P_r}{P_t} = \frac{\lambda^2}{(4\pi)^3 R^4} G_t^2 \sigma^T. \tag{9.7}$$

This equation is valid for the monostatic configuration and looks identical for the bistatic configuration if both antenna gains are equal and at the same distance from the target. However, the bistatic RCS, σ^T, is a function of various combinations of incident and observation angles. In the following sections of this chapter, we deal with monostatic measurements only.

9.2   Radar cross-section measurements

We only discuss RCS measurements in indoor test ranges. RCS characterizations are preferably performed in propagation media where other echoes than the scattered field from the target are cancelled out or maintained at a reasonably low level. There are basically two combined principles that make it possible.

First, echoes from the target must be easily separated from echoes coming from any other possible obstacles surrounding the transmitting antenna and the target. This is a function of the properties of the target, of the distance from the target and of the time resolution. The achieved time resolution depends on the available bandwidth of the source signal, which may be a pulse signal or a constant wave signal swept on a frequency band Δf. Consequently, the time resolution t_r is given by $t_r = 1/\Delta f$. Equivalently, at light speed (c) the distance resolution ΔR is defined as $c/(2\Delta f)$. A minimum requirement is to separate echoes from the target and from other sources, but a more stringent requirement may be further applied to resolve the different echoes from the source. For instance, echoes from distant obstacles may nevertheless interfere with long response time of resonating targets. Outdoor or large indoor test ranges favour distance/time discrimination of any specular reflection with regard to the travelling time of the target echoes. Since it would impose large test sites, this advocates the use of anechoic chambers where absorbing materials are adapted to the frequency range under investigation.

Second, absorbing materials have limited performance and partly attenuate those undesired echoes. Time gating further enhances the signal-to-noise ratio, rejecting the remaining undesired contributions out of the time window centred around the backscattered signals from the target [3]. The time-gating operation is performed through an inverse Fourier transform of the frequency transfer function measured over Δf. The resulting time-domain signal is truncated and transformed back in the frequency domain. It is interesting to note that this time-gating operation may be achieved with some success without the use of anechoic environment, if parasitic echoes do not overlap those of the target. Therefore, if and only if this condition is satisfied, a time-gating operation is sufficient to retrieve the target response in any kind of propagation media, including RCs. However, we show in the following part of this chapter that the statistical properties of diffuse fields in RCs may be advantageously used for such purpose.

9.3   Scattered cross section and antenna patterns in reverberation chambers

RCs are diffuse field environments. As a result, they are well adapted to sense the overall energy which is either absorbed or scattered by any object placed anywhere in the chamber (except near the boundaries [4]). The total transmitted power in the chamber is balanced with the contributions of different losses mechanisms (walls, antennas, objects). As a result, once any other losses are calibrated, the averaged total received power over (many) chamber states measured across the load impedance of an antenna characterizes the total radiated power of an unintentionally radiating object. In Chapter 6, the average absorbing cross section of objects is retrieved from the decrease of this received power which may be related to the modification of the composite Q-factor or coherence bandwidth of the chamber.

To our knowledge, the first contribution in the scientific literature dedicated to the determination of scattered cross section of targets is that of Lerosey and de Rosny [5]. The method consists in moving the target(s) in the chamber in order to cancel out the contribution of the scattered energy. Indeed, in the hypothesis of a perfectly disordered or ideally diffuse field, the average scattered field from the moving targets vanishes. In particular, this ensemble of scattered fields does not contribute anymore to the received power at the load impedance of a receiving antenna. Thus, the contribution of the scattered field may be assimilated to a loss factor and its associated Q_s-factor. The resulting average scattered cross section of the target(s), σ_s, must not be confused with the definition of σ_t in (9.1) that accounts for the specific direction of scattering fields. On the contrary, σ_s characterizes the total power scattered by the target over 4π steradian. Eventually, it is given by

$$\sigma_s = \frac{\omega V}{c Q_s}, \tag{9.8}$$

where V is the volume of the Faraday cage. There is therefore a contrast of Q-factor for moving target(s) with regard to fixed targets. In the time domain, the ratio between the averaged square electric field strength over target(s) positions and the square electric field strength for fixed target(s) position(s) is equal to $\exp(-t/\tau_s)$ with $Q_s = 2\pi \tau_s$. This provides an elegant method to retrieve σ_s from pulse excitation, assuming a roughly constant τ_s over the frequency band of interest. Experimental results are shown in [5]. A better estimation of average fields is performed using arrays of antennas. The scattering energy contrast of amplitude between fixed and moving configurations may also be enhanced using several identical targets.

Estimation of the RCS pattern remains a challenge in RC. Like the antenna pattern measurements, it requires the knowledge of magnitude, phase and polarization of the scattered field in a particular direction that may be that of the transmitter in the monostatic case or a different one in the bistatic case. This was definitely not the initial purpose of RCs but recent works contributed to use them in an unconventional way. A breakthrough was recently proposed to control incidence and polarization of waves in RCs. This proposal is known as the time reversal electromagnetic chamber as initially proposed by Cozza [6]. It uses the principle of time reversal mirror while generalizing it at multiple locations to synthesize incident-polarized plane waves with a specified angle of arrival (or several of them at a time). It is therefore possible to use such principle to illuminate a target under different angles of incidence and polarization with electronically processed signals. The calibration method remains a challenge but recent and spectacular experimental results have been obtained and published in 2017 [7]. From there, applications to RCS measurements would be an interesting subject of investigation.

A completely different approach consists in detecting a line-of-sight path between target and measurement antennas. In these conditions, the test setup is achieved in complete contradiction with standard requirements for a RC. More precisely, the measurement antenna is a directive antenna pointing at its maximum directivity in the direction of the target. In principle, the impinging wave on the target or antenna under test must appear as a locally plane wave; then the line-of-sight contribution must be quantified. Such a procedure was already used in the context of sensing antenna radiation pattern. From the complex S-parameters measured at ports of the measuring antenna and of the antenna under test, an estimation of the K-factor from a Rice probability density function (pdf) is performed [8]. Indeed, two coupled antennas in a chamber produce a Rice-distributed field due to the superposition of a diffuse field and a ballistic wave between both antennas. Therefore, the K-factor is proportional to the directivity of the antenna under test. A calibration is needed with a known antenna, and the method was confronted to standard measurements, where it appears that it is valid to identify the main beam of the radiation pattern. In [9], a clever method was introduced to extract the ballistic wave (line-of-sight) by moving the antenna under test in the direction of the measurement antenna (or vice versa). The movement is performed over a distance of some wavelengths at the considered working frequency, and S-parameters are recorded at uniform space intervals. This enables to compute the pseudo-Doppler spectrum (pseudo stands for arbitrary speed). The maximum pseudo-Doppler frequency identifies the intensity of the energy that comes from the measurement antenna in direct path to the antenna under test. Transposition of this method to RCS measurements could certainly be investigated.

In the following, we present a specific method which was initially published in 2017 [10] and further examined in [11] and in [12] using time gating as a post-processing operation.

9.4   Theory of radar cross-section pattern measurements in reverberation chambers

This theory of RCS pattern measurements deals with a monostatic RCS measurement in RC. Its generalization to a bistatic measurement is achievable and should be straightforward. The proposed theory is then illustrated through the measurements of simple targets. We mean as simple target, an object whose RCS may be assimilated to the radiation of a point-like source of backscattering in the direction of the receiver. It is, for example, the case of a target behaving as a specular reflector. The case of multiple point-like sources requires more advanced (but still not complex) signal processing. Multiple point sources occur for multiple targets or single targets with different contributing parts. A more complex problem is that of resonating structures and will deserve attention in the future. This theory is based on the measurement setup described hereafter.

9.4.1 Measurement setup

The measurement setup appears in Figure 9.2. The RC is the largest one at IETR (UMR CNRS 6164), its dimensions are 8.7 m × 3.7 m × 2.9 m. A directive antenna (a horn antenna is used in all the experiments shown in this chapter) is placed in an arbitrary location within the working volume of the RC. It is used both as a transmitting and receiving antenna. It is connected to a port (say port 1) of a vector network analyser (VNA) which records the S_{11} scattering parameter. The calibration is performed at the end of the cable supplying the horn antenna. The antenna points at the target in the direction of its maximum directivity. In Figure 9.2, the target (a rectangular metal plate on the photograph) is installed on the top of a rotating mast. It is located at the same height above ground as the transmitting/receiving antenna. The horizontal distance between the rotation axis of the mast and the aperture of the horn antenna is approximately 3.0 m on the photograph. Though the transmitting antenna does not radiate in free-space, the distance between the target and the antenna is selected so that the far-field condition is approximately satisfied. A piece of absorbing material is placed in front of the bottom of the rotating mast to minimize reflections from its metal parts. The grey electronic box, lying on the ground, controls the mast rotation. The measurement axis is in the direction of the stirrer behind the target. This choice is arbitrary, but the stirrer plays a somewhat different role compared to standard RC measurements. All measurements are based on constant wave measurements performed within some frequency band Δf.

Figure 9.2 *Test setup for RCS pattern measurement in a RC of IETR. A horn antenna is pointing at the target (a rectangular plate) installed above a polystyrene rotating mast. A piece of absorbing material prevents reflections from the metallic base of the mast*

9.4.2   Backscattered field in an empty RC

Let suppose a horn antenna transmitting a continuous wave signal at a frequency f_0 (angular frequency ω_0). This working frequency is higher than the threshold frequency for which the RC produces an ideally disordered field (i.e. an ideally diffuse field) [13,14]. It was shown from multiple studies that such conditions are met for much higher frequency than the so-called lowest usable frequency based on some limit of standard deviation of maximum electric field strength at different locations of a probe [15–17]. Such conditions are roughly met above 700 MHz in the large RC of Figure 9.2. Measurements shown in Section 9.5 of this chapter are performed in the 10 GHz frequency range.

We are interested to quantify the complex reflection coefficient S_{11} at port 1 of the VNA connected to the measurement antenna. We use the notation $S(f_0)$ for S_{11} measured at a constant wave frequency f_0. It is well known that the reflection coefficient of an antenna in a RC is totally different from the 'intrinsic' one measured when the identical antenna radiates in open space. However, we assume that the resulting reflection coefficient is an additive contribution of the free-space and backscattered contributions from the RC and other objects within the RC. This assumption is valid as long as we consider that the antenna gives birth to radiation in steady state before any electromagnetic energy is coupled back to it. This assumption holds more easily if the antenna is at distance from any reflecting structure. In the case of a resonating antenna, this condition may be more stringent due to longer transient time before reaching steady state.

For reasons which will be explained further on, we make a distinction between the backscattered diffuse field from the walls of the RC and the backscattered field in line-of-sight or in specular reflection from the stirrer. We therefore write $S(f_0)$ as

$$S(f_0) = S_{FS}(f_0) + (1 - |S_{FS}(f_0)|^2)H(f_0)\eta_{ant}$$
$$+ (1 - |S_{FS}(f_0)|^2)h_s(f_0, p_i)\eta_{ant}. \tag{9.9}$$

In this equation, S_{FS} stands for the reflection coefficient of the antenna in free-space, and η_{ant} represents its radiation efficiency that we assume to be constant over the considered frequency bandwidth, Δf. Those characteristics are intrinsic parameters of the antenna. The second term on the right-hand side of (9.9) is related to the backscattered diffuse field from the RC. In particular, $H(f_0)$ is a complex-valued transfer function describing the backscattered signal from the diffuse field towards the antenna. The last additional term of (9.9) is similar to the previous one but $H(f_0)$ is substituted for $h_s(f_0, p_i)$ that defines the line-of-sight or specular reflection from the stirrer at a given position p_i.

Before continuing our description, we discuss the rationale behind the second term on the right-hand side of (9.9). The transmitted power, P_{tr}, in the chamber is related to the output power from the VNA, P_{out}, such as

$$\frac{P_{tr}}{P_{out}} = (1 - |S_{FS}(f_0)|^2)\eta_{ant}. \tag{9.10}$$

The transfer function of the chamber $H(f_0)$ relates the input power, P_{in}, received back at the antenna port to the transmitted (radiated) power in the RC

$$\frac{P_{in}}{P_{tr}} = (1 - |S_{FS}(f_0)|^2)\eta_{ant}H^2(f_0). \tag{9.11}$$

Then, the second term on the right-hand side of (9.9) is homogeneous to $\sqrt{P_{in}/P_{out}}$.

According to perfect random field hypothesis, both real and imaginary parts of $H(f_0)$ are statistically described by random variables following a centred Gaussian distribution. This position p_i of the stirrer at frequency of excitation f_0 defines a RC state and therefore a particular realization of $H(f_0)$. Equation (9.9) defines the value of $S(f_0)$ in the absence of the target on the mast.

9.4.3 Backscattered field from the target

A target is installed on the mast. This is the only modification with regard to the previous configuration. It also means that the stirrer remains at the same position p_i. The addition of a target on the mast yields a modification of the previous equation according to two main hypothesis. We assume that the electromagnetic field in the RC is weakly perturbed by the target. The initial transfer function $H(f_0)$ is modified but follows the same Gaussian distribution with identical variance. In other words, the background electromagnetic field distribution remains statistically unchanged. The addition of the target is therefore dealt with the Born approximation [18], stating that the modification of the backscattered field is due to the only interaction between the target and the antenna. Then, the reflection coefficient is modified as follows

$$\begin{aligned}
S^T(f_0) &= S_{FS}(f_0) + C(f_0)\sqrt{\sigma^T(f_0)} \\
&\quad + (1 - |S_{FS}(f_0)|^2)H^T(f_0)\eta_{ant} + (1 - |S_{FS}(f_0)|^2)h_s(f_0,p_i)\eta_{ant} \tag{9.12}
\end{aligned}$$

The additional term $C(f_0)\sqrt{\sigma^T(f_0)}$ consists of the ballistic wave backscattered by the target towards the antenna. The term $C(f_0)$ is a backscattered coefficient. It is a complex-valued quantity that describes the wave propagation from the antenna to the target and then back to antenna at the frequency f_0. Specular reflections are neglected. Finally, σ^T at the frequency f_0 is the RCS of the target for the particular position of the target on the rotating mast. The modified transfer function is denoted $H^T(f_0)$. The last term on the right-hand side of (9.12) is not modified with respect to (9.9). Thus, according to the Born approximation, we assume that the line-of-sight or specular reflection from the stirrer is not affected by the presence of the target.

9.4.4 The backscattered coefficient

The distance R between the antenna and the target is chosen as if the measurement was carried out in free-space conditions. This distance is supposed to be greater than the Fraunhofer distance which describes an artificial boundary between near-field and far-field propagation. It is at least $2D^2/\lambda$ where D is the largest dimension of the target and λ the minimum considered wavelength, or better $4D^2/\lambda$ to account for cumulated phase difference of progressive and backscattered waves. We therefore assume the ballistic wave being a locally plane wave at the target position, despite the existence of specular reflections. The analysis performed in Section 9.1 is therefore applied. Looking at (9.12), it appears that $C(f_0)\sqrt{\sigma^T(f_0)}$ is homogeneous to the square root of the radar equation in (9.7). Thus, we may readily establish the following expression

$$|C(f_0)| = \frac{G_{ant}(f_0)\lambda_0}{(4\pi)^{3/2}R^2}(1 - |S_{FS}(f_0)|^2). \tag{9.13}$$

It comes not to a surprise that the quantity $|C(f_0)|$ evolves with the inverse of the square distance R between the antenna and the target. Moreover, it is proportional to the antenna gain $G_{ant}(f_0)$. It is important to note that the target may be represented by a set of punctual or distributed sources. The previous derivation of radar equation supposes implicitly that the target is approximated as a punctual scattering source. The distance R in the radar equation corresponds without ambiguity to the distance between the scattering source and the phase centre of the spherical wave radiated by the transmitting antenna. Accounting for the forward and backward travelling waves, we may also determine the complex expression of the backscattered coefficient including its phase

$$C(f_0) = |C(f_0)| \exp\frac{-j2\pi f_0 2R}{c} \exp(j\phi_0), \tag{9.14}$$

where ϕ_0 is an arbitrary constant phase.

9.4.5   RCS equation of the target

According to the Born approximation, we perform the difference between both measurements before and after the installation of the target on the mast. By doing so, we reduce the contribution from the background environment of the RC to random fluctuations. Substituting (9.13) into (9.14) and this last equation into (9.12), the difference response $S^T(f_0) - S(f_0)$ is then related to the RCS, σ^T, as follows

$$\begin{aligned}
S^T(f_0) - S(f_0) &= (1 - |S_{FS}(f_0)|^2)(H^T(f_0) - H(f_0))\eta_{ant} \\
&\quad + \sqrt{\sigma^T(f_0)}\frac{G_{ant}(f_0)\lambda_0}{(4\pi)^{3/2}R^2}(1 - |S_{EL}(f_0)|^2) \\
&\quad \times \exp\frac{-j2\pi f_0 2R}{c} \exp(j\phi_0).
\end{aligned} \tag{9.15}$$

The first term on the right-hand side of (9.15) represents the difference of the diffuse transfer functions of the chamber in two different states. The presence of the target acts as a more or less efficient stirrer. This perturbation must be accounted for with regard to signal processing. It is an interfering signal with regard to the backscattered wave in line-of-sight from the target. Interestingly, it is proportional to the difference of two random variables, $H^T(f_0) - H(f_0)$, both distributed according to a centred

Gaussian *pdf* with equal variance. Then, the second term on the right-hand side of (9.15) contains the magnitude of the backscattered signal from the target.

9.4.6 Extraction of RCS

We first suppose that $\sigma^T(f_0)$ remains constant over some frequency band centred around f_0. We may also assume that $G_{ant}(f_0)$ and $S_{FS}(f_0)$ are also constant over this frequency band. However, it is not a requirement, since it may be easily compensated for in (9.15). Evaluating $S^T(f) - S(f)$ for $f_0 - \Delta f/2 \leq f_0 \leq f_0 + \Delta f/2$, it appears that the square root of RCS is modulated by a sine wave signal as a function of frequency, with a periodicity (in the frequency space) $\delta f = c/2R$. Since the distance of measurement is in the range of metres, this periodicity is in the range of a few tens of MHz (e.g. 50 MHz for $R = 3$ m). The fluctuation associated with the difference of the two transfer functions in the RC varies much more rapidly with frequency, since it is associated with the coherence bandwidth of the chamber that is roughly two orders of magnitude below $\delta f = c/2R$. Figure 9.3 gives an example of the waveform of the real part of $S^T(f) - S(f)$ obtained between 9.8 and 10.2 GHz for a rectangular metal plate target of size 148×151 mm^2 at an approximate distance of 3 m. Its orientation corresponds to the maximum RCS value of the plate. This signal clearly represents an oscillating waveform with the addition of an interfering signal which, after analysis, appears to be Gaussian distributed. The standard deviation of this interference is greater than the signal amplitude. The imaginary part of the same difference is also

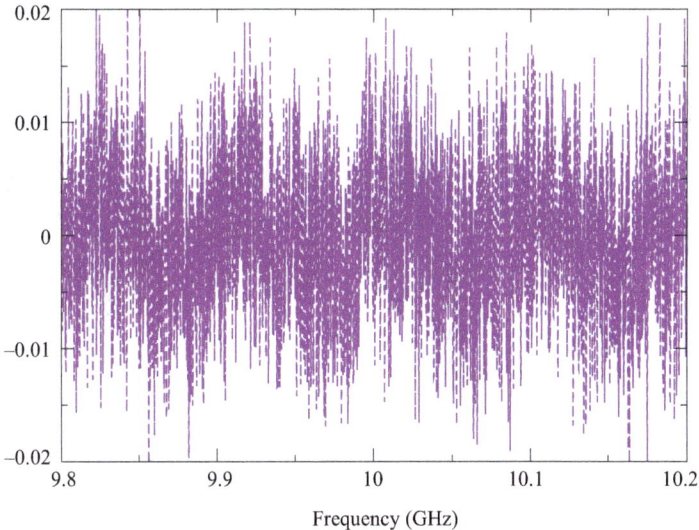

Figure 9.3 *Waveform of the real part of $S^T(f) - S(f)$ for a rectangular metal plate of size 148 mm $\times$ 151 mm in the 9.8–10.2 GHz frequency range. The plate is placed at a distance of 3 m from the measurement antenna. The stirrer remains at the same position for $S(f)$ and $S^T(f)$ measurements*

shown in Figure 9.4. It exhibits the same pattern as the real part of the difference. However, the sinusoidal waveform in Figure 9.4 appears to be in quadrature with respect to that of Figure 9.3.

There are different ways to retrieve the magnitude of $\sigma^T(f_0)$ from (9.15). These examples of experimental waveforms obtained for the real part of (9.15) in Figure 9.3 and for the imaginary part of (9.15) in Figure 9.4 suggest several options for signal processing. The superposition of a rapidly varying signal and of a smoothly varying signal suggests that a Fourier transform from frequency domain to a dual-space domain is able to separate those two contributions. This is definitely a possible way, although sensitive to low signal to interference ratio. We rather present an alternative method, well adapted, at least for the identification of a bright spot and based on a sine wave regression.

As far as this method is concerned, the determination of $\sigma^T(f_0)$ relies upon the identification of a sine wave regression curve of period δf and magnitude $A(f_0)$ from which the RCS value is determined from

$$\left|\sigma^T(f_0)\right| \approx |A(f_0)|^2 \frac{(4\pi)^3 R^4}{(1 - |S_{EL}(f_0)|^2)^2 G_{ant}^2(f_0)\lambda_0^2}, \tag{9.16}$$

where $|A(f_0)|^2$ is the square magnitude of a sine wave signal estimated from the difference of the real and imaginary part of $S^T(f) - S(f)$.

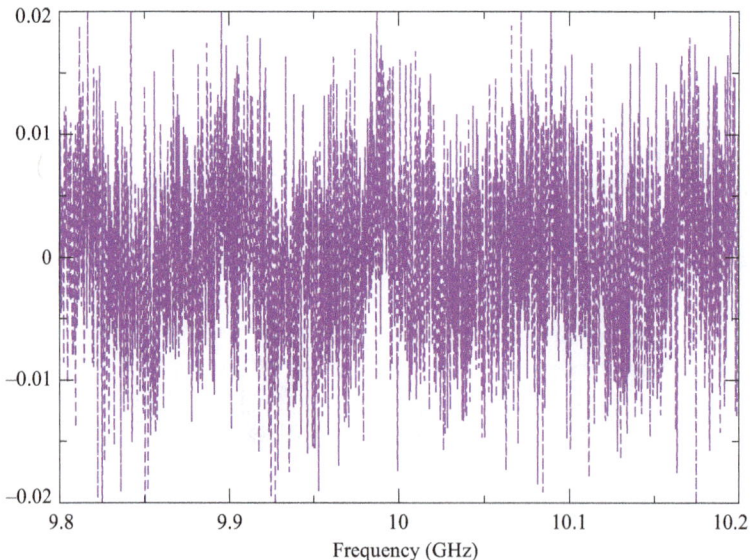

Figure 9.4   Waveform of the imaginary part of $S^T(f) - S(f)$ for a rectangular metal plate of size 148 mm × 151 mm in the 9.8–10.2 GHz frequency range. The plate is placed at a distance of 3 m from the measurement antenna. The stirrer remains at the same position for $S(f)$ and $S^T(f)$ measurements

The researched magnitude, $A(f_0)$, is that of a cosine and of a sine wave that best fits the real part and imaginary part of (9.15), according to the following minimization problems given by (9.17) and (9.18), respectively,

$$\text{argmin}_{A(f_0),\phi_0} |A(f_0) \cos \left(\frac{2\pi(f_0+f_i)2R}{c} - \phi_0 \right)$$

$$-\Re(S^T(f_0+f_i) - S(f_0+f_i))|$$

$$\text{with } f_i = i\delta f_s, \ i = -N, -N+1, \ldots, 0, \ldots, N-1, N. \tag{9.17}$$

$$\text{argmin}_{A(f_0),\phi_0} |A(f_0) \sin \left(\frac{2\pi(f_0+f_i)2R}{c} - \phi_0 \right)$$

$$-\Im(S^T(f_0+f_i) - S(f_0+f_i))|$$

$$\text{with } f_i = i\delta f_s, \ i = -N, -N+1, \ldots, 0, \ldots, N-1, N. \tag{9.18}$$

These two equations may be solved independently through a sinusoidal wave fitting involving the phase estimation ϕ_0. For the sake of brevity and clarity, we use the notation $B(f) = S^T(f) - S(f)$ and $\psi = 4\pi R/c$. Let us introduce the solutions A_1 and A_2 as the solution of (9.17) and (9.18), respectively. These solutions may be identified as

$$\Re(B(f)) = A_1 \sin(\phi_1 + \psi f) = A_1 \sin(\phi_1) \cos(\psi f) + A_1 \cos(\phi_1) \sin(\psi f),$$
$$\tag{9.19}$$

and, similarly for A_2

$$\Im(B(f)) = A_2 \sin(\phi_2 + \psi f) = A_2 \sin(\phi_2) \cos(\psi f) + A_2 \cos(\phi_2) \sin(\psi f).$$
$$\tag{9.20}$$

Two systems of equations may be built to find a solution for A_1 and A_2. We write this system for A_1 only

$$\sin(\psi \mathbf{f}) \quad A_1 \cos(\phi_1) = [\mathbf{Re}\,(\mathbf{B}\,(\mathbf{f}))]. \tag{9.21}$$

$$\cos(\psi \mathbf{f}) \quad A_1 \sin(\phi_1) = [\mathbf{Re}\,(\mathbf{B}\,(\mathbf{f}))]. \tag{9.22}$$

It means that the matrices (9.21) and (9.22) have a single column and each of their line corresponds to their values for the sequence of frequencies f_i defined in (9.17). These two equations are solved in the least square sense. The norm of A_1 is then easily extracted. The same procedure is applied to estimate A_2 from $\Im(B(f))$. Eventually, we extract A_0 from

$$|A_0| = \frac{|A_1| + |A_2|}{2}. \tag{9.23}$$

This is the value that is injected in (9.16) to find $\sigma^T(f_0)$.

9.5   Validation on simple targets

9.5.1   *Preliminary measurements in anechoic chamber*

In order to assess the performance of RCS measurements in RC, targets under analysis are first evaluated in a standard anechoic chamber dedicated to RCS measurements. The RCS measurements are performed using a rotationally controlled positioning mast and a measuring system. A VNA controls the transmitting and receiving signals. Figure 9.5 illustrates the test setup for these RCS measurements.

Two pairs of X-band horn antennas, one transmitting and the other receiving, are used. The two pairs allow to perform vertical and horizontal polarization measurements. Since receiving and transmitting antennas are different, but close from each other, this is a quasi-monostatic measurement. Antennas are at a 4-m distance from the rotating mast. Similarly, to the previous description of RCS measurements in RC, a first measurement of the empty room is carried out beforehand. The result of this initial measurement is substracted from the measurements with the targets. Then, a first step is to calibrate the measurement thanks to a reference target. The reference target considered in this chapter (both for anechoic chamber and RC measurements) is the above-mentioned metal plate of size 148 mm × 151 mm. It is rotated on the mast, so that the measurement system exhibits a maximum RCS. Then, a measurement over a wide frequency band (8–12 GHz) is carried out in order to be able to precisely position the target in distance on the temporal response obtained by inverse Fourier transform applied to the frequency measurement. In a second step, we perform a time-gating operation to only extract the response from the target and improve

Figure 9.5   Test setup for RCS pattern measurement in the anechoic chamber of IETR with the metallic plate of size 148 mm × 151 mm installed on the rotating mast

the signal-to-noise ratio. Finally, the result obtained is compared with the theoretical response of the plate. This theoretical maximum RCS is equal to $(4\pi S^2/c^2)f^2$ for a rectangular metal plate of area S under the physical optics approximation [1]. We then obtain the calibration coefficient that can be applied to the other targets to be measured.

Figure 9.6 provides a photograph of two different targets. The first one is a plate of size 148 mm × 151 mm and the second one a dihedral, consisting of two orthogonal square metal facets of size 100×100 mm^2. Figure 9.7 represents the measurement results for both targets. The two patterns are very different from each other. The plate exhibits strong fluctuations due to alternating out of phase and in phase contributions of currents induced on the metal plate according to its azimuthal orientation. The azimuth angle corresponds to the maximum RCS of the plate (the incident front wave being perpendicular to the plate plane). For any angle θ, the RCS of a rectangular plate of height a and width b is theoretically given by [1]

$$\sigma^T(\theta) = 4\pi \left(\frac{ab}{\lambda}\right)^2 \cos^2\theta \left[\frac{\sin(2\pi b/\lambda \sin\theta)}{2\pi b/\lambda \sin\theta}\right]^2. \tag{9.24}$$

Therefore, the plate should exhibit nulls of RCS for any n integer such as $\theta = \arcsin(n\lambda/(2b))$. At 10 GHz, according to the size of this plate the two first theoretical nulls appear at $\theta = \pm 5.7°$ and $\theta = \pm 11.5°$. Instead of nulls, the response of the target in the anechoic chamber exhibits local minimum at the same angles. This is due to the limited distance $R = 4$ m of measurement which introduces phase errors. Such limitation appears also in RC measurement as shown in Section 9.5.3 for the same reason. The pattern associated with the dihedral exhibits much smoother fluctuations due to the association of the two perpendicular plates forming the dihedral. One may also notice that both resulting RCS patterns are not strictly symmetric. This is mainly due to the imprecise placement of the targets on the mast and inaccurate alignment

Figure 9.6   Targets for RCS measurements. On the left-hand side, a dihedral with metal facets of size 100 mm × 100 mm. On the right-hand side a metallic plate of size 148 mm × 151 mm

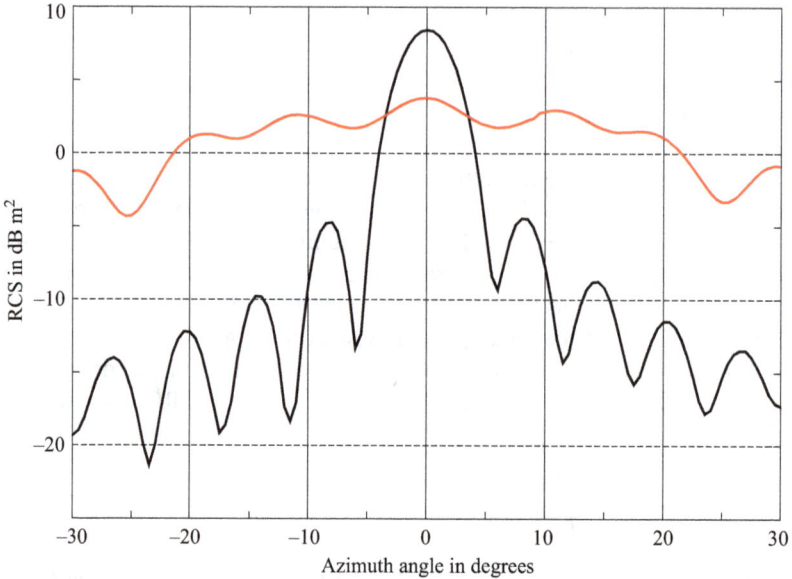

Figure 9.7   *RCS pattern measurement in the anechoic chamber of IETR of a metal plate of size 148 mm × 151 mm (black curve) and of a dihedral (red curve)*

with regard to antennas. Positioning was performed manually. These measurements were performed as basis for comparison with RC measurements discussed next.

9.5.2   Calibration in RC

Calibration in RC is performed in a similar way as in anechoic chamber, using the theoretical maximum RCS of the metal plate of size 148 mm × 151 mm as a reference. According to (9.17) or (9.18), the oscillation of the real part or imaginary part of $S^T(f) - S(f)$ oscillates with a frequency periodicity of $\delta f = c/2R$. In Section 9.4, R is defined as the distance travelled by the wave in free-space between the target and the measuring antenna. However, in practice, this periodicity is slightly different due to the transformation from radiated to guided waves (and vice versa) in horn antenna. The total propagation time until the reference plane of the VNA must account for this guided wave travelling time. The sinusoidal fit performed from measurements implies an equivalent overall distance R_{eq}, slightly longer than R. Equation (9.15) may be therefore modified such as

$$
\begin{aligned}
S^T(f_0) - S(f_0) = {}& (1 - |S_{FS}(f_0)|^2)(H^T(f_0) - H(f_0))\eta_{ant} \\
& + \sqrt{\sigma^T(f_0)} \frac{G_{ant}(f_0)\lambda_0}{(4\pi)^{3/2}R^2} (1 - |S_{EL}(f_0)|^2) \\
& \times \exp\frac{-j2\pi f_0 2R_{eq}}{c} \exp(j\phi_0).
\end{aligned}
\tag{9.25}
$$

In this last expression, $R_{eq} = R + \delta R$. This equivalent distance R_{eq} is approximated with some accuracy from the sinusoidal fit. This is not the case for R since the phase centre of the horn antenna remains unknown or only roughly determined. A first adjustment for δR was performed through analysis of the pulse response of the horn antenna at different distances from the wall of the RC. It was estimated to be 0.29 m. Therefore, a first estimation of RCS is performed, through (9.25) and (9.16) with $R = R_{eq} - 0.29$ m. At 10 GHz, the applied correction factor is about 1 dB with regard to the raw estimation of RCS.

9.5.3   Measurement of the metal plate

Once the calibration factor is applied at the position of maximum RCS, the mast is rotated over the $-30°$ to $30°$ azimuthal range, $\theta = 0$ corresponds to the maximum RCS value, the wavefront being orthogonal to the plate plane. A limited excursion of the azimuth angle was selected, the RCS of the plate being very low out of this range. The RCS pattern in the RC is then compared to that obtained in anechoic chamber. Both patterns are reported in Figure 9.8. Patterns are similar to each other. Local maximum of the RCS pattern of the first and second side lobes are well retrieved both in angular position and magnitude. According to (9.24), local minimum RCS magnitudes are nulls of RCS for an infinitely thin plate, measured at an infinite

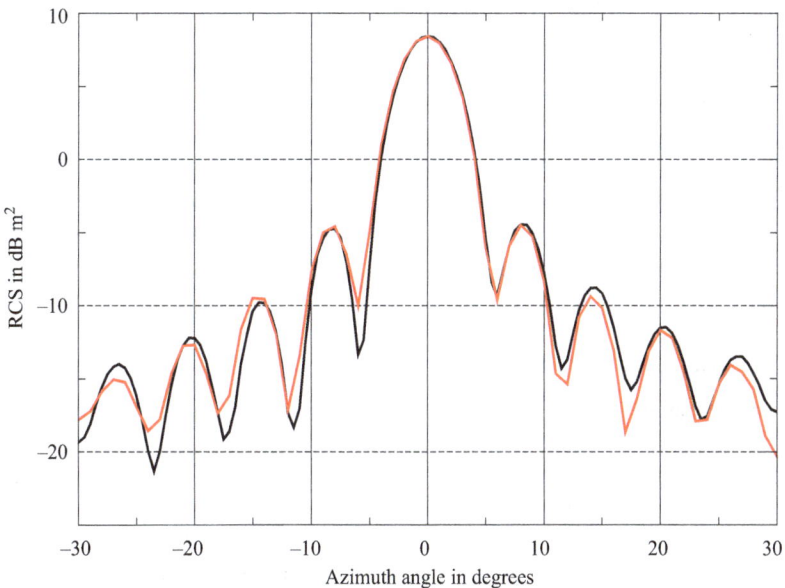

Figure 9.8   *Comparison of RCS pattern of a metal plate of size 148 mm × 151 mm in azimuth plane measured in anechoic chamber (black curve) and in reverberation chamber (red curve)*

distance. Experimental results in both test means exhibit different minimum values at levels well above noise. This is mainly explained by the distance of measurement which is about 4 m in anechoic chamber and 3 m in RC. Both distances are of the order of magnitude of far-field conditions $4D^2/\lambda$ that is about 3 m at 10 GHz. In other words, the phase relative error may reach $\pi/8$ in RC. Nulls of the RCS pattern are associated with a perfect phase compensation of the sum of all current sources on the plate and can only be resolved at much larger measurement distances that are incompatible with these test setups according to this target size and that frequency range of operation. We notice that local minimum may have (for negative angles of azimuth) lower RCS magnitudes in anechoic chamber results, since the measurement distance is greater than in RC. Both diagrams do not appear to be perfectly symmetric according to the azimuth angle, though the target is perfectly symmetric with reference to its rotation axis. This is mainly due to a lack of precision in positioning for both test setups. This implies the target position over the top of the masts and the rotation of the masts themselves.

9.5.4   Measurement of a dihedral target

The dihedral target is placed on the rotating mast in the RC instead of the rectangular metal plate. No other calibration procedure is followed. Therefore, the correction coefficient estimated from measurements with the plate still applies for this new target measurement. The mast is rotated over the $-30°$ to $30°$ azimuthal range, $\theta = 0$ corresponds to the maximum RCS value, the wavefront impinging each square facet with a $45°$ angle. Due to the shape of the target, the range of RCS magnitude is limited to no more than 9 dB. The maximum magnitude of RCS patterns estimated from anechoic chamber and RC measurements are 4.2 and 3.8 dBm^2, respectively. The overall oscillating shapes of the RCS patterns are similar to one another. Some deviations, including imperfect symmetry, are observed and may be partly attributed to the positioning method already mentioned. Nevertheless, this result demonstrates the proof of concept of realization of RCS measurements in RC. Different paths may be followed to improve the quality of measurements and this is currently under investigation in several research teams. In particular, the role of the stirrer represents an interesting topic of investigation.

9.5.5   The role of the stirrer

The stirrer may play a role with respect to the accuracy of the measurements. So far, this role has not been much discussed with regard to the theory of RCS measurements in RC. It was only taken into account through the presence of a term proportional to $h_s(f_0, p_i)$ in expressions of the scattering parameter in the empty RC (Equation (9.9)) and the target-loaded RC (Equation (9.12)). Once the difference of these two scattering parameters is performed, this term was supposed to cancel out.

Let recall that $h_s(f_0, p_i)$ stands for the line-of-sight or specular reflection from the stirrer at a given position p_i. A sufficient condition for this term to cancel out is to perform the measurement of the empty RC and of the target-loaded RC at the same stirrer position. Measurements presented in Figures 9.8 and 9.9 were performed this

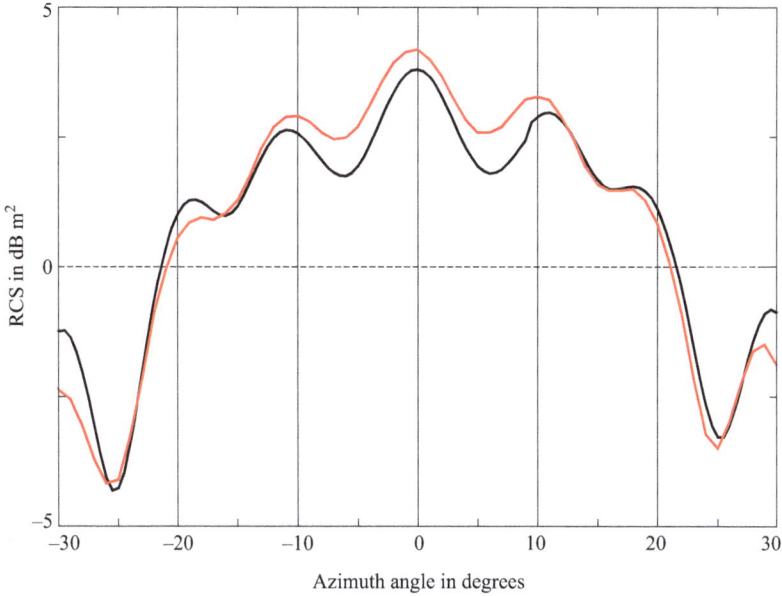

Figure 9.9 *Comparison of RCS patterns of a dihedral in azimuth plane measured in anechoic chamber (black curve) and in reverberation chamber (red curve)*

way at an arbitrary position of the stirrer. However, if the line-of-sight or specular reflection from the stirrer is weak enough with respect to the contribution of the target at its specific position, this is no longer required. We illustrate this fact looking at a movement of the stirrer in between both measurements.

9.5.5.1 Changing the stirrer position after empty RC measurement

As an example, a RCS pattern estimation is performed for the metal plate of size 148 mm × 151 mm once the stirrer is rotated by a few degrees with respect to the initial position where an empty measurement was recorded. The result of Figure 9.10 confirms the role of $h_s(f_0, p_i)$ for small values of RCS. In that figure, the plots of the result obtained from identical and different stirrer positions are plotted together. In the latter case, the RCS estimation is performed for two rotations of the stirrer of 3.6° and 7.2° from the initial position corresponding to the empty RC measurement. Deviation from the original results are very limited for the 3.6° stirrer rotation and are more apparent for side lobes beyond $\theta = \pm 20°$. It is also clear that results for the 7.2° stirrer rotation are much more perturbed with a clear impact on the shape of the side lobes. However, the main scattering lobe of the RCS pattern remains nearly identical to the initial result. Magnitudes of RCS in that direction are probably high enough to remain unperturbed by line-of-sight or specular reflections from the stirrer.

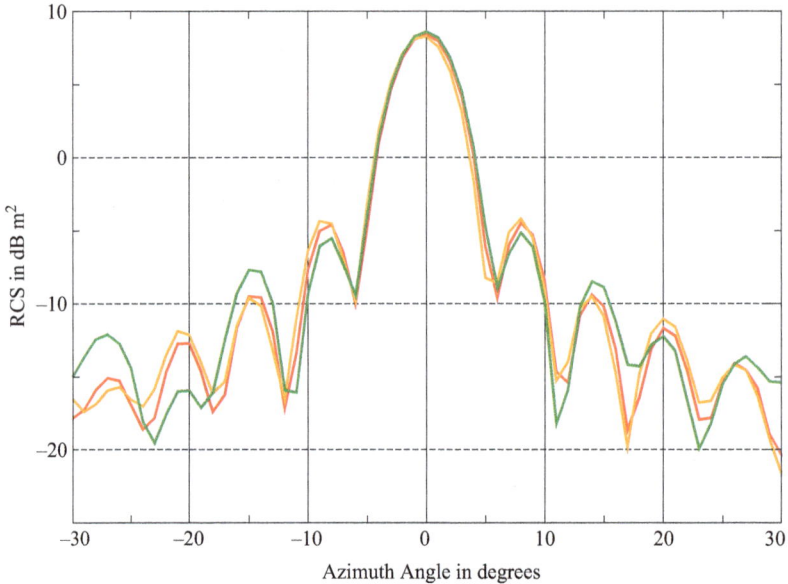

*Figure 9.10   Estimation of RCS pattern in RC of the metal plate of size
148 mm × 151 mm: (i) without changing stirrer position for empty RC
and target measurement (red curve as of Figure 9.8), (ii) with a 3.6°
rotation of stirrer position between both measurements (orange
curve), (iii) with a 7.2° rotation of stirrer position between both
measurements (green curve)*

9.5.5.2   Multiple RCS measurements at several stirrer positions

Results of Figure 9.10 indicate that it is highly recommended to keep an identical stirrer position between both phases of the RCS measurement in RC. We may therefore choose not to use the stirrer at all and opt for other alternative solutions such as time gating that was investigated in [12]. On the other hand, it is also relevant to have a look at what the stirrer could bring for the benefit of the measurement quality.

In a first phase, we have performed several RCS pattern estimations of the same metal plate at different and arbitrary positions of the stirrer. We selected five consecutive step positions of 3.6°. For each of these five stirrer positions, an empty chamber and a loaded chamber measurements are performed. Therefore, each couple of measurements yields to a RCS estimation attached to a particular stirrer position. It appears clearly from Figure 9.11 that reproducibility of the results is not fully guaranteed over the complete range of azimuth angles. More specifically, fluctuations among all results become significant for RCS values below $\approx -10\ dBm^2$. According to the theory presented in Section 9.4, extraction of RCS is made possible thanks to the distinction of slowly and rapidly oscillating signals in the measured frequency span. The rapidly oscillating signals are moreover Gaussian distributed. The slow

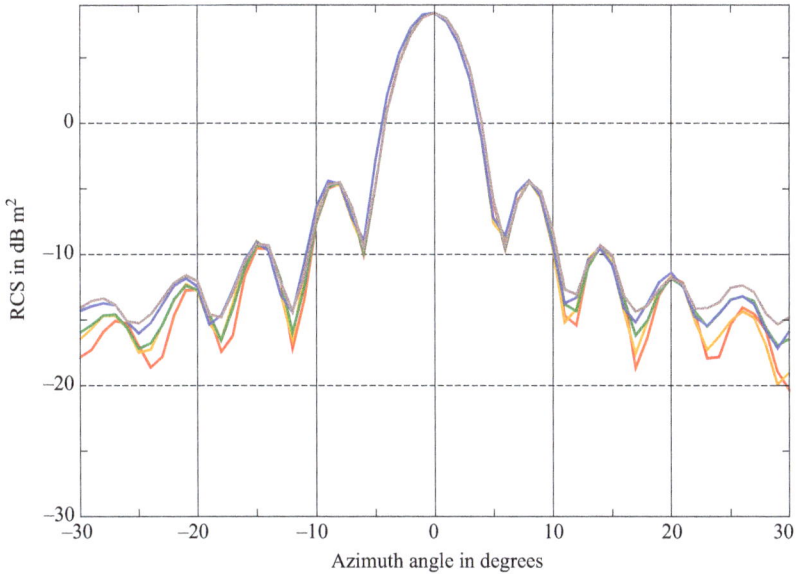

Figure 9.11   Comparison of RCS patterns in the RC of IETR for a metal plate of size 148 mm × 151 mm at five different stirrer positions. At each of these positions, a RCS pattern is estimated from the measurement of the empty chamber and the measurement of the target-loaded chamber. Each coloured curve corresponds to one of these positions

signal extraction is performed with a sinusoidal regression technique that is sensitive to high level of signal-to-noise interference, where interference stands for the power of the diffuse field in RC. A more detailed, still preliminary, analysis was provided in [11].

The same analysis may be performed with the above-mentioned dihedral target. Another set of five RCS patterns is obtained in the same conditions as the plate. However, Figure 9.12 shows that RCS patterns are quasi-identical from one measurement to another. Fluctuations may be noticeable but reach no more than a fraction of a dBm^2. It is the order of magnitude of RCS values that explains this result, typically for this measurement setup, well above -10 dBm^2. Note that this approximative threshold is not at all a universal feature but depends on the measurement distance between the radar antenna and the target as well as on the quality factor of the specific RC used for these measurements. Adding a limited amount of absorbing materials in the RC may enhance the sensitivity of the measurements.

Eventually, results of Figure 9.11 indicate that averaging over several stirrer positions (or several chamber states) seems to be a promising technique to improve the sensitivity of RCS measurements in RC. This question is currently investigated by the authors of [11] in collaboration with the author of this chapter.

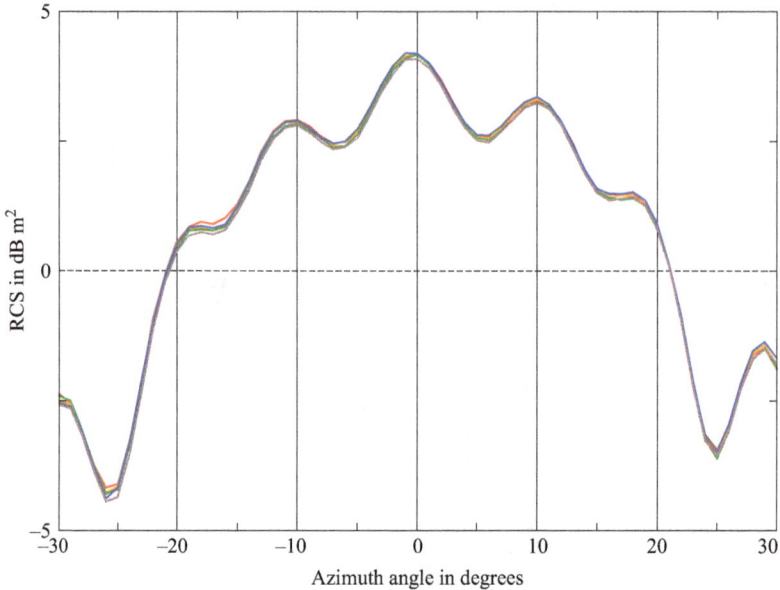

Figure 9.12   Comparison of RCS pattern measurements in RC for a dihedral target at five different stirrer positions. Each coloured curve corresponds to one of these positions

9.6   Discussion

This chapter was dedicated to advanced RCS measurements in RC. More specifically, we focused on a technique that provides the RCS pattern of a target installed on a rotating mast. Any direct echo from the target in the direction of the illuminating radar antenna is analysed through the observation of the complex scattering parameter over a pre-determined frequency span. This scattering parameter is composed of two major contributions: *(i)* the real and imaginary parts of the complex transfer function of the considered ideal RC with random fluctuations described by centred Gaussian distributions; *(ii)* the direct echo from the target for which the complex *S*-parameters oscillates according to the antenna–target distance. According to the setup configuration, these two contributions may have very different signatures according to their rate of fluctuations with frequency. In other words, they have different spectra in the dual space of the frequency variable. This provides possible strategies to retrieve the direct echo alone (time gating, Fourier transform, regression, etc.). Among these, we selected a simple sinusoidal regression and showed how it allows to retrieve some target responses. This post-processing technique is sufficient for targets behaving as single-point scatterers. Moreover, it takes advantage of the centred Gaussian interference provided by the diffuse field in the chamber.

Finally, the role of the stirrer within the RC was also discussed. At a first glance, the measurement technique itself does not take the stirrer into account, since it is based on a preliminary estimation of the empty chamber. In these conditions, the stirrer may be considered as part of the background as any electromagnetic scatterer in the RC such as walls. Modifying the stirrer position between both measurements may have the same detrimental impact as moving a scatterer in a conventional anechoic chamber. This impacts mainly the estimation of low levels of target reflectivity. Looking more carefully at the measurement technique, the stirrer may play a far more consistent role to improve the sensitivity of measurements through averaging the level of interference. Research works are currently in progress with regard to this aspect.

There are many more aspects to deal with and there is a large avenue for further researches in that recent subject. Detection of more complex or multiple targets, phase resolution are among important aspects to be treated in the future, for which more intense signal processing should come into play.

References

[1] Ruck GT, Barrick DE, Stuar WD, *et al.* Radar cross-section handbook. Plenum Press, New York; 1970.

[2] Friis HT. A note on a simple transmission formula. In: Proceedings of the I.R.E and Waves and Electrons; May 1946;34(5):254–256.

[3] Loredo S, Pino MR, Las-Heras F, *et al.* Echo identification and cancellation techniques for antenna measurement in non-anechoic test sites. IEEE Trans Antennas Propag. 2004;46(Pt 1):100–107.

[4] Hill D. Boundary fields in reverberation chambers. IEEE Trans Electromagn Compat. 2005;47(Pt 2):281–290.

[5] Lerosey G and de Rosny J. Scattering cross-section measurement in reverberation chamber. IEEE Trans Electromagn Compat. 2007;49(Pt 2):280–284.

[6] Cozza A. Emulating an anechoic environment in a wave-diffusive medium through an extended time-reversal approach. IEEE Trans Antennas Propag. 2012;54(Pt 8):3838–3852.

[7] Cozza A and Monsef F. Steering focusing waves in a reverberation chamber with generalized time reversal. IEEE Trans Antennas Propag. 2017;65(Pt 3):1349–1356.

[8] Lemoine C, Amador E, Besnier P, *et al.* Antenna directivity measurement in reverberation chamber from Rician K-factor estimation. IEEE Trans Antennas Propag. 2013;61(Pt 10):5307–5310.

[9] Garcia-Fernandez M, Carsenat D, and Decroze C. Antenna gain and radiation pattern measurements in reverberation chamber using Doppler Effect. IEEE Trans Antennas Propag. 2014;62(Pt 10):5389–5394.

[10] Besnier P, Sol J, and Méric S. Estimating radar cross-section of canonical targets in reverberation chamber. In: International Symposium on Electromagnetic Compatibility, EMC EUROPE; 2017 Sep; Angers, France. IEEE; 2017. p. 1–5.

[11] Reis A, Sarrazin F, Richalot E, *et al.* Mode-stirring impact in radar cross section evaluation in reverberation chamber. In: International Symposium on Electromagnetic Compatibility, EMC EUROPE; 2018 Sep; Amsterdam, The Netherlands. IEEE; 2018. p. 875–878.

[12] Soltane A, Andrieu G, and Reineix A. Monostatic radar cross-section estimation of canonical targets in reverberating room using time-gating technique. In: International Symposium on Electromagnetic Compatibility, EMC EUROPE; 2018 Sep; Amsterdam, The Netherlands. IEEE; 2018. p. 355–359.

[13] Besnier P and Demoulin B. Electromagnetic reverberation chambers. ISTE, UK and Wiley & Sons, Hoboken, NJ, USA; 2011.

[14] Hill D. Electromagnetic fields in cavities: deterministic and statistical theories. IEEE Press Wiley, Hoboken, NJ, USA; 2009.

[15] Lemoine C, Besnier P, and Drissi M. Investigation of reverberation chamber through high-power goodness-of-fit tests. IEEE Trans Electromagn Compat. 2007;49(Pt 4):745–755.

[16] Cozza A. The role of losses in the definition of the overmoded condition for reverberation chambers and their statistics. IEEE Trans Electromagn Compat. 2011;53(Pt 1):296–307.

[17] Adardour A, Andrieu G, and Reineix A. On the low-frequency optimization of reverberation chambers. IEEE Trans Electromagn Compat. 2014;56(Pt 2): 266–276.

[18] Born P and Wolf E. Principles of optics: electromagnetic theory of propagation, interference and diffraction of light. Elsevier, Amsterdam, The Netherlands; 2013.

Conclusion

Guillaume Andrieu[1]

As a conclusion for this book, we have tried to identify some important open issues the community of RC users will have to face in the future. This list is not exhaustive, we are sure that original applications not mentioned here will appear soon. However, the less we can say is that all the challenges mentioned below are quite exciting!

C.1 OTA test for new protocols

RCs are right now extensively used for over-the-air (OTA) testing of wireless devices as discussed in Chapter 5. However, this domain is subject to a rapid evolution for instance due to the continuous increase in the number of connected devices, related in particular to the exponential development of Internet-of-Things (IoT) devices. This is compounded by the appearance of new protocols and frequency bands requiring users to update continuously the OTA testing techniques (including experimental setups and post-processing programs) in order to match the characteristics of each protocol. As an example, in the case of recent narrow-band protocols (such as NB-IoT for instance) when the bandwidth of the modulated signal may be smaller than the RC coherence bandwidth even when the RC is unloaded, it will probably be unnecessary to load the chamber. In this framework, it will be interesting to investigate the trade-offs on uncertainty of using a high Q-factor RC, finite (nonzero) coherence bandwidth and number of samples on the realization of such OTA tests.

C.2 "Anechoic" measurements

We are referring here to measurements (for instance antenna radiation pattern or radar cross-section measurements as discussed, respectively, in Chapters 8 and 9) traditionally performed in an anechoic chamber but performed in a Faraday cage equipped of a mode stirrer (or not). As discussed in this book, these measurements requiring to discard the influence of the multipath environment are limited, to date, to the case of non resonant objects, the sensitivity of the obtained results being also lower than the measurements performed in an anechoic chamber. Future progress on this topic will for sure concern the improvement of the measurement sensitivity by

[1]XLIM Laboratory, UMR7252, University of Limoges, Limoges, France

optimizing the experimental setups and the post-processing techniques. This could help to characterize with a reasonable measurement time ultrawide-band antennas or resonating targets in order to identify their bright spots.

C.3 Chaotic cavities

A controversial topic in the field of RCs concerns the positive effect (if any) of inserting diffracting curved objects in an RC, the stirring process being still ensured by a mode stirrer. Indeed, this so-called "chaotic" RC is assumed to obtain better performance from the point of view of the quality of the electromagnetic field generated inside.

There is in our opinion a basic (but still open) question to solve first before drawing definitive conclusions which is "what is the real definition of a chaotic chamber?" Figure C.1 presents four kinds of RCs that are candidates to be considered as a chaotic RC (at different degrees). A common idea considers that the multipaths

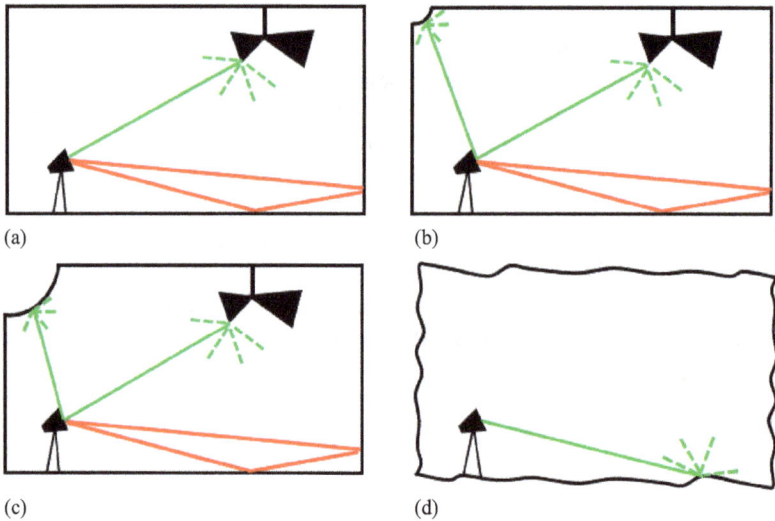

Figure C.1 *Four different types of "chaotic" RCs: (a) the traditional RC: a parallelepipedic Faraday cage with a rotating mode stirrer, (b) the same with a small curved diffractor inserted in, (c) the same with a larger curved diffractor and (d) a VIRC where the entire geometry of the tent is potentially in movement (i.e., the VIRC is not lying on the ground). An example of unstirred path is represented in red in the first three. Some examples of diffracted paths are represented in green.*

are more randomly spatially distributed in a chaotic cavity (as in a Sinai billiard) than in a parallelepipedic cavity, this having the effect to prevent the appearance of "regular" modes and potentially to improve the isotropy and the homogeneity of the field.

However, a parallelepipedic chamber equipped with a rotating mode stirrer (cavity "a") can be assumed to be a chaotic cavity in the sense that the emitted paths are diffracted in all the directions when encountering the mode stirrer (of random geometry). It is therefore impossible to predict analytically the routing of all the different paths after they hit the mode stirrer, this being theoretically possible in a parallelepipedic cavity without any mode stirrer. It is clear that cavities "b" and "c," modified by the insertion of curved objects expected to increase the degree of chaoticity of the RC, do not prevent the appearance of unstirred paths. However, we can notice that the larger size of the curved object in the cavity "c" increases the probability of these paths to encounter the diffracting object. Finally, only the last cavity "d," which refers to a VIRC potentially entirely movable (i.e., not lying on a ground), is the only one able to (theoretically) discard the appearance of any unstirred path.

In most of the studies available in the literature on this important topic, a potential bias can occur when the performance of an RC with and without the presence of curved diffractors is compared. Indeed, and whatever the metrics used, these two conditions are not directly comparable in reason of the (slight) decrease of the cavity Q-factor due to the additional metallic losses related to the diffractors (leading also to a reduction of the RC total volume). An unbiased study should compare the RC performance (for instance using the "well-stirred" condition method) with and without the diffractors at the same Q-factor. It means that a small piece of absorber has to be inserted without the presence of the diffractors in order to compensate their inherent losses when they are inserted in the chamber.

C.4 VIRCs

VIRCs have made their appearance at the beginning of this century following the proposal of M. Frank Leferink. Despite several nonnegligible advantages discussed in Chapter 4 (low-cost, movable, mountable/dismountable among others), the scientific literature about VIRCs is, to date, not particularly abundant especially in scientific journals. This is why we encourage here the community of RC users to investigate the influence of some specific characteristics of VIRCs which could have interesting properties, in particular with respect to "classical" RCs. In addition to the possibility mentioned previously to discard the influence of any unstirred path (and therefore reduce the RC K-factor) and then to obtain a perfectly (or fully) chaotic RC, a VIRC could also be helpful in order to reduce the uncertainty budget of RC measurements (antenna efficiency, OTA tests for instance). This would require to be able to determine the number of independent samples obtained in a VIRC over a large observation time. Finally, the numerical simulation of a VIRC with a (commercial or homemade) electromagnetic software is also quite challenging.

C.5   Future EMC strategy in the transportation industries

Nowadays, transportation industries, highly concerned by EMC validations, face major changes. For instance, the appearance of electric, autonomous, connected or lightened vehicles or aircrafts raises fundamental challenges on how to ensure a satisfying EMC certification process in the future. In this framework, current EMC certification processes may become unreasonably time-consuming, irrelevant and obsolete in the near future. Therefore, innovative certification methods dealing with these new constraints but involving reasonable testing times will have to be proposed. For sure, RCs will have an important role to play in the highest frequency range.

Index

absorbers 2–3, 7, 20–1, 25, 27, 30–4, 36–7, 41, 46, 48, 52–3, 60, 65–7, 73, 83, 85–6, 88–90, 104, 108, 113, 115–16, 124, 132, 134, 137, 144, 176, 187–91, 201, 205, 222, 239, 247–8, 250, 253, 261, 287

absorbing area cross section (AACS) 3, 175–6, 186–90, 193, 201

aeronautics 75

agar phantom 193–5

Anderson-Darling (AD) 13–15, 88, 147

anechoic chamber 3, 10, 125, 151–4, 156, 205, 239, 242, 245, 247–50, 252–3, 258, 264, 274–9, 283, 285

anechoic measurements 285–6

animal exposure 3, 176, 191–3, 201

anisotropy 7, 146–8

antenna 8–10, 12, 21–2, 24, 26–8, 30–1, 33–4, 39–41, 45, 48, 50, 55, 63–5, 67–8, 76, 83, 85, 90, 92–4, 106, 112–15, 126, 142–3, 146, 151–3, 165, 178–80, 184, 205, 207–10, 214–15, 217, 219–20, 222, 224, 245, 266–70, 282

  efficiency 3, 85, 97, 99, 127, 151–2, 167, 179, 205–35, 287

  network 67–8

  radiation efficiency 205–24, 235

  radiation pattern 3, 166, 239–58, 266, 285

  total efficiency 224

  ultra wide band antenna 205, 216, 220, 223, 225, 235

  under test 206, 239, 266

arbitrary waveform generator (AWG) 216–17, 222, 225

autocorrelation 18, 82, 92, 94–5, 137–8, 242

autocorrelation coefficient 16–18, 71–3, 75

  first-order autocorrelation coefficient 16–17, 57–8, 91

autocovariance 139, 143

automotive 192

average mode bandwidth 50–1, 57, 62, 85, 113, 136, 248

backscattered coefficient 269–70

  enhanced 183, 186, 200

backscattered field 268–9

bandwidth 21, 25, 28, 41, 50–1, 55, 61–2, 68, 74–6, 126, 135, 137, 152, 156, 176, 179, 220, 245–6, 248, 254, 258, 261, 264, 285

  coherence bandwidth (CBW) 7, 92, 126, 128, 135–9, 175, 187, 265, 271, 285

  resolution bandwidth 21, 87

base-station emulator (BSE) 127–8, 151–4

bioelectromagnetics 3, 176

bit-error-rate (BER) 153

calibration 2, 74, 76, 104, 106, 192–3, 201, 210, 222, 265, 267, 275–8

  alternative calibration procedure 36–40

  procedure 8, 74, 176, 192, 201, 278

cavity modes 287

certification testing 135

channel 124–6, 129, 131, 135–6, 151–2, 157–8, 208

  emulation 125, 157–9

  isotropic channel 125, 158

  multipath channel 125, 131, 149, 156–9, 168

chaotic cavity 287

chaotic chamber 286
chaotic properties 177
correlation 16, 59, 64, 67, 71–2, 88–9, 93, 95, 124, 127, 132, 135, 139–40, 142–3, 162–3, 228, 231
  cross correlation 139, 143–6
  frequency correlation 135, 138, 150
  sample correlation 7, 9, 15–20, 25, 36, 57, 60, 65, 68, 71, 88–9, 92, 107
  spatial correlation 139–40, 143–4, 150, 160, 162

decorrelation time 82, 92–4, 96, 98
degrees-of-freedom (DoF) 9, 103, 107–9, 145, 147, 161–2, 167–8, 229–30, 232
delay spread 157
  RMS delay spread 21, 131, 138–9, 158
density function
  cumulative density function (CDF) 13–15, 18, 148
  probability density function (PDF) 18, 103, 105–6, 224, 226, 228–9, 266, 271
device under test (DUT) 94, 96, 123, 126–30, 134, 136, 143–4, 149, 151–4, 156, 162, 164–7, 178
directivity 2, 101, 109, 205, 241, 266–7
distribution 13, 15, 71, 108, 127, 131, 147–8, 187, 226
  electromagnetic field distributions 269
  exponential distribution 114, 226
  Gaussian distribution 9, 105, 131–2, 146, 181, 261, 269, 282
  Rayleigh distribution 12–15, 71, 94–6, 103–4
  Weibull distribution 15
Doppler effect 242
Doppler frequency 242–4
Doppler spectrum 242–5, 258
dosimetry 3, 175–201

efficiency 2–3, 7, 39, 46, 48, 51, 76, 88, 126–7, 129, 151–2, 154, 160, 175, 190–1, 205–6, 208, 235, 239
electric field 9, 12, 31, 37–9, 56, 103–4, 106, 111, 131, 146, 181, 228, 262–3, 265, 268

triaxial electric field probe 37, 40, 56, 102–4
electromagnetic compatibility (EMC) 1–2, 8, 37, 62, 101–2, 109–11, 116, 119, 123–6, 168, 177, 288
electromagnetic distributions 7–9, 13–15, 25, 34, 36, 39, 56, 60, 65, 103, 269
electromagnetic field 3, 7–8, 62, 64, 67, 82, 102, 176, 188, 192, 196, 269, 286
electromagnetic field strength 26, 39, 75, 101, 103–4, 111
equipment under test (EUT) 9, 12, 30–2, 39, 41, 51, 62, 101–4, 106, 108–21
estimator 209, 227, 230, 233, 235
  biased estimator 227, 229, 234
  unbiased estimator 227–9, 233, 235
extreme values 15, 93, 106

failure 102
  probability of 102, 104, 111
far-field 246, 262–3, 267, 269, 278
flexible tent 81
Fourier 138, 219–20
Fourier transform (FT) 198, 242, 272, 282
  fast Fourier transform (FFT) 220
  inverse Fourier transform (IFT) 51, 130, 179, 198, 210, 218, 245, 247, 249, 252, 264, 274
free space 3, 10, 26, 54–5, 68–9, 74, 87–8, 129, 151–2, 178–82, 189, 207–8, 250, 267–9, 276
frequency-domain 82, 86–93, 95, 98, 208–9, 216–17, 220, 224, 245–7, 252, 256
fundamental mode 47
fundamental resonance 20, 75, 83, 103

Gaussian noise 61
generic equipment 112–14
goodness-of-fit (GOF) test 13–14, 95, 123, 147

heat equation 195–6, 198, 201
heat transfer 193–9
homogeneity 287

impedance mismatch 21, 26, 28, 55, 151, 153–4, 209

inhomogeneity 7
Internet-of-Things (IoT) 285
isotropy 8, 30, 37, 123, 127, 146–8, 168, 287

K-factor 7, 88, 127, 130, 134, 149–50, 165–7, 240–1, 266
  method 240–2, 258
  Rician K-factor 134, 149–50
Kolmogorov–Smirnov (KS) 15

level of confidence 19, 22
level of significance 13–15, 22
line-of-sight (LOS), 30–1, 149, 178, 239, 241, 247–51, 253, 258, 261, 266, 268–70, 278–9
line-of-sight path 239, 241, 243–6, 253
loaded chamber 140, 142, 162, 195, 280
  heavily loaded chamber 128, 131–2, 134, 142, 144, 160
  unloaded chamber 124, 132, 140, 142, 158, 162–3, 165
loading 15, 21–5, 34–6, 53, 60, 65, 74–5, 93, 107, 124, 126, 132, 134–5, 139, 152, 157, 232–5, 251, 253, 255
loss 39, 82, 127, 163, 184, 229, 265
  antenna loss 209, 222
  metallic loss 178, 287
lossless 50, 176, 208
lowest usable frequency (LUF) 36, 39

maximum likelihood method 13
metallized tent 2, 82–3, 85–6, 92, 98
minimum susceptibility level 119–21
modal density 43, 50, 89, 99, 103, 175, 177, 261
modal overlap 25, 175, 177, 261
modal overlapping 25, 39, 60, 88
mode stirrer 2, 8–12, 16–17, 20–1, 33–4, 36, 45–8, 52–3, 56, 86, 96, 112, 114, 241, 287
  dual-mode stirrer 47–8
  mode stirrer positions 10, 12, 16, 52, 56, 101, 240–1
  rotating mode stirrer 2, 19–20, 36–8, 46, 48–9, 60, 63–6, 75, 92, 96–7, 112, 242, 247, 286–7

mode-stirring sample 126, 128, 130–2, 134–5, 137, 139–40, 142, 145–7, 149–51, 154, 160–4
mode-stirring sequence 123, 125–6, 129–32, 136, 139–43, 145–6, 149–51, 154–5, 160, 162–3, 167
Monte Carlo 19, 107, 109, 116–17, 120, 148, 228

near-field 269
non line-of-sight path 240, 245–6, 248–51, 258

overmoded condition 8
overtesting 101, 116–19, 121
overtesting coefficient 116–20
over-the-air (OTA) 97, 99, 125, 127, 150, 152, 160–8, 285
over-the-air test 97, 99, 123–68

periodicity 92–4, 271, 276
plane wave 8, 177, 181–2, 187–8, 262–3, 265–6, 270
polynomial curve-fitting 15, 28–30, 33, 41, 57, 71, 88, 90
power-delay profile (PDP) 7, 52, 125, 130–1, 158, 168, 248
probabilistic model 2, 101–21

quality factor 2, 136, 138, 175, 177–9, 183, 185, 187, 190, 201, 206, 208–9, 241, 281
  average composite quality factor 7, 12, 50, 75, 106, 113, 248

radar cross-section 3, 239, 261–83, 285
radar equation 262–4, 270
radiated susceptibility test 1–2, 12, 51, 62, 94, 101–21

sample 7, 9, 13, 15–16, 60, 82, 105, 114, 139, 142, 150, 152, 154, 181
  correlated samples 9, 126
  independent samples 17, 48, 61, 63–4, 76, 82, 92, 94, 96–8, 112, 124, 139, 146, 212, 214, 221, 228, 230–3, 235, 242, 287
  uncorrelated samples 3, 19, 92, 132, 139–40, 142, 160

scattered cross-section 264–6
scattering measurements 51, 240, 242
scattering parameters (S-parameters) 7,
    51, 75, 112, 127, 129, 131, 140, 143,
    162, 176, 181, 200, 206, 216, 221–2,
    229, 240, 245, 266, 278, 282
signal-to-noise ratio (SNR) 217, 222,
    247, 264, 275
skin depth 187, 189–90, 193–4, 197, 201
slope 51, 53
space industry 4
spatial resolution 246
specific absorption rate (SAR) 192, 200
standard deviation 92, 106–7, 110, 130,
    134, 145, 164, 166, 200, 224, 268,
    271
standards 7, 82, 92, 98, 103, 106–7, 110,
    130, 134, 145, 164–5, 188, 200, 206,
    208, 224, 227, 235, 271, 274
stirred contribution 56
stirred paths 10, 88, 239
stirred power 241
stirring 2, 7, 45–77, 81–3, 85, 90, 92–4,
    96, 98, 124–30, 132, 135, 139–40,
    144–7, 150, 152, 154, 157, 160–1,
    163–5, 167, 177, 180, 207, 241, 286
stirring conditions 8–9, 101–21, 241
stirring strategy 83–5
stirring technique 2, 46, 47, 49, 57, 68,
    74–6, 232
    antenna moving stirring technique 67
    antenna network stirring technique
        67–8
    frequency stepping stirring technique
        50–61, 76
    hybrid stirring technique 68–75
    mechanical stirring technique 45–50,
        60, 63–4
    network antenna stirring technique 67
    platform stirring technique 63–4, 126,
        129, 139, 149, 163, 165
susceptibility level 101, 108–9, 111, 115,
    117, 121

target 3, 261–7, 269–71, 274–6, 278–9,
    281–3

time-domain 2, 7, 51–3, 81–99, 131, 208,
    216, 219–20, 224–5, 235, 245, 255,
    264
time-domain resolution 246, 255
time-gating 3, 240, 245–58, 264, 274
time-gating window 246–7, 249, 252–3,
    255, 257
time-reversal 3
total isotropic sensitivity (TIS) 123–5,
    127, 135, 145, 151–6, 160, 162,
    167–8
total radiated power (TRP) 1–2, 125, 127,
    135, 145, 151–2, 154, 160, 168, 265

uncertainty 3, 28, 63, 76, 96, 98, 107,
    110–11, 118–19, 121, 124, 126, 130,
    140, 142, 145, 154–5, 160–4, 167,
    214, 216, 221–2, 241
    combined uncertainty 146, 162
    expanded uncertainty 168
    level of uncertainty 45, 53, 107, 111,
        139
    measurement uncertainty 82, 85,
        125–6, 151, 166, 212, 214, 216,
        220–35
uncertainty budget 3, 97–8, 287
uniformity 7–8, 37
    spatial uniformity 123–8, 130, 132,
        134, 139–40, 142, 144–5, 150,
        160–4, 167–8, 213–14
unstirred contribution 8, 26, 87
unstirred paths 10, 241, 245, 258, 286–7
unstirred power 241

vibrating intrinsic reverberation chambers
    (VIRC) 2, 46, 62, 81–99, 286–7

well-stirred condition 8–9, 24, 36, 41, 57,
    73, 106–7, 121, 123–4, 126, 128,
    131
well-stirred condition method 7–41,
    45–77, 82, 86, 88, 90, 98, 287
windowing function 245, 247, 254
wireless 123, 125–6, 129, 143, 149, 152,
    167
wireless devices 2, 97, 99, 123–68, 285

www.ingramcontent.com/pod-product-compliance
Lightning Source LLC
Chambersburg PA
CBHW060247230326
41458CB00094B/1500